Landscape Series

Volume 25

Aims and Scope

Springer's innovative Landscape Series is committed to publishing high-quality manuscripts that approach the concept of landscape from a broad range of perspectives. Encouraging contributions on theory development, as well as more applied studies, the series attracts outstanding research from the natural and social sciences, and from the humanities and the arts. It also provides a leading forum for publications from interdisciplinary and transdisciplinary teams.

Drawing on, and synthesising, this integrative approach the Springer Landscape Series aims to add new and innovative insights into the multidimensional nature of landscapes. Landscapes provide homes and livelihoods to diverse peoples; they house historic—and prehistoric—artefacts; and they comprise complex physical, chemical and biological systems. They are also shaped and governed by human societies who base their existence on the use of the natural resources; people enjoy the aesthetic qualities and recreational facilities of landscapes, and people design new landscapes.

As interested in identifying best practice as it is in progressing landscape theory, the Landscape Series particularly welcomes problem-solving approaches and contributions to landscape management and planning. The ultimate goal is to facilitate both the application of landscape research to practice, and the feedback from practice into research.

More information about this series at http://www.springer.com/series/6211

Todd R. Lookingbill • Peter D. Smallwood
Editors

Collateral Values

The Natural Capital Created by Landscapes of War

Springer

Editors
Todd R. Lookingbill
Department of Geography
and the Environment
University of Richmond
Richmond, VA, USA

Peter D. Smallwood
Department of Biology
University of Richmond
Richmond, VA, USA

ISSN 1572-7742 ISSN 1875-1210 (electronic)
Landscape Series
ISBN 978-3-030-18990-7 ISBN 978-3-030-18991-4 (eBook)
https://doi.org/10.1007/978-3-030-18991-4

This Springer imprint is published by the registered company Springer Nature Switzerland AG.
The registered company address is: Gewerbestrasse 11, 6330 Cham, Switzerland

Foreword

In June 1970, during the height of the US-Vietnam War, Motown singer Edwin Starr re-recorded the Temptations album track "War" and issued it as a single. It quickly went to #1 on the Billboard charts. With its blasting, syncopated horn section and Starr's deep voice soul-shouting the unadorned lyrics, the song cut through the AM radios of America's youth as both protest anthem and pop song anguish:

War! huh, yeah
What is it good for?
Absolutely nothing!

If only such clarity were true. Historians of empire, revolution, and independence, chroniclers of medicine and technological innovation, soldier poets, and political leaders all struggle with Starr's question: War, *what is it good for?*

Editors Todd R. Lookingbill and Peter D. Smallwood, and the varied set of scholars who authored the individual chapters, grapple with this question in terms of conservation, ecosystem services, and cultural landscapes. Edwin Starr's song echoes throughout the book, even in its potent title *Collateral Values*.

Wisely, the editors partitioned the book into two sections: Battlefields and Borderlands. In the first half of the book, the authors visit the US, English, and French battlefields, crossing historical periods from medieval England to World War I. Jason Julian's chapter on the Southern Plains of Texas and Oklahoma examines the relationship between frontiers, forts, and front-country conservation. Carly Sibilia and her coauthors, Lookingbill and Geoffrey Carter, conduct a similar analysis, this time exploring the history of warfare in early English history and introducing the concept of "constructive conservation." Rémi de Matos Machado and J.P. Hupy shift to the larger-scale landscape of the 10,000 hectare Verdun Forest (site of one of the biggest battles of World War I), describing its ecosystem services and its service as a "place of memory." Lookingbill and coauthors, Emily Minor and Lisa Wainger, focus on a different kind of invasion than in most military strategic studies, in this case the invasion of non-native plants. They warn that such invasions can catch managers "unaware of dangerous thresholds in long term resource management of battlefield landscapes."

The second half of the book is even more global, while maintaining its focus on localized case studies from the Korean War Demilitarized Zone to the conflicts of Latin America, the Cold War's creation of European borderlands, and the unfinished future of Afghanistan's painful wars and the fate of the US Guantanamo Bay Naval Station in Cuba. (Full disclosure: In 2013, working as science advisor to the director of the National Park Service, I wrote an earnest white paper on the creation of an international peace park at Guantanamo and how such a designation could solve several diplomatic, national political, and environmental problems. It disappeared in the yaw of bureaucracy). Lisa Brady examines the environmental and political history of the Korean Demilitarized Zone and translates that history into a trajectory of conservation possibilities—even going so far as to describe it as potentially a "green ribbon of hope." Saleem Ali provides a similar analysis of the Cordillera del Condor corridor, buffering the conflicted boundaries of Peru and Ecuador, and reminds us of the importance of post-conflict strategies that can make conservation workable and effective in the long term. David Havlick looks at the Cold War and "the transition of militarized landscapes into new land uses dedicated to conservation." His discussion of "recasting" military landscapes is both a process for peace and a potential policy of merit. Joe Roman's chapter on Guantanamo is both a historical review and a creative presentation of a model for conservation as a diplomatic tool. Smallwood and Chris Shank's chapter on Afghanistan's Wakhan National Park, larger than Yellowstone and home to 17,000 persons and iconic wildlife, also focuses on transition—what they call "the long journey from geopolitical buffer zone to national park."

The diversity of cases, methodological strategies, and insights gives *Collateral Values: The Natural Capital Created by Landscapes of War* a robustness of importance. This is a place-based book with sweeping consequences.

As cultural sites, these places are crucial for memory—they are physical remembrances that neither necessarily reify nor reject events but can preserve their recognition for future generations. As protected areas that provide measurable (and often significant) ecosystem services, they are (or can be) transcendent landscapes—the destruction of war transformed in peace as described in Isaiah 2:3–4:

> …and they shall beat their swords into ploughshares, and their spears into pruning hooks: nation shall not lift up sword against nation, neither shall they learn war any more.

The authors (both editors and chapter writers) wisely avoid the moral myopia of describing the "benefits" of war. They write, "The central premise of this book is that these sites represent potential opportunities for environmental conservation and restoration." I would go further: They have described, as my colleague and teacher William R. Burch, Jr., has argued, a form of conservation practice that redeems the faults of prior and present generations—what he has called "redemption ecology." And in doing so, they provide a way forward that is both visionary and practical—for scientific and scholarly research; for conservation planning, policy, and practice; and for civic action.

We may not be able to yet answer Edwin Starr's piercing question, but the book you are about to read is a stepping stone on the redemptive path.

Gary E. Machlis
University Professor of Environmental Sustainability
Clemson University
Clemson, SC, USA
Former Science Advisor to Director of the National Park Service
Washington, DC, USA

Acknowledgments

It took the efforts of a great many people to produce this book, and we are deeply grateful to them all. This volume represents many years of thinking about the intersection of warfare and conservation. We are first thankful to all of our chapter authors, who helped expand and extrapolate the concept of "collateral values" to the many time frames and locations represented in the book. The intellectual framework benefited from early conversations with Gary Machlis and colleagues at the National Park Service, including Shawn Carter, Peter Sharpe, Christine Arnott, Kristen Allen, Dorothy Geyer, and Tim Blumenschine. Kimberley Browne, Taylor Holden, Amanda Waggoner, and others in the University of Richmond's Spatial Analysis Lab assisted with map creation for several of the chapters. We are also indebted to Savannah Wilson, who assisted with copy editing and formatting as part of the last push across the finish line. Finally, we appreciate the support, advice, and guidance provided by Springer and the editors of this series, especially Janet Silbernagel.

Contents

Part I
Introduction

Chapter 1
Collateral Values: The Natural Capital Created by Landscapes of War

Todd R. Lookingbill and Peter D. Smallwood

Abstract Warfare and related military activities have intensive, long-term, negative impacts on the environment. But sites of past human conflict also present potential opportunities for conservation and restoration. We provide an approach to valuing military landscapes based on the ecosystem services that they provide. These services are often underappreciated because the benefits gained from the functioning of the ecosystems are not the primary reason for the protection of the sites. We describe these services as collateral values, drawing on the military concept of collateral damages. In this book, we provide examples from across the globe, reflecting conflicts stretching over hundreds of years. Landscapes considered include military battlefields, demilitarized borderlands, and potential peace parks. Specific conflicts include the American Revolutionary and Civil Wars, the Great European Wars, and recent engagements in Cuba, Afghanistan, and the Korean Peninsula. Examples are set within the conceptual framework of warfare ecology with a focus on post-war activities. They address conservation issues including land preservation, protection of biodiversity and water resources, and sustainable tourism. Where possible and appropriate, lessons learned from historical landscape trajectories are discussed in the context of their potential application to the future management of ecosystems still engaged in conflict. We suggest that the increased recognition of the multiple values of these unique landscapes should lead to their increased protection and careful management to preserve and promote the diverse services that they provide.

T. R. Lookingbill (✉)
Department of Geography and the Environment, University of Richmond,
Richmond, VA, USA
e-mail: tlooking@richmond.edu

P. D. Smallwood
Department of Biology, University of Richmond, Richmond, VA, USA

T. R. Lookingbill, P. D. Smallwood (eds.), *Collateral Values*, Landscape Series 25,
https://doi.org/10.1007/978-3-030-18991-4_1

1.1 A Proposal

In 2016, the United States National Park Service (NPS) celebrated its 100-year anniversary. As part of its Call to Action for the centennial celebration, a new vision was articulated for a second century of national park management (NPS 2015). The grand idea of national parks has been evolving since the inception of the NPS to protect remote areas of pristine wilderness to, more recently, also incorporate lands representing Bill Cronon's sense of "wildness in our own backyards" (Cronon 1995). The centennial vision continues this evolution and includes a greater emphasis on promoting visitation in close proximity to diverse population centers that have not always been so thoughtfully served (Weber and Sultana 2013). Explicit within the Call to Action is the charge of "improving urban residents' awareness of and access to outdoor and cultural experiences close to home by promoting national parks in urban areas" (NPS 2015). Consistent with this evolving model of parks, the Science Committee of the NPS Advisory Board proposed a paradigm for the next century of NPS management that avoids the false dichotomy of labeling parks as either "natural" or "cultural" units and instead embraces all parks as coupled human-natural systems (NPS Advisory Board 2012).

At nearly the same time (from 2011–2015), the U.S. was also commemorating the 150-year anniversary of its Civil War. Benchmark anniversaries like this one have been shown to drive up visitation and attention to battlefield sites (e.g., van der Merwe 2014), and the number of people visiting U.S. battlefield parks nearly doubled in the 1960s as part of the centennial recognition (Madron and Tilton, Chap. 2). During the five-year period of the Civil War sesquicentennial, millions of private, local, state, and federal dollars were allocated to battlefield preservation efforts. The catalyzing effect of the war's anniversary led to a total preservation of more than 18,000 hectares of battlefield land by the non-profit Civil War Trust alone (American Battlefield Trust 2018). At the federal level, nearly $35 million were allocated by the American Battlefield Protection Program to help state and local governments to preserve all or parts of 62 Civil War battlefield sites. Planning for additional preservation efforts extended to another 55 sites (NPS 2018).

The concurrence of these two anniversaries provided a prime conservation opportunity. The history and geographic distribution of U.S. Civil War battlefield sites lend themselves to the new NPS directives of the Call to Action. The U.S. Civil War Sites Advisory Commission has identified 384 sites of military significance (NPS 1993). Strategically, many of the sites are located in urban and suburban settings in close proximity to large population centers, as the historical campaigns were fought on the outskirts of cities, including the Capital of the Confederacy (Richmond, VA) and the Nation's Capital (Washington, DC). Ownership of these lands varies from completely private to a mix of different government and non-profit organizations, but more than 70 units of the NPS already preserve spaces associated with the Civil War. The NPS is busy working on new ways to interpret these landscapes that tell a richer story of the multifaceted aspects of the parks (NPS 2017; Rudy 2017). This initiative is consistent with the broader movement toward recognizing the value of multifunctional landscapes (Fischer et al. 2017), including sacred sites such as cemeteries

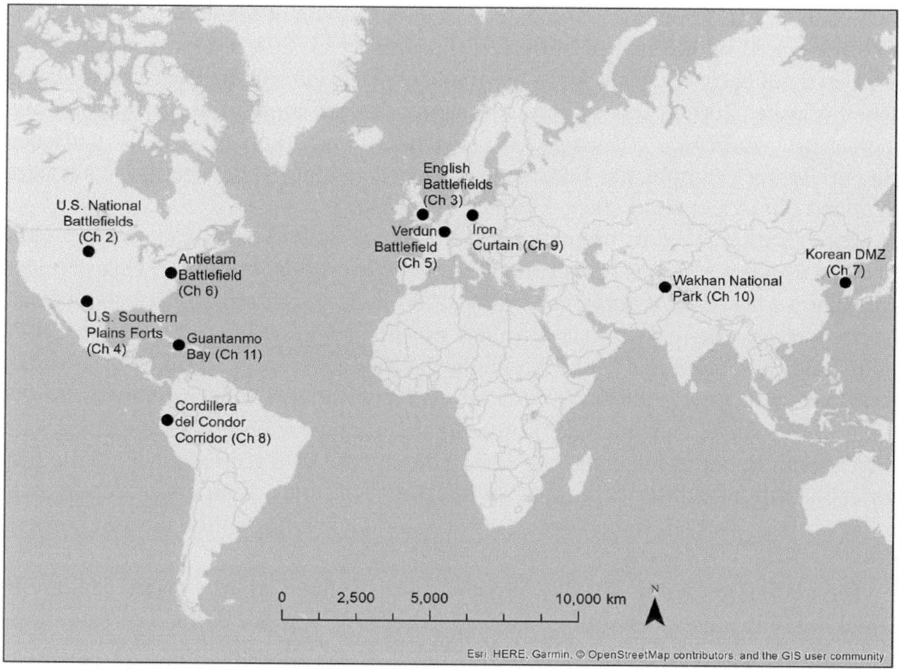

Fig. 1.1 Overview of case study sites presented in this book

(Kowarik et al. 2016), church forests (Reynolds et al. 2017), and battlefields (Pearson et al. 2010). The chapters in this book expound upon the benefits of battlefield conservation and landscape restoration through global examples representing military conflicts dating back more than a thousand years (Fig. 1.1).

Geography, as a discipline, has long been concerned with different aspects of military activity (Sheppard and Tyner 2016). The book provides an approach to valuing military landscapes based on the ecosystem services that they provide. Ecosystem services have been defined in numerous ways, but a common definition is that they are the outputs of natural systems from which humans may derive benefits (National Research Council 2005; Boyd and Banzhaf 2007). By this definition, ecosystem services require use or appreciation by people, but can incorporate any concerns people may have, including values for preserving the integrity of natural systems. Services may include recreational opportunities, carbon sequestration, aesthetic enjoyment, and improved quality of natural resources like water (de Groot et al. 2002; Fisher et al. 2009).

The central premise of this book is that sites of past warfare and related military activity represent potential opportunities for future ecosystem services and warrant environmental conservation. For the most part, these sites are initially recognized for their cultural and historical value. The immediate spiritual benefits provided by these sites to former combatants and their families are quickly followed by recreational and economic benefits as visitation to these sites increases (Miles 2014). In this sense, battlefield tourism comprises a significant sub-discipline of war tourism (which itself can be considered a branch of dark tourism or thanotourism, i.e., tourism associated with death or suffering; Fonseca et al. 2016). Docummented war tourism dates back at least as far as the Battle of Waterloo and the Boer Wars (Seaton 1999; Moeller 2005). The economic benefits of dark tourism are substantial and increasing worldwide (Zhang 2010; Madden and Shipley 2012). Culloden Battlefield, for example, has become the most visited tourist destination in northern Scotland (Gold and Gold 2003; McLean et al. 2007).

Once memorialized as protected landscapes, battlefield parks can provide additional ecosystem services that extend beyond their original mission. These services are what we describe as "collateral values," drawing on the military concept of collateral damage (the term collateral values was first suggested to us by Gary Machlis in a conversation we had with him about the book: we are grateful to him for his imagination in giving the book such an evocative title). Collateral values are the focus of this book. They include promoting the future existence of rare species, safe recreational opportunities, and aesthetic enjoyment. As a consequence of the historical setting of military conflicts on the fringe of cities of strategic importance, many of these parks have become oases of natural amenities within the encroaching urban and suburban matrix (Lookingbill et al. 2014).

The book is set within the existing conceptual framework of warfare ecology in which three distinct phases of ecological impacts of war are defined: preparations for war, war itself, and post-war activities (Machlis and Hanson 2008). The growing literature within this field begins with an acknowledgement that the impacts of warfare are overwhelmingly negative, to society and also to the environment (Dudley et al. 2002; Lawrence et al. 2015; Gaynor et al. 2016; Daskin and Pringle 2018). The ultimate objective should be the elimination of these destructive behaviors. Direct and indirect environmental impacts include pollution and contamination, habitat destruction, poaching and resource extraction, cratering, fires, noise, and suspension or lack of enforcement of existing environmental regulations and laws (Hanson 2018). While these impacts occur at all three stages of warfare, the case studies in this book focus on the post-war period. This period can be characterized by all of the same pressures and adverse effects of a disrupted society and environment.

However, there are also certain opportunities created by warfare. For example, active military training areas may create heterogeneous disturbance regimes that enhance biodiversity by creating habitat in varied successional stages (Warren et al. 2007; Zentelis and Lindenmayer 2014). Post-war, decisions need to be made about how to manage lands that may have been abandoned during conflict or that remain unsafe or that are considered hallowed. The days and years immediately following the cessation of military activities are a critical period for landscape planning, management, and potentially conservation (Negret et al. 2017). Proactive environmental stewardship can be a fulcrum of the peace process. Central to the stories conveyed in this book, positive environmental outcomes can be accomplished relatively inexpensively as "collateral values" when combined with other social, cultural, and political goals. Within this context, the dedication of the resources required to study, plan, and create ecological conservation areas is minor compared to the overall expenses associated with military warfare. The stories provided in this book provide powerful arguments for the creation of post-war conservation areas.

1.2 Battlefields

The book is divided into two halves, with the first half focused on landscapes recognized primarily for their historical value as sites of significant military battles (Fig. 1.2). These case studies represent over a millennium of warfare including England's long military history, frontier battles associated with early European colonization of the American Southern Plains, the American Revolutionary and Civil Wars, and the First World War. The second half of the book focuses on more recent engagements in borderlands of Europe, Afghanistan, Cuba, South America, and the Korean Peninsula. Examples of the collateral values created by landscapes of war are thus drawn from around the world, and each chapter begins with a full-page map of the focal study site.

Dating back to the designation of Yellowstone as the world's first national park in 1872, the U.S. NPS has been a thought leader in protected lands management. In Chap. 2, Madron and Tilton describe the evolution of the agency's conservation strategy as it has progressed to being more inclusive of battlefield landscapes. The chapter provides a history of how battlefield parks were first added to the U.S. NPS portfolio of protected lands over a century ago, and how they have become, through

Fig. 1.2 High Bridge State Park, Virginia. Virginia contains the largest number of Civil War battlefields in the United States. Landscapes commemorating historical battlefields, like this one memorializing the conflict from over 150 years ago, provide a mix of ecosystem types (grasslands, forests, and the Appomattox River crossed by the High Bridge itself) and services (hiking, bicycling, horseback riding, water regulation, wildlife habitat, and education)

time, some of the most visited sites in the national park system. Through textual analysis of the enabling legislation of national parks, this chapter provides evidence of how the rationale for creating battlefield parks has changed over the past 100 years. A major emphasis of the book is on the temporal changes in attitudes about battlefield conservation. A related emphasis is on the post-war changes in the physical landscapes themselves. Chapter 3 provides a proposed trajectory for post-war landscapes leading to their management for multiple, layered assets through principles of constructive conservation.

Chapters 3, 4, and 5 offer examples of the ecosystem services provided by battlefield parks in a chronology from sites preserved to commemorate the Battle of Maldon in 991 to the Battle of Verdun in 1916. In addition to the overall theoretical conservation framework laid out by Sibilia and her co-authors, Carter and Lookingbill, Chap. 3 provides a formal definition of a "battlefield" and outlines a brief history of English domestic warfare, which has created over 500 known battlefield sites. Forty-six of these sites have been formally recognized on the Register of Historic Battlefields. In their chapter, the authors describe the ecosystem services provided by three English battlefield landscapes that vary along a management and land-use gradient. The chapter concludes with an analysis of the patterns of land-cover change in and around the case study sites.

In Chap. 4, Julian picks up with a description of the battles fought between European settlers and various Tribal Nations in the U.S. Southern Plains states of Oklahoma and Texas between 1821 and 1890. Of the many forts that were built during these conflicts, 33 have been designed as publicly accessible, conserved spaces. This chapter provides a detailed inventory of the ecosystem services that have evolved at these old fort sites. The study provides an example of a broad category of militarized landscapes where forts act as the first inroads of an invading people into a new land. One could argue that the Southern Plains forts were a continuation of a process that had been going on for more than a century to encourage European settlement into the New World, and similar networks of protected lands commemorating fort installations can be found elsewhere in the Eastern U.S. For example, many of the 42 forts created as part of the Third System of coastal fortification in the 1800s are also publicly accessible today (Weaver 2001).

Jumping forward more than a century, Machado and Hupy (Chap. 5) describe the evolution of the forested landscape of Verdun from its pre-war condition in the early twentieth century, through the First World War, and into the present day, in which it attracts nearly 250,000 visitors per year. The chapter details the many functions provided by the present-day landscape and provides an example of how the natural resources of battlefield sites that may be difficult or dangerous to access can be mapped and assessed using remotely sensed imagery. It concludes with a listing of some of the contempory conservation challenges confronting the site.

As time passes, many of these landscapes share common threats to the ecosystem services that they provide. These threats include habitat loss and isolation, invasive species, and regional stressors over which local management has limited control. Challenges to managing these landscapes include incomplete ownership of core habitat areas, busy roads dissecting park interior areas, air quality problems, climate change, and mixed mandates to preserve cultural and natural resources that are sometimes at odds. For example, of the 385 battlefields identified by the U.S. Civil War Sites Advisory Commission, over 70 sites were considered completely lost to residential and commercial construction, and an additional 161 were experiencing moderate to high levels of threat (NPS 1993). Chapter 6 describes in more detail the types of threats encountered by battlefield parks as their surrounding landscapes evolve. Lookingbill, Minor, and Wainger then quantify the impact of one of these stressors, exotic plant invasions, to the ecosystem services provided at Antietam National Battlefield, the site of one of the bloodiest and most significant battles of the U.S. Civil War. The first half of the book ends with this consideration of the long-term management of battlefield parks.

Although the most familiar form of collateral values may be represented within battlefield parks, exemplified by Gettysburg National Military Park (e.g., Williams and Patterson 1999) or Antietam National Battlefield (e.g., Landsman and Bowman 2017), the landscapes considered here are not restricted solely to traditional battlefields. The attention is on any areas associated with historical, military conflict. Existing battlefield parks provide perhaps the longest-standing examples of the evolution of landscapes of war towards generators of multiple ecosystem benefits. By comparison, borderlands between countries provide perhaps the greatest potential for future ecological restoration as collateral value landscapes.

1.3 Borderlands

Borderlands as referenced in this book may, in some circumstances, be peace parks but only where these parks have been associated with prior armed conflicts. We do not include the over 100 transboundary conservation areas (TCAs), like the Waterton-Glacier International Peace Park on the border of the U.S. and Canada or the Great Limpopo Transfrontier between South Africa, Mozambique, and Zimbabwe, that were established primarily for biodiversity conservation. The value of connecting conservation efforts across international borders has gained attention recently, for example, along the U.S.-Mexico border (Fowler et al. 2018); however, while TCAs have potential for providing valuable ecosystem services, their creation is fraught with close ties to concepts such as economic neoliberalism and can come at the expense of human populations living in the region (King and Wilcox 2008). The ecosystem values described in this book are secondary to the primary values associated with establishing or maintaining military peace and security.

Moving from battlefield parks that, in some cases, have not seen warfare for hundreds of years, the second half of the book examines landscape of more recent conflict (Fig. 1.3). While the first half of the book has a backward-looking focus, paying close attention to the histories of the sites and how those histories inform their current character and values, the second half of the book is more prospective, considering the future collateral values that could be attained by establishing parks at conflict sites along international borders. In many cases, these conflict zones are

Fig. 1.3 The Wakhan corridor, Afghanistan. The Wakhan corridor became part of Afghanistan in the late nineteenth century as a buffer zone between the Russian empire and the British Raj in India. Already remote, it was further isolated through the twentieth century by cold war hostilities and the civil war that erupted after the Soviet Union withdrew its troops in 1989. In 2014, it was designated as Afghanistan's second national park

already providing ecosystem services. With more fences and walls being constructed along borders around the world, especially following the September 11, 2001 attacks in New York and Washington, D.C. (Vallet 2014), it is worth stepping back and considering the habitats and ecosystem services that are impacted by these barriers and the opportunity costs of not protecting them (see for example Lasky et al. 2007; Ogden 2017).

Perhaps the most dramatic and dangerous border in the world is the Demilitarized Zone (DMZ) between North and South Korea. The Korean War was a bitter, lethal conflict. In three years, hundreds of thousands of military personnel were killed, and estimates of civilian deaths range over two million (DeBruyne 2018). The DMZ was established in the summer of 1953 as a mechanism to stop the fighting, albeit there has never been a treaty to formally end the war. The leaders of the countries could not have anticipated that the DMZ would still be in place more than six decades later, or that it could end up playing such an important role in preserving Korean biodiversity, as well as important historical and cultural sites. Brady outlines the ecosystem services now provided by the DMZ, as well as the important cultural functions. In Chap. 7, she describes the risks of losing the benefits that have developed from the DMZ, and a vision for conserving them. Developing the DMZ into a transboundary protected area would be especially challenging given the tremendous disparity in standards of living between North and South Korea. Nevertheless, Brady finds reason to hope that the DMZ's beneficial functions can be conserved.

In Chap. 8, Ali describes a case where potential collateral values from a demilitarized border were not only planned for, but leveraged for peacebuilding. Peru and Ecuador had fought repeatedly over a disputed border. Tensions rose again in 1998, and both sides feared a resumption of military conflict. The presidents of Peru and Ecuador met with the American President Clinton, asking him to lead the process to develop a proposal to settle the dispute. A key factor in the proposal was an agreement that both sides would establish ecological parks on their sides of the disputed border. A treaty was eventually signed, the peace has held, and Cordillera del Condor was widely lauded as an example of environmental peace-making. Unfortunately, the two sides have not protected the natural resources of the Cordillera del Condor. As Ali demonstrates, this park has succeeded in keeping the peace, but not in conserving the area's natural capital.

Stretching over 12,500 km, the Iron Curtain between the Warsaw Pact nations allied with the USSR and the North Atlantic Treaty Organization (NATO) nations allied with the U.S. may have been the longest border between conflicting parties. Havlick (Chap. 9) describes the European Green Belt initiative to create a system of protected areas, parks, trails, and bike paths along the entire length of the old Iron Curtain. Along with a personal narrative drawn from his experiences crossing the border during the Cold War, and later biking 1200 km of the developing Green Belt, Havlick explores the tensions inherent in ecological conservation on militarized landscapes. While we should not forget the brutality of the conflicts that gave rise to these landscapes, Havlick also explores the fears that working to develop their collateral values may also legitimize the original militarization of these landscapes.

In 2014, Afghanistan declared the whole of the Wakhan Corridor as its second national park. Although its status as a national park is very recent, it became defined as a region and attached to Afghanistan long ago in the latter half of the nineteenth century. The Russian Empire and British Raj in India were the superpowers of the Victorian age, competing for territory and power in Asia. They were at risk of colliding in a poorly mapped region between Afghanistan and China. Smallwood and Shank recount the story of how Afghanistan's borders were drawn in the Great Game by those superpowers, and how the Wakhan came into being as a demilitarized buffer zone between them. Chapter 10 describes what is currently known of the natural capital of the park (primarily biodiversity, including some rare, charismatic species, such as snow leopards), and the challenges that lie ahead of efforts to sustainably manage that capital.

Roman offers a proposal to create a peace park on another disputed border region, the US naval base on Guantanamo Bay, Cuba. While the US maintains that treaties give it the right to maintain the naval base in perpetuity, it is no longer necessary as a navy base, and instead serves as an impediment to improved relations with a rapidly changing Cuba. In Chap. 11, Roman proposes to turn the base into a peace park, dedicated to the conservation and study of biodiversity. While initial reactions to such a proposal have been mixed, Roman makes a compelling case for the many benefits that would be provided by an internationally operated park, research center, and scientific training facility at Guantanamo.

The evolution of warfare suggests that current and future wars are unlikely to be restricted to well-defined battlefields that can later be set aside as protected landscapes. However, while the future of warfare is changing, it is an unfortunate reality that there will always be conflict zones. Chapter 12 concludes the book by looking to the future of warfare and border disputes. Smallwood and Lookingbill argue that tensions along borderlands will continue to create new landscapes that can acquire collateral values. The book concludes with a description of how a remarkable group of diverse stakeholders has forged new natural capital from national tragedy, by creating a 2200-hectare National Memorial at the site where UAL Flight 93 crashed into Pennsylvania coal country on 9/11. In this example and throughout the book, the authors express the hope that a recognition of the multiple values of these unique landscapes of war will lead to their increased protection and their careful management to preserve and promote the diverse services that they provide.

References

American Battlefield Trust. (2018). https://www.battlefields.org/. Accessed 12 Feb 2018.

Boyd, J., & Banzhaf, S. (2007). What are ecosystem services? *Ecological Economics, 63*, 616–626.

Cronon, W. (1995). The trouble with wilderness; or, getting back to the wrong nature. In W. Cronon (Ed.), *Uncommon ground: Rethinking the human place in nature* (pp. 69–90). New York: W. W. Norton & Co.

Daskin, J. H., & Pringle, R. M. (2018). Warfare and wildlife declines in Africa's protected areas. *Nature, 553*, 328–332.

de Groot, R. S., Wilson, M. A., & Boumans, R. M. J. (2002). A typology for the classification, description, and evaluation of ecosystem functions, goods, and services. *Ecological Economics, 41*, 393–408.

DeBruyne, N. (2018). *American war and military operations casualties: Lists and statistics*. Congressional Research Service, RL32492. Accessed 14 Nov 2018.

Dudley, J. P., Ginsberg, J. R., Plumptre, A. J., Hart, J. A., & Campos, L. C. (2002). Effects of war and civil strife on wildlife and wildlife habitats. *Conservation Biology, 16*, 319–329.

Fischer, J., Meacham, M., & Queiroza, C. (2017). Plea for multifunctional landscapes. *Frontiers in Ecology and the Environment, 15*, 59–59.

Fisher, B., Turner, R. K., & Morling, P. (2009). Defining and classifying ecosystem services for decision making. *Ecological Economics, 68*, 643–653.

Fonseca, A. P., Seabra, C., & Silva, C. (2016). Dark tourism: Concepts, typologies and sites. *Journal of Research Tourism & Hospitality*, S2–S002.

Fowler, N., Keitt, T., Schmidt, O., Terry, M., & Trout, K. (2018). Border wall: Bad for biodiversity. *Frontiers in Ecology and the Environment, 16*, 137–138.

Gaynor, K. M., Fiorella, K. J., Gegory, G. H., Kurz, D. J., Seto, K. L., Withey, L. S., & Brashares, J. S. (2016). War and wildlife: Linking armed conflict to conservation. *Frontiers in Ecology and the Environment, 14*, 533–542.

Gold, J. R., & Gold, M. M. (2003). Representing Culloden: Social memory, battlefield heritage and landscape of regrets. In S. P. Hanna & V. J. Del Casino Jr. (Eds.), *Mapping tourism*. Minneapolis: University of Minnesota.

Hanson, T. (2018). Biodiversity conservation and armed conflict: A warfare ecology perspective. *Annals of the New York Academy of Sciences, 1429*, 50–65.

King, B., & Wilcox, S. (2008). Peace Parks and jaguar trails: Transboundary conservation in a globalizing world. *GeoJournal, 71*, 221–231.

Kowarik, I., Buchholz, S., von der Lippe, M., & Seitz, B. (2016). Biodiversity functions of urban cemeteries: Evidence from one of the largest Jewish cemeteries in Europe. *Urban Forestry & Urban Greening, 19*, 68–78.

Landsman, A. P., & Bowman, J. L. (2017). Discordant response of spider communities to forests disturbed by deer herbivory and changes in prey availability. *Ecosphere, 8*, e01703.

Lasky, J. R., Jetz, W., & Keitt, T. H. (2007). Conservation biogeography of the US–Mexico border: A transcontinental risk assessment of barriers to animal dispersal. *Diversity and Distributions, 17*, 673–687.

Lawrence, M. J., Stemberger, H. L. J., Zolderdo, A. J., Struthers, D. P., & Cooke, S. J. (2015). The effects of modern war and military activities on biodiversity and the environment. *Environmental Reviews, 23*(4), 443–460.

Lookingbill, T., Schmit, J., Tessel, S., Suarez-Rubio, M., & Hilderbrand, R. (2014). Assessing national park resource condition along an urban-rural gradient in and around Washington, D.C., USA. *Ecological Indicators, 42*, 147–159.

Machlis, G. E., & Hanson, T. (2008). Warfare ecology. *Bioscience, 58*, 729–736.

Madden, M., & Shipley, R. (2012). An analysis of the literature at the nexus of heritage, tourism, and local economic development. *Journal of Heritage Tourism, 7*, 103–112.

McLean, F., Garden, M.-C., & Urquhart, G. (2007). Romanticizing tragedy: Culloden site in Scotland. In C. Ryan (Ed.), *Battlefield tourism: History, place and interpretation* (Advances in tourism research) (pp. 222–234). New York: Elsevier.

Miles, S. (2014). Battlefield sites as dark tourism attractions: An analysis of experience. *Journal of Heritage Tourism, 9*, 137–147.

Moeller, M. (2005). *Battlefield tourism in South Africa with special reference to Isandlwana and Rorke's drift KwaZulu-Natal*. Dissertation, University of Pretoria.

National Park Service. (1993). *Report on the Nation's civil war battlefields*. Washington, DC: Civil War Sites Advisory Commission c/o National Park Service. https://www.nps.gov/abpp/cwsac/cws0-1.html. Accessed 16 April 2015.

National Park Service. (2015). *A call to action: Preparing for a second century of stewardship and engagement*. https://www.nps.gov/calltoaction/PDF/C2A_2015.pdf. Accessed 11 Jan 2018.

National Park Service. (2017). *NPS civil war to civil rights commemoration summary report*. https://www.nps.gov/civilwar/upload/civil-war-to-civil-rights-summary-report-1-v2.pdf. Accessed 14 Dec 2017.

National Park Service. (2018). *American battlefield protection program*. https://www.nps.gov/abpp/index.htm. Accessed 12 Feb 2018.

National Park System Advisory Board Science Committee. (2012). *Revisiting Leopold: Resource stewardship in the national parks*. https://mylearning.nps.gov/wp-content/uploads/2016/08/Revisiting-Leopold_Resource-Stewardship-in-the-National-Parks-2012.pdf. Accessed 5 Nov 2017.

National Research Council. (2005). *Valuing ecosystem services: Toward better environmental decision-making*. Washington, DC: The National Academies Press.

Negret, P. J., Allan, J., Braczkowski, A., Marone, M., & Watson, J. E. M. (2017). Need for conservation planning in postconflict Colombia. *Conservation Biology, 31*, 499–500.

Ogden, L. E. (2017). Border walls and biodiversity: New barriers, new horizons. *Bioscience, 67*, 498–505.

Pearson, C., Coats, P., & Cole, T. (Eds.). (2010). *Militarized landscapes: From Gettysburg to Salisbury Plain*. New York: Continuum.

Reynolds, T., Collins, C., Eshete, A. W., Liang, J., Briggs, W., Sisay, T., Lowman, M., & Adamu, E. (2017). Sacred natural sites as mensurative fragmentation experiments in long-inhabited multifunctional landscapes. *Ecography, 40*, 144–157.

Rudy, J. (2017). From tokenism to true partnership: The National Park Service's shifting interpretation at the Civil War's sesquicentennial. In K. M. Levin (Ed.), *Interpreting the Civil War at museums and historic sites* (pp. 61–76). Lanham: Rowman & Littlefield Publishers.

Seaton, A. V. (1999). War and thanatourism: Waterloo 1815–1914. *Annals of Tourism Research, 26*, 130–158.

Sheppard, E., & Tyner, J. (2016). Forum on geography and militarism: An introduction. *Annals of the American Association of Geographers, 106*, 503–505.

Vallet, E. (Ed.). (2014). *Borders, fences and walls: State of insecurity?* (Border Regions Series). Burlington: Ashgate Publishing.

van der Merwe, C. (2014). Battlefields tourism: The status of heritage tourism in Dundee, South Africa. *Bulletin of Geography, 26*, 121–129.

Warren, S. D., Holbrook, S. W., Dale, D. A., Whelan, N. L., Elyn, M., Grimm, W., & Jentsch, A. (2007). Biodiversity and the heterogeneous disturbance regime on military training lands. *Restoration Ecology, 15*, 606–612.

Weaver, J. R., II. (2001). *A legacy in brick and stone: American coastal defense forts of the Third System, 1816–1867*. McLean: Redoubt Press.

Weber, J., & Sultana, S. (2013). Why do so few minority people visit national parks? Visitation and the accessibility of "America's Best Idea". *Annals of the Association of American Geographers, 103*, 437–464.

Williams, D. R., & Patterson, M. E. (1999). Environmental psychology: Mapping landscape meanings for ecosystem management. In H. K. Cordell & J. C. Bergstrom (Eds.), *Integrating social sciences and ecosystem management: Human dimensions in assessment, policy and management* (pp. 141–160). Champaign: Sagamore Press.

Zentelis, R., & Lindenmayer, D. (2014). Bombing for biodiversity—Enhancing conservation values of military training areas. *Conservation Letters, 8*, 299–305.

Zhang, J. J. (2010). Of Kaoliang, bullets and knives: Local entrepreneurs and the battlefield tourism enterprise in Kinmen (Quemoy), Taiwan. *Tourism Geographies, 12*, 395–411.

Part II
Battlefields

Chapter 2
Rethinking Resources in Our National Battlefields

Justin Madron and Lauren Tilton

Abstract The importance of protecting battlefield landscapes has been recognized by the National Park Service for over a century. As we seek strategies to protect these lands, a look to their fraught histories offers insight into how these lands came to be seen as valuable. Specifically, we argue that the rationale for their preservation conceals conservation opportunities for battlefields. Assessing the language used in the enabling legislation of battlefield parks as a subset of all national parks reveals how the value of cultural and natural resources differs among the units. The rationale for creating battlefield parks also has changed through time. While terms like "landscape" have appeared in the earliest of these founding documents, there was a shift to highlighting the unique historical importance of these sites. More recent designations have been more likely to include terms indicating the importance of natural resource values as justification for park establishment. These urban greenspaces provide crucial habitat for wildlife while serving as outdoor enjoyment and recreational opportunities. Recognizing the significance of the natural resources alongside the cultural and historical importance of battlefield parks results in missed opportunities for these parks to provide ecosystem services. This is important considering that most national battlefields lie within urban environments and are in close proximity to some of the most populated metropolitan areas of the United States. Explicit consideration of the ecosystem value of battlefield parks could also benefit future discussions of the political and cultural functions of these sites.

J. Madron (✉)
Digital Scholarship Lab, Boatwright Memorial Library, University of Richmond, Richmond, VA, USA
e-mail: jmadron@richmond.edu

L. Tilton
Department of Rhetoric and Communication Studies and Digital Scholarship Lab, University of Richmond, Richmond, VA, USA

T. R. Lookingbill, P. D. Smallwood (eds.), *Collateral Values*, Landscape Series 25,
https://doi.org/10.1007/978-3-030-18991-4_2

2.1 Battlefields' Natural Capital

On August 25, 2016, the National Park Service (NPS) turned 100. The yearlong celebration offered the federal agency an opportunity to promote the nation's parks through thousands of events across the nation. The occasion affirmed NPS's role as stewards of conservation, historic preservation, and recreation on America's public lands. Yet, how did these three values – conservation, historic preservation, and recreation – come to define the mission of the NPS; how are these values applied to national battlefields versus parks; and, how might they inform policy strategies for preserving natural resources in the future?

We pose these three questions in order to address how public lands can be protected from and benefit urban growth. While urban areas continue to expand, national lands have been identified as a particularly important site for ecological benefits. As we seek strategies to protect these lands, a look into their fraught histories offers insight into how these lands have come to be seen as valuable. These histories combined with acknowledging these lands as a site for natural resources offers strategies for protecting these lands for the future. Along with highlighting the benefits of the battlegrounds as sites for cultural and historical interpretation, we can also argue that battlegrounds are a source of natural resources such as clean air and water, which can help protect the ecological benefits that these lands provide.

The debates over public lands, specifically battlefields, in the United States is also indicative of a shift in conservation debates that this book illuminates. Subsequent chapters detail the many ecosystem services provided by battlefield landscapes around the world. Debates over the historical and cultural interpretations continue, but like with the U.S. national parks, the benefits of these sites' ecosystem services has often been secondary or absent. The history traced in this chapter and the argument to at least acknowledge, if not shift, our emphasis to the ecological benefits of these sites are central to this book.

Our chapter focuses on how historical and cultural interpretations have shaped debates over battlefields by focusing on the history of battlefield conservation within the U.S. National Park Service (Fig. 2.1). This larger context is beneficial when examining the subsequent chapters.

2.2 Identifying Priorities: Why Protect Public Lands

An initial look at the enabling legislation of military and battlefield parks versus the national parks offers insight into the motives for the creation of national parks and battlefields. The language used in the purpose statements for United States national parks alongside military and battlefield parks reveals two trends.[1] Analyzing word

[1] Text analysis was used to calculate word frequency count in the enabling legislation. The legislation was coded as either a battlefield or national park. For battlefields, we included the eleven

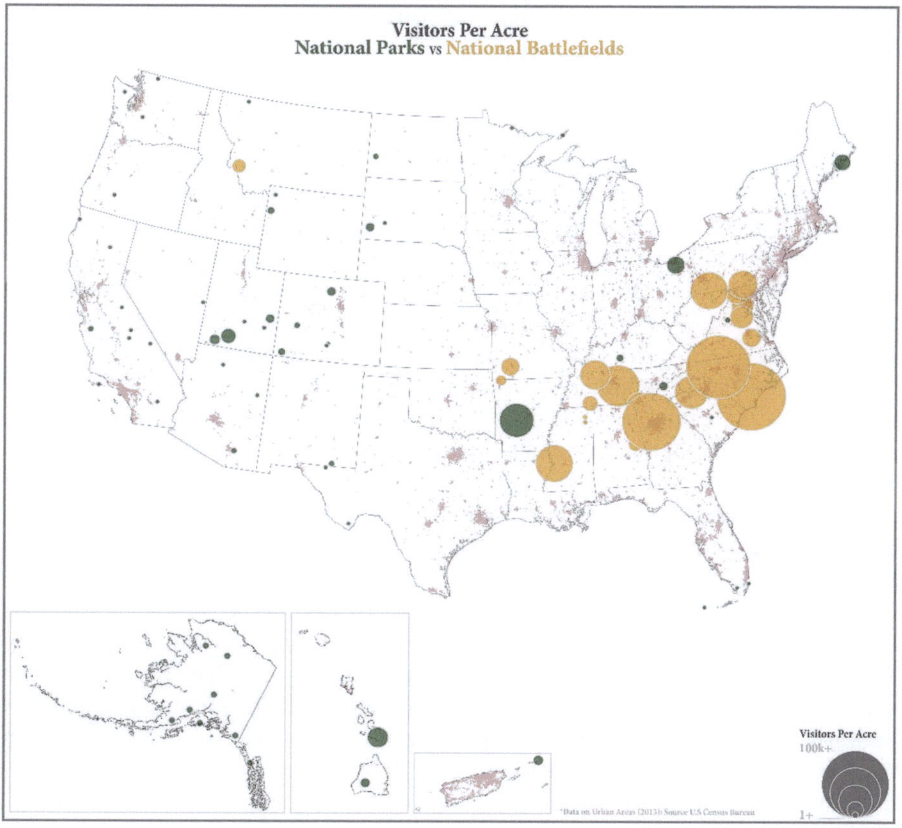

Fig. 2.1 National Battlefields and National Parks highlighting visitors per acre and proximity to urban areas (Data: NPS Stats and U.S. Census Bureau)

frequencies shows that national parks are continually framed as "wilderness," "natural," and "scenic," while national battlefields are commonly associated with the terms "American," "history," and "interpret." The word choice highlights how two competing values — natural and cultural resources — undergirded federal and local initiatives to protect and preserve public lands. While national parks were a place to protect the wilderness for tourists eager to experience pristine, untouched nature, national battlefields served as a site to preserve and interpret American history. We now track how these cultural values have been assigned to different kinds of public lands since their formation.

National Battlefields, nine National Military Parks, and four National Battlefield Parks. We used a stoplist to remove common words and then calculated the word frequency to assess the most common words describing national parks versus battlefields.

2.3 The Early Years of Preservation: Battlefields for Historical Preservation

Questions about whether and how to protect public lands and for whom have enlivened debates about the practice of conservation and preservation for over a century. The concept of national battlefields and parks emerged in the late nineteenth century as anxieties about the nation's history and values played out in ideas about the land and nature. From the start, battlefields were central to debates about how to preserve and interpret American history. The practice of preserving battlefields began after the Civil War. As the efforts of radical reconstructionists to create a more equitable racial and economic order were rapidly overturned, champions of reunion and reconciliation looked to battlefields as a means to celebrate both sides' fallen soldiers and obscure the centrality of slavery. While this effort was certainly not uncontested, the reconciliationists found a crucial and powerful ally in the United States War Department.

The government's imperialist ambitions throughout the late nineteenth century entangled the country in wars, such as the Spanish-American War and the Ghost Dance War. The War Department sought to use battlefields as a place to study strategy for current military initiatives and to train officers. At the same time, veterans both outside and within the department opposed altering the battlefields, hoping they would be a site of Civil War memory (Smith 2008). The War Department won and took control of the lands. The use of the battlefields for military training was minimal and short-lived, leaving minor alterations to the landscape, a fortuitous turn for advocates of battlefield preservation. Because veterans were advocating for saving battlefields, Congress listened and in the 1890s designated five Civil War battlefields – Antietam, Chickamauga and Chattanooga, Shiloh, Gettysburg, and Vicksburg – as national parks under the War Department (Table 2.1). The inclusion of Chickamauga, a Confederate victory in September 1863 that resulted in extensive casualties second only to Gettysburg, indicated that national reunion would involve honoring the Confederacy, an approach that would help obscure the role of race and slavery in the war and therefore curtail efforts to address racial inequality (Blight 2003). The rapid commitment to and establishment of these parks earned the era the reputation as "the Golden Age" of battlefield park preservation (Smith 2008). Yet, the battle over the contours of preservation, including which battlefields to preserve and histories to tell, had just begun.

Table 2.1 First five designated Civil War battlefields

Park	Current designation	Established date
Chickamauga and Chattanooga National Military Park	National Battlefield	1890-08-19
Antietam National Battlefield	National Military Park	1890-08-30
Shiloh National Military Park	National Military Park	1894-12-27
Gettysburg National Military Park	National Military Park	1895-02-11
Vicksburg National Military Park	National Military Park	1899-02-21

2.4 The Early Years of Preservation: National Parks for Conservation and Recreation

While the debate over battlefields ensued, the idea of creating national parks gained momentum among bureaucrats, industrialists, and environmentalists who sought to preserve lands acquired during westward expansion. The government and white Americans together constructed the myth of the West as vacant lands unoccupied by Native American communities to justify colonization (Spence 1999; Blackhawk 2008; Duval 2007). The rapid expansion produced anxiety that the great, untouched natural wonders of the nation would no longer be available to the nation's populace. National pride and romanticization of America's natural beauty also drove calls for preservation as artists and writers alike celebrated the landscape as a site for national regeneration and aesthetic pleasure (Demars 1990).

Led by industrialists as well as environmentalists like John Muir, the national park emerged as the strategy to preserve large swaths of the West as a natural playground for Americans (Sellars 1999). Tensions flared immediately over what was to be protected, for whom and towards what ends. These interlocking issues centered on a core question: Are the parks for tourism or preserving nature? The answer changed which lands to acquire, who could have access and when, and what land management strategies to employ. Those committed to preservation debated how to define conservation and resource protection, while businessmen called for an investment in tourism in the national parks (Sellars 1999). How the landscape would look depended on whether the government let a park naturally evolve unabated, be altered intentionally for tourism, or be preserved as it must have looked at a particular historical moment.

The founding legislation for Yellowstone, established as the first national park in 1872, deemed it "a public park or pleasuring-ground for the benefit and enjoyment of the people" (Dilsaver 1994). Advocacy for Yellowstone came from the Northern Pacific Railroad Company, which sought a tourist destination accessible exclusively by its trains (Sellars 1999). Infrastructure including roads and hotels followed as tourists flocked to the 2.2 million-acre park. The tourism industry boomed as the country set aside parks throughout the late nineteenth century and early twentieth century (Table 2.2). For those worried the government might wane in its commitment to the parks, tourism proved they were a value to the public and therefore deserved sustained federal support and growth.

Table 2.2 National Parks designated prior to 1920

Park	Current designation	Established date
Hot Springs Reservation (NP in 1921)	National Park	1832-04-20
Yellowstone National Park	National Park	1872-03-01
Kings Canyon National Park	National Park	1890-10-01
Sequoia National Park	National Park	1890-09-25
Yosemite National Park	National Park	1890-10-01

The excitement over development was not universal. Naturalists and environmentalists like John Muir were concerned (Sellars 1999). They were invested in a wilderness composed of pristine, untouched nature. The parks were to protect this "virgin land," an ideology inculcated in gendered relationships to the natural world. The federal government should protect the innocent purity of nature for the pleasure of a powerful and strong nation, these activists argued. The result was legislation that often included land in excess of the amount necessary for tourism at the time (Sellars 1999).

2.5 National Battlefields: Renewed Attention During the Civil War's Semicentennial

As tensions rose over conservation and recreation within the national parks, the purpose of the nation's battlefields shifted in the early twentieth century. The 50th anniversary of the war and the passing of veterans brought renewed attention to the nation's battlefields, which had stagnated since the 1890s. Advocates suggested preserving small pieces of land where the battles could be commemorated and remembered as opposed to large swaths of land that were difficult to maintain and, particularly in urban areas, risked conflicting with industrialization and urbanization (Smith 2008). Thirteen battlefields were designated from 1926 to 1936, including ten from the Civil War, two from the Revolutionary War, and one from the French and Indian War (Table 2.3). The result was a push to designate battlefields as memorials.

Simultaneously, the Civil War battlefields lost value as sites for contemporary military strategy, a major impetus behind their initial preservation. Veterans were not present to designate field lines, which were necessary if the military was to learn strategy from the war (Smith 2008). Disinterest in the lands was exacerbated by increasing U.S. isolationism following World War I and the economic and social turmoil of the Great Depression (Keith 2004). Having minimal strategic value, the battlefields were shuffled between bureaus and buried in the federal bureaucracy. Seeking a proper administrative home for the battlefields, FDR signed an executive order in summer 1933 to move them to the NPS, formed in 1916, from the War Department. With this transfer, the battlefields became primarily a site for historic preservation and public history (Lee 1973). However, the national parks would remain NPS's priority; only seven of the 24 extant national battlefield parks were created after 1933 (Fig. 2.2).

Table 2.3 National Battlefields designated under the War Department

Battlefield	Current designation	Established date
Big Hole National Battlefield[a]	National Battlefield	1910-06-23
Guilford Courthouse National Military Park[b]	National Military Park	1917-03-02
Kennesaw Mountain National Battlefield Park[c]	National Battlefield Park	1917-02-18
Moores Creek National Battlefield[b]	National Battlefield	1926-06-02
Petersburg National Battlefield[c]	National Battlefield	1926-07-03
Fredericksburg and Spotsylvania National Military Park[c]	National Military Park	1927-02-14
Stones River National Battlefield[c]	National Battlefield	1927-03-03
Brices Cross Roads National Battlefield Site[c]	National Battlefield Site	1929-02-21
Cowpens National Battlefield[c]	National Battlefield	1929-03-04
Fort Necessity National Battlefield[d]	National Battlefield	1931-03-04
Fort Donelson National Battlefield[c]	National Battlefield	1933-08-10
Tupelo National Battlefield[c]	National Battlefield	1933-08-10
Kings Mountain National Military Park[c]	National Military Park	1933-08-10
Monocacy National Battlefield[c]	National Battlefield	1934-06-21
Manassas National Battlefield Park[c]	National Battlefield Park	1936-11-14
Richmond National Battlefield Park[c]	National Battlefield Park	1936-03-02

[a]Nez Perce War
[b]Revolutionary War
[c]The American Civil War
[d]French and Indian War

2.6 National Parks: Negotiating Two World Wars

World War I brought new pressures as the military sought natural resources, particularly timber, for the war effort (Keiter 2013). Unlike the national parks, forest lands could be mined for their natural resources. The United States Forest Service, established in 1881 as the Division of Forestry, clashed immediately with the military over whether and when to extract (Sellars 1999). Park advocates worried the military would extend its reach and responded by lobbying Congress to pass the Organic Act of 1916. The act called for the creation of a National Park Service under the Department of the Interior charged with conserving "the scenery and the natural and historic objects and the wildlife therein and to provide for the enjoyment of the same in such manner and by such means as will leave them unimpaired for the enjoyment of future generations" (NPS 2017c). The bill solidified the nation's commitment to preserving the national parks, including outlining how to penalize those who violated the law (NPS 2017a).

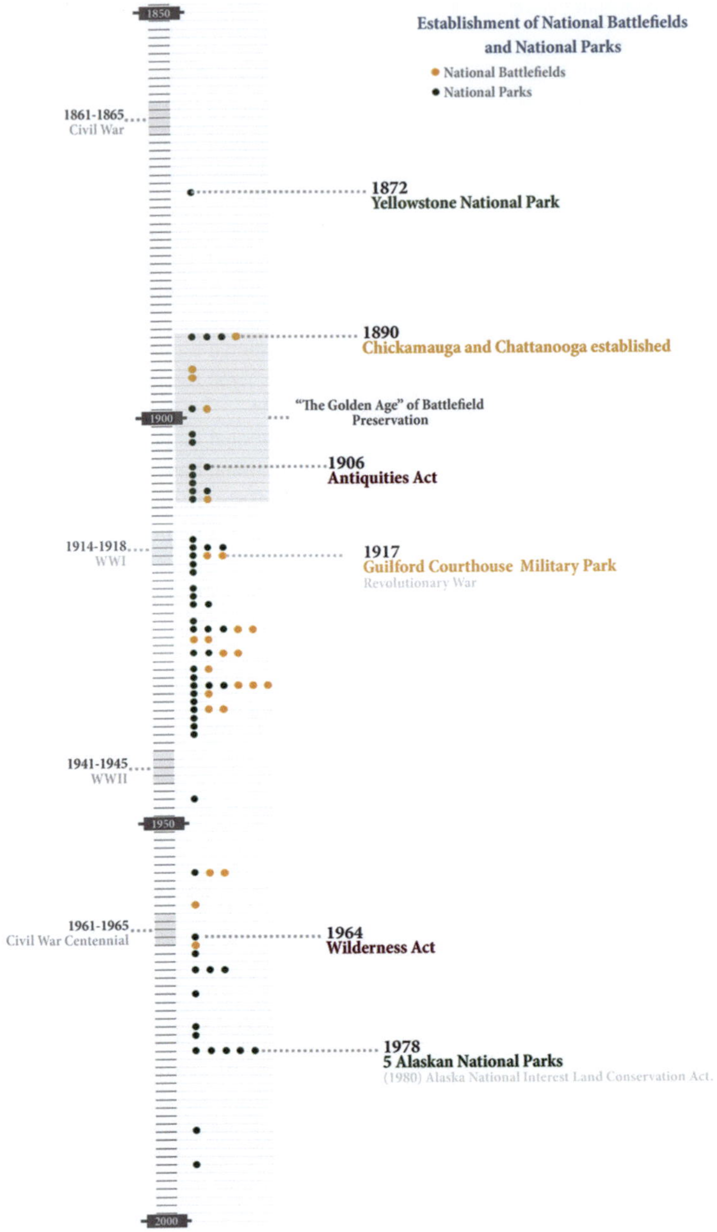

Fig. 2.2 Timeline of establishment of National Parks and National Battlefields

The dual goals of conservation and leisure became a feature of the national parks. Writing to NPS Director Stephen Mather in 1918, Secretary of the Interior Franklin Lane outlined the three goals of the nascent government organization. In what has become known as the "Lane Letter," he wrote:

> First that the national parks must be maintained in absolutely unimpaired form for the use of Future generations as well as those of our own time; second, that they are set apart for the use, observation, health, and pleasure of the people; and third, that the national interest must dictate all decisions affecting public or private enterprise in the parks (Dilsaver 1994).

To maintain an "absolutely unimpaired" park that facilitated "use, observation, health, and pleasure" at first reads as a paradox. Providing access for the American people meant modifying the park. However, rather than understanding this as a contradiction, Lane understood that the aesthetic and physical pleasure garnered from the parks should be unimpaired, even if this meant adjusting the parks (Keiter 2013). Infrastructure projects were not only permitted but encouraged as long as the engineers possessed "knowledge of landscape architecture or have a proper appreciation of the esthetic value of park lands" (Dilsaver 1994). Educational and recreational use were exuberantly encouraged as these were already established activities that characterized, not devastated, a national park. Lane believed adjustment to the land could be done by "harmonizing... these improvements with the landscape" while "improv[ing] the scenic features" of the parks (Dilsaver 1994). Conceptualized this way, protecting the parks, facilitating access, and supporting recreation and education were mutually reinforcing goals.

Under Lane's leadership, NPS established several parks and catered to tourism throughout the 1920s and early 1930s (Fig. 2.3). The park service built lodges, visitor facilities, and roads for a nation on the move thanks to a newly affordable technology for middle-class Americans – the automobile (Keiter 2013; Sellars 1999). The federal government emphasis on tourism meant the need for park infrastructure that justified its jobs programs during the Great Depression. For example, the Civilian Conservation Corps (CCC) employed unmarried men who built over 2000 miles of roads through the nation's parks (Keiter 2013). At the same time, the rapid expansion alarmed conservationists who worried this buildup was destroying the natural landscape (Keiter 2013).

World War II brought new pressures. The War Department again sought the parks' natural resources (Keiter 2013). The NPS protected the lands as visitation plummeted. For example, visits to Shenandoah dropped from 1,071,000 in 1941 to less than 42,000 in 1943 (Fig. 2.4). Millions of Americans were occupied either at home or abroad as leisure gave way to the necessities of a nation at war. Yet, with the end of the war, the parks would see unprecedented visitation as Americans enjoyed economic prosperity.

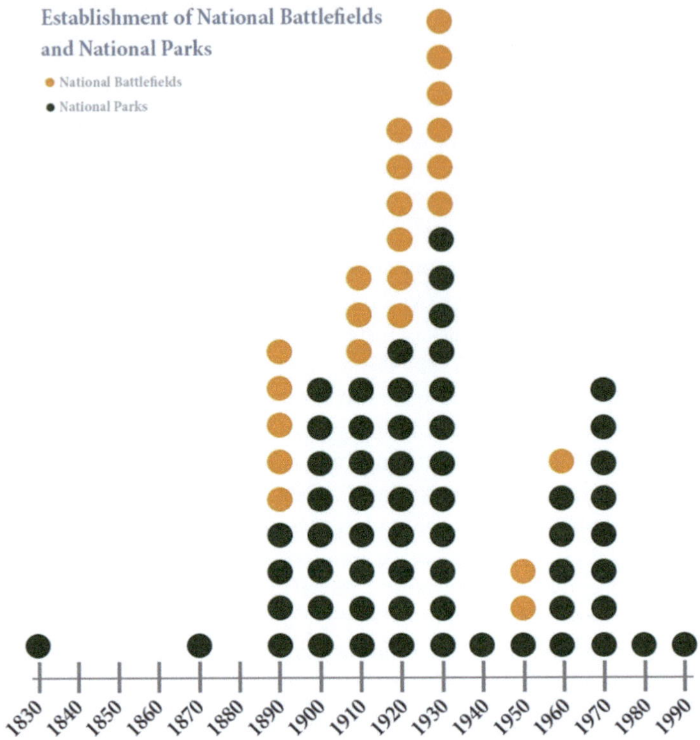

Fig. 2.3 National Battlefields and National Parks designated by decade

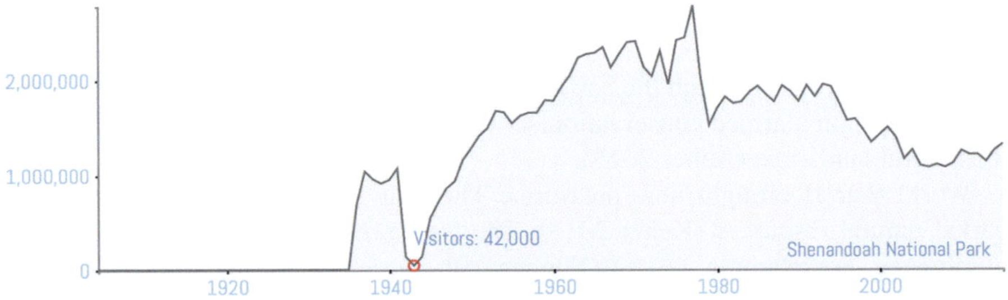

Fig. 2.4 Shenandoah National Park visitation numbers (Data: NPS Stats (https://irma.nps.gov/Stats/Reports/Park))

2.7 The Civil War Centennial: The Battle Over Historical Memory Intensifies

As the national parks enjoyed renewed interest, so too did the nation's battlefields. The centennial of the Civil War brought renewed attention in the late 1950s and 1960s to the battlefields amid a nation in great flux. Communities of color organized for civil rights as racial tensions escalated across the nation and revealed the degree to which white Americans, often with assistance from the state, would work to shore up de jure and de facto inequality (Sugrue 2008; Payne 2007). How to remember the Civil War became a battleground over the memory and meaning of the war. Initially, those seeking a Civil War narrative that celebrated the heroism of both sides were anxious that commemorative events would disparage the South by focusing on the critical role of slavery. To alleviate these fears, local interest groups and Civil War enthusiasts set out to plan the commemorations. History enthusiasts and historians came together in 1953 to form the Civil War Centennial Association. This was not without opposition, in part from the National Park Service, which argued the centennial should be overseen by a federal agency (Cook 2007). President Eisenhower responded by signing into law the creation of the U.S. Civil War Centennial Commission in 1957, placing the commemoration in federal hands (United States 1968). However, the commission would do little to change the often conflicting narratives of the war. Rather, the centennial became a flashpoint for those struggling for and those looking to curtail civil rights. While George Wallace of Alabama revived the reconciliationist narrative of the war in order to celebrate the Confederacy and affirm segregationist agendas, civil rights leaders hoped to revive emancipationist narratives that would recenter the critical role of slavery in the war and reveal yet the unfinished promises of full racial equality (Cook 2007).

As tensions flared, the NPS invested in the parks. The battlefields enjoyed renewed interest with the centennial. Visitors to national battlefields nearly doubled to 6.5 million during the anniversary (Fig. 2.5). Battlefield advocates used the

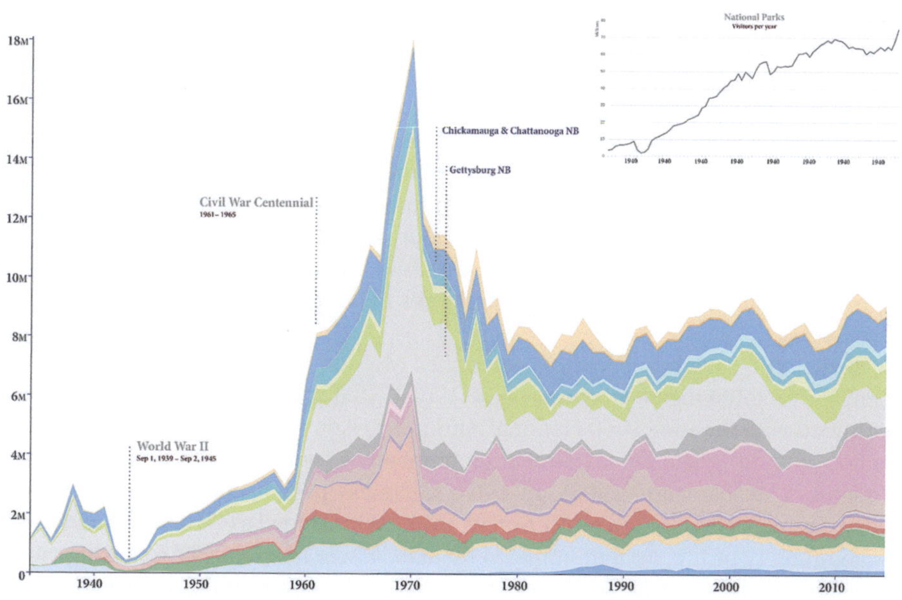

Fig. 2.5 National Parks and National Battlefields visitation statistics (Data: NPS Stats (https://irma.nps.gov/Stats/Reports/Park))

Table 2.4 National Battlefields designated during the 100-year anniversary of the Civil War

Park	Current designation	Established date
Horseshoe Bend National Military Park[a]	National Military Park	1956-07-25
Pea Ridge National Military Park[b]	National Military Park	1956-07-20
Wilson's Creek National Battlefield[b]	National Battlefield	1960-04-22

[a]Creek War
[b]The American Civil War

opportunity to support the creation of Pea Ridge (1956) and Wilson's Creek (1960) as a military park and battlefield, respectively (Table 2.4). The choice to preserve sites of Union and Confederate victory showed that efforts to valorize both sides persisted. With the passing of the centennial, public interest in the battlefields diminished in the following decades. The growth of the park system alongside a recession and rising conservative movement meant minor federal investment in the sites in the decades to follow. At the same time, environmentalists increased pressure on the NPS to manage the park system with more careful attention to science and ecology. The national resource value of these public lands was coming into sharper focus.

2.8 A Consumer's Republic: Tourism and Its Discontents

By the 1950s, the economy boomed for white, middle-class Americans who enjoyed new access to leisure facilitated by paid vacation and car ownership (Keiter 2013). A flourishing economy alongside population growth in the West paired with a baby boom put increasing pressure on the park system. Millions of Americans flocked to the great outdoors, wearing down already precarious infrastructure that had received minimal maintenance since the New Deal (NPS 1965). In response, NPS Director Conrad Wirth used the fiftieth anniversary of the NPS in 1966 to launch Mission 66. Through the program, he procured $500 million in federal funding to improve and expand park infrastructure. The improvements led to over 100% growth in visitation over the decade, with the number of visitors increasing to 121 million in 1965. This increased visitation shaped the parks for the decades to follow (Keiter 2013; Carr 2007).

While NPS leaders and visitors celebrated Mission 66, the environmental movement and its lobbying efforts grew in the 1960s. Strategies for protecting lands were expanded with the June 1960 National Antiquities Act, which gave the president the authority to designate lands national monuments, a power previously reserved for Congress (McManamon 2000). In 1964, Congress passed the Wilderness Act of 1964. The act called for the protection of wilderness defined as "where the earth and its community of life are untrammeled by man, where man himself is a visitor who does not remain" in order "to assure that an increasing population, accompanied by expanding settlement and growing mechanization, does not occupy and modify all areas within the United States" (Dilsaver 1994). The act created a National

Wilderness Preservation System. In order to identify which lands from the NPS would comprise this system, the act ordered the Secretary of the Interior to "review every roadless area of five thousand contiguous acres or more in the national parks, monuments and other units of the national park system and ... [to] report to the President his recommendation as to the suitability or non-suitability of each such area or island for preservation as wilderness" (Dilsaver 1994).

As the legislative and executive branches continued to vie for control over setting land policy, debates continued about how to define wilderness and which government agency would have oversight. Regardless of whether lands were under the NPS or Forest Service, environmentalists agreed that protecting lands meant curtailing invasive infrastructure, especially roads and the cars that followed. The NPS was less interested in the environmentalists' position as tourism remained a priority. NPS leaders took their time on their wilderness report, fearing they would lose management over their lands (Keiter 2013).

While the NPS dillydallied, the "Wildlife Management in the National Parks" report, better known as the Leopold Report, gained traction. Written in 1963, the report laid out a plan for ecosystem management that would return the parks to "the condition that prevailed when the area was first visited by the white man. A national park should represent a vignette of primitive America" (Dilsaver 1994). In order to accomplish this goal, the report laid out a process for park management that involved restoring them to their "primitive" state. Controlled fires, animal population control, and introducing or removing plants were all permitted as long as the reasoning and approach were based on scientific research.

Importantly, the report also argued that ecosystem management and tourism were not mutually exclusive. It acknowledged the importance of tourism but called for limits. For example, the report suggested that overcrowded parks develop road plans that limit the number of visitors rather than add more roads (Dilsaver 1994). The report's emphasis on plant and animal management suggested the secondary role of tourism, a goal heralded by the growing environmental movement, which advocated for protecting nature and reducing human interference (Kline 2001; Taylor 2016).

Activism for environmental causes enjoyed further success in the 1970s. Environmentalists drew on the strategies and gains of the civil rights and liberation movements, which demonstrated that change required changing hearts and minds as well as the law. For example, the movement created Earth Day in 1970 in which millions of Americans organized for environmental protection. A slew of environmental laws that enjoyed bipartisan support followed, building off of the Wilderness, Clean Air and Clean Water Acts (Keiter 2013). The National Environmental Policy Act required federal agencies to develop environmental assessments and impact statements, while the Endangered Species Act gave two federal agencies the ability to protect endangered species and their habitats (Dilsaver 1994). These environmental gains redirected national policy toward conservation and preservation as concerns increased about the nation's resources within and beyond the parks.

Yet, the gains of the environmental movement would slow as the support they enjoyed from the Nixon administration faded and the conservative backlash of the

Reagan administration took hold (Turner 2009). A final major gain of this movement was the 1980 Alaska National Interest Land Conservation Act. The process had begun in 1968 when President Johnson protected the lands under the Antiquities Act of 1906 at the urging of Secretary of the Interior Stewart Udall (NPS 2015a, b).[2] It would take over a decade for Congress to pass the 1980 omnibus law, which designated over 150 million acres for federal protection under a series of designations including forests, monuments, and parks (Dilsaver 1994). Five new parks were added to the NPS, which in 2016 totaled nearly 30 million acres. However, with the Reagan administration's embrace of free enterprise and deregulation, environmentalists would find themselves spending much of the next three decades defending the gains of the 1960s and 1970s.[3]

2.9 NPS Changing Course

While expansion of the national park system came to a near halt, the 75th anniversary of the parks offered a moment to reflect and look toward the future. The 1993 Vail Agenda noted the NPS's inconsistent embrace of scientific approaches to ecological management, calling on the agency to reaffirm its commitment in order to better manage the park's natural resources (Sellars 1999). In part this was a response to the challenges of managing overcrowding and infrastructure in the parks. Science could help the NPS understand and maintain its ecosystems while justifying visitor regulations. It was not that the Vail Agenda vilified tourism. Rather, the report's objectives included recommendations for how to improve access, education, and enjoyment in the parks, which included improving existing facilities and catering to visitors' needs. Like with the Lane Letter in 1933, the goal was to find a better balance between tourism, which on average grew by about a million visitors a year in the nineties, and conservation.[4] Accordingly, NPS looked to shift away from leisure activities that had a high impact on the natural environment, particularly those that encouraged roads and vehicles (Keiter 2013).

NPS's effort to find a balance resulted in an embrace of ecotourism throughout the 1990s. This approach was the subject of much debate including what practices fell under its rubric (Higham 2007). While their views were contested, advocates agreed that there should be a specific tourism sector catering to natural environments and conservation. In other words, those committed to ecotourism shared the

[2] The NPS lists the founding of the parks as 1978, for this is when President Jimmy Carter designated the land national monuments. Two years later Congress would follow by passing the necessary legislation.

[3] The only new land added after Reagan was the National Park of American Samoa in 1988.

The Dry Tortugas National Park was designated a national monument under the Antiquities Act in 1935. It was then named a national park in 1992. See https://www.nps.gov/drto/learn/historyculture/index.htm

[4] 58,929,580 people visited in 1990 compared to 66,593,439 in 2000. For the data, see Keiter 2013.

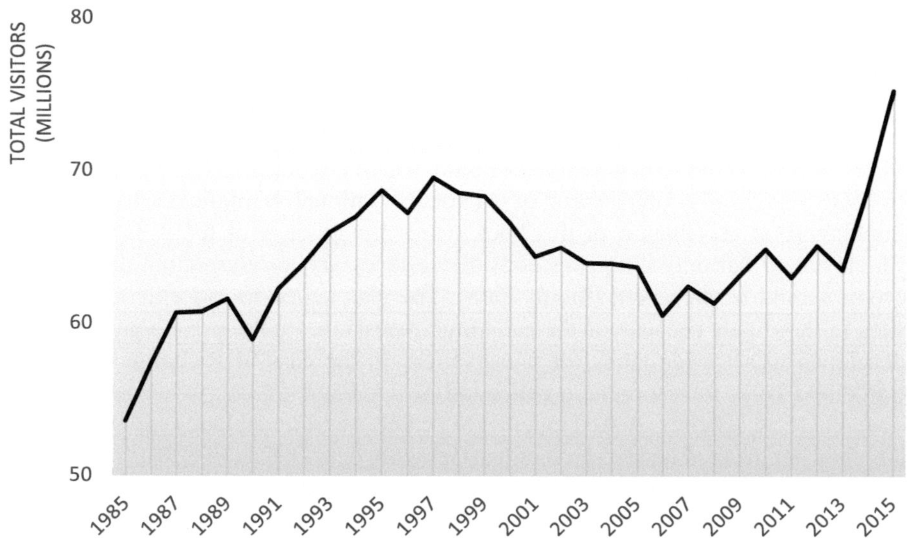

Fig. 2.6 National Parks visitation numbers highlighting increase in the 1990s (Data: NPS Stats (https://irma.nps.gov/Stats/Reports/Park))

desire to create a nature-focused experience geared toward learning about the ecosystem and nature while simultaneously protecting the landscape and nature.

One major impetus for NPS's embrace of ecotourism was the growing market share for this type of leisure. Visitation numbers soared in the 1990s with a 10.6-million-visitor increase between 1990 and 1997 (Fig. 2.6). The park service devised and applied a framework called Visitor Experience and Resource Protection (VERP), which served to protect natural resources while applying minimal restrictions on the visitor experience (NPS 2015a, b). Accordingly, NPS set out to identify best practices for assessing the carrying capacity of parks, in other words the type and number of visitors the parks could manage without adversely affecting park resources, and handling ecotourism (NPS 1997). NPS's embrace of ecotourism offered opportunities for nature observation, nature-centered outdoor activities, and extensive interpretive programs. This corresponded with an approach toward infrastructure development where facilities were designed to blend into the natural landscape (Department of State 2002). To track impact, NPS created a central database that would help managers compile indicators and standards for visitor impact in their parks. This helped set best practices for tourism and visitor use at the park-specific level (Rees et al. 2006). And interest in ecotourism is not letting up. It is the fastest growing sector of the tourism industry today (Juutinen et al. 2011).

2.10 Battlefields: Historic Preservation Meets Environmental Protection

While the development of ecotourism remained primary, the NPS continued to adjust its approaches to historic preservation, which continues to impact battlefields. In 1980 with the establishment of Antietam National Battlefield (discussed in Lookingbill et al. Chap. 6 of this book) came the passage of the Antietam Plan, which stated that only core sites considered necessary to interpreting the historic events should be preserved (Smith 2008). The plan called for prioritizing specific sites in large part because of the economic recession. Public-private partnerships developed to set aside lands for preservation. In the case of Richmond National Battlefield Park, for example, a collection of interested private parties along with state and local officials worked together to preserve additional lands around the city that they determined needed protection (Geraghty et al. 1993; NPS 2017b).

By the late 1980s and early 1990s, battlefields came back into vogue with an emerging heritage tourism industry (Zeppel and Hall 1991). The existing sites received increased investment as tourism officials, historians, and advocacy organizations partnered. Begun in 1991 by the Secretary of the Interior and authorized by Congress in 1996, the American Battlefield Protection Program continues today to identify battlefields in the United States and promote their preservation. The program's priority is focused "primarily on land use, cultural resource and site management planning, and public education" (NPS 2017a). The government has not approved new national sites, but over 14,000 acres have been added to existing battlefields (Fig. 2.7). Richmond National Battlefield Park and Shiloh National

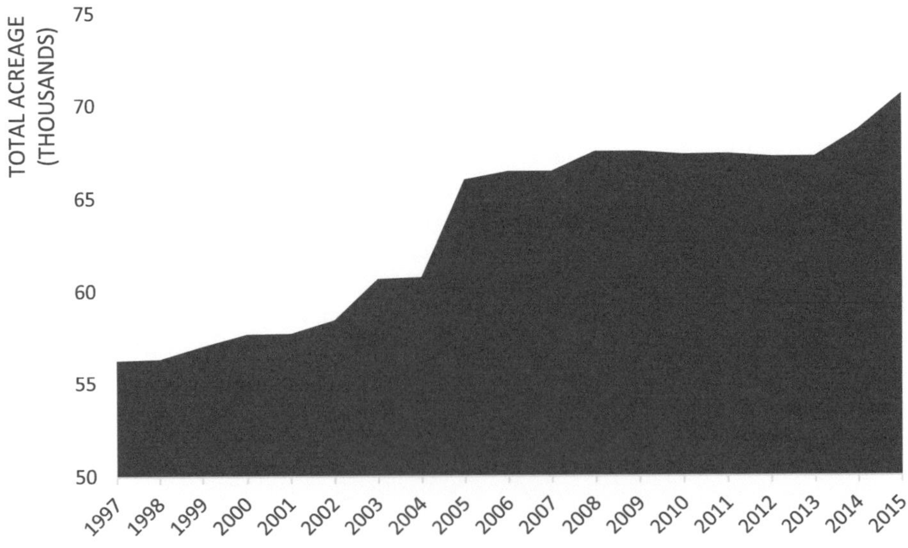

Fig. 2.7 National Battlefields acreage of established parks through time (Data: NPS Stats (https://irma.nps.gov/Stats/Reports/Park))

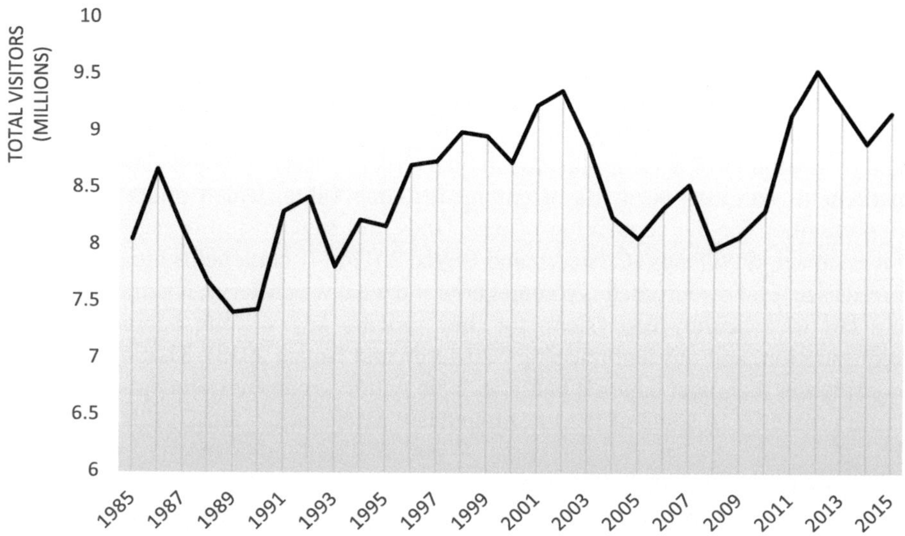

Fig. 2.8 National Battlefields visitation numbers increasing as of 2009 (Data: NPS Stats (https://irma.nps.gov/Stats/Reports/Park))

Military Park constitute the largest additions with 7111 and 3949 acres respectively. Most of the activity has centered around Civil War sites; more than two out of three of the battlefields were from the Civil War. Special interest groups such as the Civil War Trust have stepped in and became a driving force in battlefield preservation along with other grassroots organizations; the Civil War Trust alone has protected 45,839 acres in 23 states with a focus on American Revolution, Civil War and War of 1812 sites Civil War Trust. (n.d.). The focus on battlefields has resulted in increased visitation in recent years (Fig. 2.8).

A reframing of these sites as worthy of environmental protection has also come as visitation increases and urban areas continue to expand and threaten to destroy battlefields that used to be protected because of their rural geography. Urban battlefield parks such as Richmond National Battlefield are often comprised of small parcels of land and deeply impacted by surrounding lands that are out of the parks' control (DuVal 2007). The increased human development around these sites presents a multitude of environmental issues for battlefield parks such as water pollution, invasive species, and sediment runoff. Environmentalists have turned their attention to these sites as places for environmental preservation, supported by an increasing consensus that these parks are a critical safe haven for biodiversity and the ecosystem (Smith 2008; Geraghty et al. 1993; NPS 2009). Attention to such issues has renewed efforts to protect larger swaths of land and brought attention to the ecoservices offered by battlefields alongside their cultural value.

2.11 The Future

We argue that the rationale for their preservation risks concealing conservation opportunities for battlefields. The ideas about conservation, historic preservation, and recreation have been applied to national parks versus battlefields obscures how battlefields can also be a site of natural resource value, which offers promising opportunities for the acquisition of new lands and stewardship of ecosystem services on acquired lands (Cristescu and Boyce 2013).[5] If battlefields are also recognized as natural resources, this changes how and what we preserve. Natural resources are key to a healthy ecosystem, for they provide key wildlife habitats, carbon sequestration, and nutrient trapping (Cristescu and Boyce 2013). This is important considering that most national battlefields lie within urban environments and are in close proximity to some of the most populated metropolitan areas. Although small in size — compared to national parks — national battlefields received over half a million visitors per acre in 2015.

Protecting natural resources in urban parks requires continued resource management and regular inventory of resources. It also requires that the communities in charge of surrounding lands work with the NPS to understand how their land use policies can have an impact on these public lands and vice versa (NPS Advisory Board 2001). These urban greenspaces provide crucial habitat for wildlife while acting as outdoor enjoyment and recreational opportunities. Failure to recognize the significance that these battlefield parks have for our environment could exacerbate the ecological degradation already occurring in urban ecosystems. Furthermore, thinking about these sites and how they play into future political climates is an important consideration. If legislation like the Antiquities Act is rolled back, this could have dire ecological consequences. Understanding the importance of these lands' historical meaning alongside their ecosystem services can help protect these lands and help contemporary urban communities thrive.

Acknowledgements We would like to thank our colleagues, friends, and family for their support. This chapter would not have been possible without the encouragement of Rob Nelson, Director of the Digital Scholarship Lab. He generously provided crucial feedback on the drafts throughout the process. Nathaniel Ayers also helped develop an innovative design for a digital project based on this chapter, and to which we are grateful. The University of Richmond's DSL is a special place to be a part of. We also want to thank Tiffany Madron and Taylor Arnold who patiently listened as we talked through ideas. Finally, a special thanks to Todd R. Lookingbill and Peter D. Smallwood for organizing these efforts and asking us to be a part of this book.

[5] For an example of this approach as applied to the national parks, see "National Resource Condition Assessment" NPS. 2013. https://nature.nps.gov/water/ProgramBriefs/NRCAProgramBrief2013.pdf

References

Blackhawk, N. (2008). *Violence over the land: Indians and empires in the early American West*. Cambridge, MA: Harvard University Press.

Blight, D. W. (2003). *Race and reunion: The Civil War in American memory*. ACLS Humanities E-Book.

Carr, E. (2007). *Mission 66: Modernism and the national park dilemma*. Amherst: University of Massachusetts Press.

Civil War Trust. (n.d.). *The land before the National Park Service began*. https://www.civilwar.org/learn/articles/land-national-park-service-began?referrer=https%3A//www.google.com/. Accessed 25 Aug 2017.

Cook, R. (2007). *Troubled commemoration: The American Civil War centennial, 1961–1965* (Making the modern South). Baton Rouge: Louisiana State University Press.

Cristescu, B., & Boyce, M. S. (2013). Focusing ecological research for conservation. *Ambio, 42*, 805–815. https://doi.org/10.1007/s13280-013-0410-x.

Demars, S. E. (1990). Romanticism and American national parks. *Journal of Cultural Geography, 11*, 17–24. https://doi.org/10.1080/08873639009478434.

Department of State. The Office of Electronic Information, B. of P.A. (2002). *Overview of sustainable ecotourism in the United States of America*. https://2001-2009.state.gov/g/oes/rls/or/19412.htm. Accessed 25 Aug 2017.

Dilsaver, L. M. (Ed.). (1994). *America's national park system: The critical documents, second edition*. Lanham: Rowman & Littlefield.

DuVal, K. (2007). *The native ground: Indians and colonists in the heart of the continent* (Early American studies). Philadelphia: University of Pennsylvania Press.

Geraghty, G. B., Boge, M., & McPherson, J. M. (1993). *Paving over the past: A history and guide to Civil War battlefield preservation*. Washington, DC: Island Press.

Higham, J. E. S. (Ed.). (2007). *Critical issues in ecotourism: Understanding a complex tourism phenomenon*. Amsterdam/Boston: Butterworth-Heinemann.

Juutinen, A., Mitani, Y., Mäntymaa, E., Shoji, Y., Siikamäki, P., & Svento, R. (2011). Combining ecological and recreational aspects in national park management: A choice experiment application. *Ecological Economics, 70*, 1231–1239. https://doi.org/10.1016/j.ecolecon.2011.02.006.

Keiter, R. B. (2013). *To conserve unimpaired: The evolution of the national park idea*. Washington, DC: Island Press.

Keith, J. (2004). *Rich man's war, poor man's fight: Race, class, and power in the rural South during the first world war*. Chapel Hill: University of North Carolina Press.

Kline, B. (2001). *First along the river: A brief history of the U.S. environmental movement* (4th ed.). Lanham: Rowman & Littlefield Publishers.

Lee, R. F. (1973). *The origin and evolution of the National Military Park idea*. National Park Service. https://www.nps.gov/parkhistory/online_books/history_military/nmpidea2.htm. Accessed 25 Aug 2017.

McManamon, F. P. (2000). *NPS archeology program: Antiquities Act of 1906*. https://www.nps.gov/archeology/tools/laws/antact.htm. Accessed 25 Aug 2017.

National Park Service. (1965). *Mission 66 visitor centers (introduction)*. https://www.nps.gov/parkhistory/online_books/allaback/vc0.htm. Accessed 25 Aug 2017.

National Park Service. (1997). *VERP: The visitor experience and resources profession framework a handbook for planners and managers*. https://www.fs.fed.us/cdt/carrying_capacity/verphandbook_1997.pdf. Accessed 31 Aug 2017.

National Park Service. (2009). *Strategic plan for natural resource inventories: FY 2008-FY 2012*. Natural Resource Report NPS/NRPC/NRR — 2009/094. National Park Service, Fort Collins, CO.

National Park Service. (2015a). *Creation of Gates of the Arctic National Park & Preserve – Gates of the Arctic National Park & Preserve*. https://www.nps.gov/gaar/learn/historyculture/creation-of-gates-of-the-arctic-national-park-and-preserve.htm. Accessed 25 Aug 2017.

National Park Service. (2015b). *Sustainable ecotourism and visitor capacity – Big Cypress National Preserve*. https://www.nps.gov/bicy/learn/management/sustainable-ecotourism-and-user-capacity-2.htm. Accessed 25 Aug 2017.
National Park Service. (2017a). *About us – American Battlefield Protection Program*. https://www.nps.gov/abpp/aboutus.htm. Accessed 11 Sept 2017.
National Park Service. (2017b). *Environmental management – Richmond National Battlefield Park*. https://www.nps.gov/rich/learn/nature/environmental-management.htm. Accessed 25 Aug 2017.
National Park Service. (2017c). *Organic Act of 1916 – Great Basin National Park*. https://www.nps.gov/grba/learn/management/organic-act-of-1916.htm. Accessed 25 Aug 2017.
National Park Service Advisory Board. (2001). *Rethinking the national parks for the 21st century*. https://www.nps.gov/policy/report.htm. Accessed 25 Aug 2017.
Payne, C. M. (2007). *I've got the light of freedom: The organizing tradition and the Mississippi freedom struggle* (2nd ed.). Berkeley: University of California Press.
Rees, M., Cahill, K., Safford, M., & Rice, H. (2006). VERP, LAC, VIM, VAMP: A database that compiles user-capacity indicators and standards on the Web. *Park Science, 24*, 38–40.
Sellars, R. W. (1999). *Preserving nature in the national parks: A history*. New Haven/London: Yale University Press.
Smith, T. B. (2008). *The golden age of battlefield preservation: The decade of the 1890s and the establishment of America's first five military parks*. Knoxville: University of Tennessee Press.
Spence, M. D. (1999). *Dispossessing the wilderness: Indian removal and the making of the national parks*. New York: Oxford University Press.
Sugrue, T. J. (2008). *Sweet land of liberty: The forgotten struggle for civil rights in the North*. New York: Random House.
Taylor, D. E. (2016). *The rise of the American conservation movement: Power, privilege, and environmental protection*. Durham: Duke University Press Books.
Turner, J. M. (2009). "The specter of environmentalism": Wilderness, environmental politics, and the evolution of the new right. *The Journal of American History, 96*, 123–148. https://doi.org/10.2307/27694734.
U.S. Department of the Interior. (n.d.). *Park Reports – NPS Stats*. https://irma.nps.gov/Stats/Reports/Park. Accessed 25 Aug 2017.
United States (Ed.). (1968). *The civil war centennial: A report to the congress*. Washington, DC: U. S. Govt. Print. Off.
Zeppel, H., & Hall, C. M. (1991). Selling art and history: Cultural heritage and tourism. *Journal of Tourism Studies, 2*, 29–45.

Chapter 3
Registering English Battlefields: The Constructive Conservation of Historic Environments

Carly D. Sibilia, Geoffrey Carter, and Todd R. Lookingbill

Abstract The extensive history of warfare in England offers abundant opportunities for the conservation of historic land and its associated heritage and natural capital. Since 1995, the Register of Historic Battlefields has recognized English battlefields of exceptional historical value. To date, 46 out of at least 500 known English battlefield sites have been formally registered. While registered status provides a degree of extra protection against development, interpretation of undefined terms in planning legislation results in varied levels of conservation. Such challenges emphasize the importance of effective management, as registered sites can make significant contributions to local ecosystems as well as neighboring communities. In this study, we provide a brief history of English battlefield designation and conservation. We describe specific ecosystem services provided by battlefield landscapes through three case studies that vary along a management and land use gradient. For each location, we describe the types of ecosystem services provided, the management methods and goals, and the persisting threats. We also examine the potential differences in land cover for the registered battlefields and the surrounding landscapes. Despite differences in time of relevance, area, and habitat type, Naseby, Bosworth, and Maldon Battlefield each demonstrates the benefits of identifying and recognizing land historically associated with warfare. While formal recognition of the 46 registered battlefields conveys a certain level of appreciation and awareness, maintaining and improving the ecological condition of these sites is necessary to support regional biodiversity as well as surrounding communities and local economies.

C. D. Sibilia (✉)
Department of Biology, University of Richmond, Richmond, VA, USA
e-mail: carly.sibilia@richmond.edu

G. Carter
Battlefields Trust, London, UK

T. R. Lookingbill
Department of Geography and the Environment, University of Richmond, Richmond, VA, USA

T. R. Lookingbill, P. D. Smallwood (eds.), *Collateral Values*, Landscape Series 25,
https://doi.org/10.1007/978-3-030-18991-4_3

3.1 Introduction

As illustrated throughout this book, battlefield sites are often located in highly productive environments. These complex landscapes blend social and ecological histories, and through their careful stewardship, can provide varied ecosystem functions that serve to benefit local communities as well as regional biodiversity. Revealing the layered character of naturalized military landscapes emphasizes the historical and present-day narratives that embed these sites with meanings and values (Hourdequin and Havlick 2016). Recognizing these stories plays an important role in developing successful conservation strategies. In this chapter, we describe an approach to managing battlefields that pays tribute to their past and paves the way for their sustainable future. Constructive conservation is an adaptive and flexible approach to management that aims to remain faithful to history while carefully considering the diverse values of the land (Drury and McPherson 2008; Catling 2013). The principles of constructive conservation, established by Historic England and practiced by the organization since 2008, guide the sustainable management of historic environments throughout the country (Bruce-Lockhard 2008).

England is home to over 500 recognized battlefields, stretching from the Roman occupation of Britain to World War II (Fig. 3.1), that together exemplify the country's rich and expansive military history (Rayner 2007). Undoubtedly, additional skirmishes occurred for which no accounts survived. Nevertheless, the high density of battlefield sites scattered throughout the English countryside is extraordinary. This long record of warfare is a function of England's turbulent and well-documented history, which can be considered in several, distinct phases.

3.2 A Brief Overview of England's Military History

The surviving documentation of English military activity begins with the invasion and occupation of Britain by the Romans in the first century. Although many accounts of battle exist for the Roman period, extending into the fifth century, they are mostly insufficiently detailed to permit the reliable geolocation of the battlefield sites. The lack of adequate spatial data continues through the Early Middle Ages (410–1065), which were a period of almost continuous warfare dictated by waves of Anglo-Saxon and Viking invaders from Europe, alongside internal struggles for supremacy as the nation of England unified and came into existence. Accounts of these conflicts are usually inadequate to determine even the general location of the battlefields. Where identification has been attempted, these locations remain controversial and the subject of ongoing debate amongst historians. For example, the location of the battle of Maldon (991 A.D.), one of three case studies presented below, has been widely accepted but is not without challengers (Bessinger 1963; Foard 2003). Unfortunately, a robust battlefield archaeology, which could be used to verify battlefield locations, is absent for this period.

Fig. 3.1 England's 46 Registered Battlefields. Stars indicate case study sites. Data from Historic England (2017a): https://historicengland.org.uk/listing/the-list/data-downloads/. Accessed 13 Dec 2016

Beginning around the time of the conquest of England by the Normans under William the Conqueror in 1066 and extending into the Middle Ages, the written evidence for battles grows much more detailed and the identification of sites becomes considerably easier. The years following William's death in 1087 were characterized by sporadic rebellions and civil conflict as different factions fought to gain control of the

Crown, and as the leading families sought to define the limits of the monarch's power. English military activities eventually expanded to other parts of the island of Britain. Edward I set about conquering Wales with an invasion in 1276, and this was completed with the installation of his son (later Edward II) as the first English Prince of Wales in 1301. However, none of the battles of this campaign occurred on England soil.

Edward I also was determined to conquer Scotland. Much of the fighting from this conflict took place in England as the Scottish were frequent invaders of northern England, and several of the most well-known English battlefield sites (e.g., Halidon Hill 1333, Neville's Cross 1346, Otterburn 1388, Homildon Hill 1402, and Flodden 1513) relate to this long, drawn-out struggle, which began in earnest at the end of the thirteenth century and did not reach its conclusion until the middle of the eighteenth century.

For much of the medieval period, England fought many of its major battles on the continent of Europe, primarily in France as part of the Hundred Years War. This conflict stretched from 1347 to 1453 as successive English kings attempted to lay claim to the throne of France. At home, the fifteenth century saw a major domestic conflict, the Wars of the Roses. These wars, which were essentially a series of internecine blood feuds over who had the better claim to the throne of England, included a number of major battles, beginning with the first battle of St. Albans in 1455 and effectively concluding with the defeat of Richard III at the battle of Bosworth in 1485. Bosworth, another of our three case-study sites, represents to many historians the end of the Middle Ages in England and led to the rise of a new dynasty, the Tudors. It is worth noting that the location of almost no battlefield from the Middle Ages period is without suggestions of alternative sites. Even a battle as pivotal as Bosworth was recently found to be incorrectly located in its original designation (Foard and Curry 2013).

A few years earlier, the battle of Towton in 1461 had the dubious distinction of being the bloodiest battle fought on English soil with a death toll estimated at 28,000 out of 75,000 combatants. Towton also is significant in that it is the first English battle for which there is any substantial archaeological evidence that could be used to geolocate the site. A mass grave was discovered in 1996 containing the remains of the soldiers who died on the battlefield.

During the Early Modern Period that followed, there were few battles fought other than a number of doomed attempts to displace the newly established Tudors and occasional reoccurrences of Anglo-Scottish border warfare. Major hostilities resumed in the middle of the seventeenth century with the British Civil Wars.

The period from 1639 to 1651 saw the outbreak of armed conflict between the Stuart king, Charles I, and the forces of Parliament leading to the execution of Charles in 1649, the abolition of the monarchy, and England's only period as a republic, which lasted until 1660 when Charles' son was restored as Charles II. The exact number of battles fought during this short period depends on the particular definition of a battle used but by most accounts was well in excess of 50 (Brooks 2005). Many of these were relatively small, localized affairs. Of the major engagements, the battle of Naseby in 1645 was the pivotal moment after which the king's defeat became inevitable. The sites of these battles were generally well-recorded and changes in the nature of warfare with the use of firearms has provided useful battlefield archaeology to support the written record. This comes both in the form of recovered artifacts and changes to the landscape associated with warfare, such as the construction of earthworks.

With the end of the Civil Wars, the number of battles fought on English soil decreased sharply, concluding with the two failed uprisings in 1715 and 1745 by the surviving members of the Stuart dynasty. The last battle fought on English soil is often taken to be the battle of Clifton Moor in 1745 between the retreating Prince Charles Edward Stuart ("Bonnie Prince Charlie") and the English forces of the Duke of Cumberland.

Neither of the major World Wars of the twentieth century saw any fighting on English soil. Neither the Battle of Britain in 1940, fought in the skies over southern England, nor the Blitz, the German bombing of London and other major cities, were traditional ground campaigns, despite being significant military actions. A number of other activities associated with these wars, such as building trenches for training in World War I and the construction of airfields and defensive structures during World War II, were later memorialized, but these actions were not strictly battlefields. Smallwood and Lookingbill (Chap. 12 in this book) describe the evolution of warfare away from these formalized battlefields and comment upon the implications for future conservation efforts.

3.3 Conservation Policy and the Realization of Battlefield Significance

The English government began formally recognizing monuments and structures particularly representative of England's heritage in 1882 through the Ancient Monuments Protection Act. The Act provided the first state protection for ancient monuments in the UK; however, it was argued that for the legislation to be effective, an inventory of significant ancient and historical monuments and constructions must be compiled and sustained (Murray 2015; Brown 1905). The Royal Commission on the Historical Monuments of England, a government advisory body, was therefore established in 1908 to supplement the 1882 Act by documenting buildings, monuments, and landscapes of archaeological, architectural, and historical importance throughout England (Sargent 2001).

By 1933, the collection consisted of 273 sites all recognized for qualities particularly illustrative of England's extensive history (English Heritage 2016). The original list, compiled with the primary objectives of recognition and preservation, was at first confined to landscapes and ancient structures such as Stonehenge and Rievaulx Abbey. It would later come to include great stately houses and castles, but did not yet include any areas of conflict.

The conservation and protection of battlefields began in the mid-1990s and is marked by projects such as the Towton Battlefield Archaeological Survey, the founding of the Battlefields Trust, and the development of the English Heritage Register of Historic Battlefields (Ferguson 2013). Although battlefield landscapes were often acknowledged implicitly alongside ancient structures, buildings, and monuments, these new actions signified a formal effort to identify and protect English battlefields for the first time. The creation of the Battlefield Register and the advocacy for these sites by the Battlefields Trust represents an important transition in the government's relationship with these historic landscapes.

In 1999, English Heritage, then a statutory advisor on all aspects of the historic environment, merged with the Royal Commission on the Historical Monuments of England with the goal of improving management and increasing visitation to historic sites. By the mid-2000s, funds raised from visitors began to contribute to the maintenance and conservation of battlefield sites, and by 2011 the collection of sites produced an operational surplus (English Heritage 2016). In response to this success, the government agreed to transfer the responsibilities of the national heritage collection to a charitable trust. On April 1, 2015, English Heritage separated into two distinct entities: a charity that looks after the collection of sites open to the public, which retained the name English Heritage; and Historic England, a non-departmental public body that runs the statutory protection system, advises the government, and provides various support to a range of bodies and groups.

Despite the significant progress that has been made in identifying and managing England's battlefields, conservation efforts are regularly constrained by equivocal planning policies that continue to impede the effectiveness of battlefield protection.

3.3.1 The Ambiguities of Battlefield Protection

In England, the National Planning Policy Framework sets forth the government's economic, environmental, and social planning policies. The framework provides a system through which local people and their associated councils can construct regional and neighborhood procedures that reflect the needs and priorities of their communities (Department for Communities and Local Government 2012). Land management, including the development of historical landscapes, is in fact referenced in the current framework; however, vague and ambiguous language often prevents the continual protection of battlefields and other "designated heritage assets" (DCLG 2012). It is outlined in the planning framework that "great weight should be given" to the conservation of such assets when considering the impact of proposed development. Section 12 of the framework establishes that "significance can be harmed or lost through alteration or destruction of the heritage asset or development within settings," and that "substantial harm or loss" should require "clear and convincing justification" (DCLG 2012). The standards that define these categories of "harm" and "loss" are never clearly presented, and protection cases presented in favor of battlefield conservation are often lost. The Battlefields Trust's Statement on Planning directly expresses this concern:

> Unfortunately, the NPPF does not spell out what constitutes harm or substantial harm, particularly in the case of battlefields where development impacts on sight lines or micro ter-

> rain are vital to obtaining an understanding of the battle ... planning authorities do not always appreciate this, leading them sometimes to underplay the extent of harm resulting from development (Battlefields Trust 2012).

The position stands in reference to registered and unregistered battlefields. Each time a case arises, the Battlefields Trust and other invested entities must argue against poorly defined language that is subject to interpretation. While registered battlefields are technically listed as heritage assets that require consideration through planning procedures, the impact of development is often miscalculated or misunderstood. Current policy is therefore insufficient in providing the necessary protection for these important areas of historical and environmental significance.

3.3.2 Constructive Conservation

England's heritage is paramount to its success as a nation. Historical landscapes, specifically areas of significant conflict, engender a sense of identity, history, and place (Magnus 2015). Historic England maintains that such legacies provide a "powerful stimulus" for domestic and international tourism, attract businesses and worldwide enterprise, "inspire creative industries," and add value to the "national brand" (Magnus 2015). An appreciation of heritage has a profound effect on the way individuals within a society feel and behave, as heritage influences interpretations of the past and aspirations for the future. The physical evidence of warfare evident at English battlefield sites provides a looking glass into past human behaviors and forms a correspondingly important part of the historic environment (Foard 2008). There is substantial public and educational interest in battlefields and similar landscapes of conflict (Pollard 2003); however, "if the battlefield is to have more than commemorative value, than the character of its landscape at the time must be understood" (Foard 2008). Practicing and promoting the principle of constructive conservation, which supports the idea that heritage can act as a dynamic cultural force, has in turn facilitated the development of a society whose support and enthusiasm for the historic environment has never been greater (Magnus 2015).

Constructive conservation is a flexible approach to management that uses change to reinforce the unique features of a site, rather than diminish them. Centered on the core concepts of "repair, care, and sustainability," the aim of constructive conservation is to balance the preservation and adaptation of landscapes in a way that promotes their function and interpretation (Coskun 2015). While skeptics have expressed fear that this method has potential to weaken the protection of historic nature, the active and collaborative creation of new, intensely managed ecosystems instead maximizes land-use in an ecologically viable fashion (Technische Universität Darmstadt 2013). This concept of change that accommodates for historical significance has had a valuable role in the development of conservation philosophy. The emphasis that constructive conservation places on innovative schemes that protect and enhance historical landscapes signifies their potential in terms of ecological, economic, educational, and cultural services.

3.4 Registration Criteria and Battlefield Designation

Battlefield sites are selected for designation based on a set of criteria that seeks to determine which battlefields best exemplify English heritage. The requirements outlined by the English Heritage Register of Historic Battlefields defined what battlefields were, as well as those factors that impact the level of special historic interest (English Heritage 2012). According to the selection guidelines, a site may be considered a battle only if it hosted "wholly or largely formed bodies of armed men, normally deployed and engaged under formal command" (English Heritage 2012). More specifically, the action must involve military forces present on each side in numbers comprising battalion strength (i.e. totaling c.1000 or more), organized in formal battle array (Foard 2008). The areas in which the troops deployed and fought while in battle formation define the boundaries of the battlefield. For management purposes, the immediate context of the battle must be properly delineated to include critical elements of subsidiary action alongside the main action. The Battlefields Registration Selection Guide (Historic England 2017b) includes the further specifications that siege sites, events of civil unrest or rioting, sites of aerial or naval bombardment, aerodromes, and bomb sites are not included in the Register of Historic Battlefields, but may be recognized through other designations.

Sites that qualify as battlefields are then assessed based on two primary criteria: historical significance and securely identified location. If the site of a battle is to merit registration, it has to have been an engagement of national significance that is capable of close definition on the ground (Historic England 2016). Although battlefields have frequently been the setting for crucial turning points in English history, most battlefields pose a challenge of how to locate events within their contemporary landscape. Only when the conflict has been accurately located can the terrain be used to understand the event itself. It is therefore necessary that the battle be placed accurately in context using the written and archaeological record (Foard 2008). However, secure and substantial archaeological evidence has yet to be retrieved from any English battlefield before the fifteenth century, and the comparative scarcity of knowledge of earlier periods of conflict is reflected in the Register entries (English Heritage 2012).

While historical significance and secured location are the key factors in determining a battlefield's merit for designation, other present features may add to the likelihood of registration, or grant increased significance to particular aspects of a site. For instance, *topographic integrity*, the survival of the character of the landscape at the time of the battle, is highly valued for its importance in interpreting the site in its historical context. The *archaeological potential* for the discovery of graves, structures, projectile scars or assemblages of bullets, arrowheads and personal effects, may also be considered, and archaeological evidence presents a unique perspective on the course and location of events. *Research potential* for further interpretation, especially in older sites, is also derived from documentation, including contemporary chronicles and preserved state papers, as well as letters, memoirs, accounts, biographic associations, and commemorations, all of which enhance the historical significance of a site by supplementing explanations of the actions that took place (Historic England 2017b).

From 1994 to 1995, English Heritage assessed 71 actions for inclusion on the Register; 43 qualified for inclusion based on the criteria, and were subsequently added to the Register. Eight sites were denied inclusion due to condition, while five failed to meet sufficient location accuracy. The remaining locations were discarded, as they did not meet the classification qualifications, and could not properly be labeled as battles (Foard and Morris 2012). Since the original evaluations in 1995, three more sites, including Edgcote and two at Lostwithiel, were added to reach the current total of 46 Registered English Battlefields.

The establishment of the Battlefields Register in 1995 was an important step in the conservation of English battlefields; its purpose is to offer protection through the planning system and to promote a better understanding of battlefield significance alongside public enjoyment (Historic England 2016). Unfortunately, ambiguous planning guidelines often inhibit the effective conservation of these sites, and many of the battlefields are still vulnerable to threats including development, contamination, and cultivation (Marsh 2016). However, when preserved and managed in a constructive fashion, English battlefields offer valuable ecological services on local and national scales.

In the sections below, we describe the approach to constructive conservation as it is applied to three historic English battlefields that vary in management focus (Table 3.1). For each site, we describe the ecosystems present and document the range of ecosystem services provided. We illustrate the different management strategies present at each site to highlight the adaptability of the constructive conservation methodology. We address environmental threats and concerns associated with each battlefield and quantify the changes in land cover type within and around each site over a 17-year period from 1990 to 2007. We end with some general lessons for landscape management in an effort to provide guidance for future conservation of historical landscapes throughout England and the world.

Land cover data for the analyses were acquired through the Centre for Ecology and Hydrology for the years 1990 (Fuller et al. 1994), 2000 (Fuller et al. 2002), and 2007 (Morton et al. 2011). Each land cover map comprised a digital dataset providing classification of land cover types at a 25 m resolution. The three land cover datasets varied in number of classes, class names, and projections. The data were reprojected in ArcGIS 10.4 to the original 1990 projection, and then the descriptions of the different land cover classes for the 3 years were compared in order to create aggregate classes. Aggregate class categories (see Table 3.2) were derived from those provided in Table 2 of the "Countryside Survey: Land Cover Map 2007 Dataset Documentation" (Morton et al. 2011). Three-kilometer buffers were then created around the three case study sites (Fig. 3.2), and land cover change was quantified for each case study battlefield and buffer for the 17-year interval.

Table 3.1 Attribute comparison of the three case study sites

	Naseby	Bosworth	Maldon
Size	452 ha	1072 ha	42 ha
Focus	Education	Community	Ecology
Site Features	Viewpoints Interpretive panels Battlefield trail	Battlefield center Battlefield trail Viewpoints	Battlefield trail Bird watching huts
Valuable Ecosystems	Hedges Wetlands	Grasslands Woodlands Wetlands	Mudflats Salt marshes Estuary Intertidal zone
Leading Management	The Naseby Battlefield Project	Leicestershire County Council	The National Trust
Persisting Threats	A14 roadway Lack of hedge maintenance Windfarm construction	Landowner intentions Modern agriculture Development	Land claim Waste pollution Sea level rise

Table 3.2 Aggregation of land cover classes across the three Land Cover Map (LCM) datasets (1990, 2000, 2007). Aggregate classes derived from LCM2007 Dataset Documentation (Morton et al. 2011)

Aggregate Class	LCM 1990	LCM 2000	LCM 2007
Broadleaf Woodland	Broadleaf Woodland	Broadleaf Woodland	Broadleaf Woodland Felled Forest
Coniferous Woodland	Coniferous Woodland	Coniferous Woodland	Coniferous Woodland
Agriculture	Arable and Horticulture	Arable Cereals Arable Horticulure Arable Non-rotational	Tilled Land Scrub/Orchard Mown/Grazed Turf
Grassland	Improved Grassland Calcareous Grassland Neutral Grassland Rough Grassland Acid Grassland Fen, Marsh, Swamp	Improved Grassland Calcareous Grassland Neutral Grassland Setaside Grassland Acid Grassland Fen, Marsh, Swamp	Moorland Grass Meadow/Verge Meadow Rough/Marsh Grass Grass Heath
Saltwater	Saltwater	Saltwater	Saltwater
Freshwater	Freshwater	Water (Inland)	Inland Water
Bare Ground	Inland Rock	Inland Bare Ground	Inland Bare Ground Ruderal Weed
Coastal	Littoral Rock Littoral Sediment Supra-littoral Rock Supra-littoral Sediment Salt Marsh	Littoral Rock Littoral Sediment Supra-littoral Rock Supra-littoral Sediment Salt Marsh	Beach and Coastal Bare Salt Marsh
Developed	Suburban Urban	Suburban/Rural Development Continuous Urban	Suburban/rural Development Continuous Urban

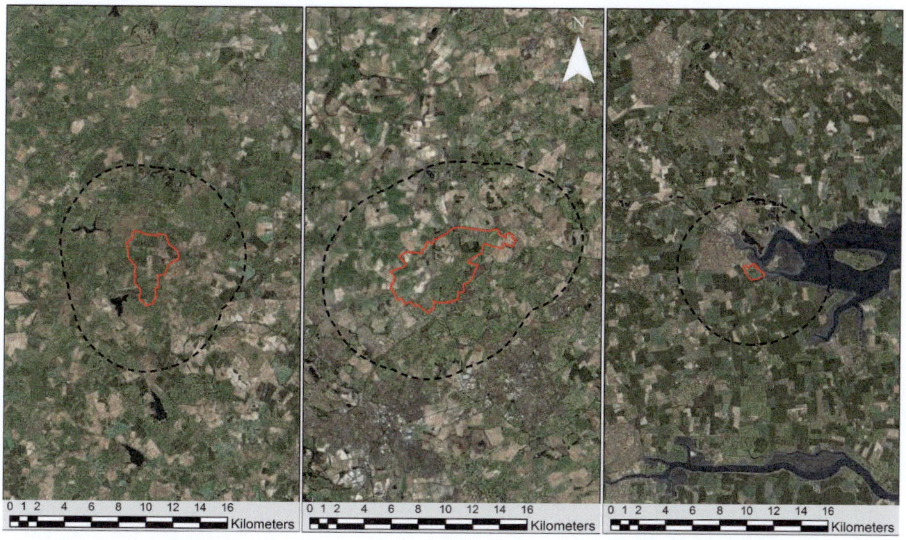

Fig. 3.2 Landscape heterogeneity in and around case study sites: Naseby, Bosworth, and Maldon (shown from left to right). Red boundaries indicate the Registered Battlefield border. Dotted lines represent a 3-km buffer. Boundary lines from the Battlefields Trust and imagery from Google Earth (1 October 2016)

3.5 Naseby Battlefield

Naseby Battlefield (52°24′31″ N, 0°59′54″ W), which marks the most important battle of the First Civil War, is replete with interpretive features that allow visitors to understand the relationship between the landscape and its history. Fought on June 14, 1645, the Battle of Naseby was the decisive engagement of the British Civil Wars between the main Royalist army of King Charles I and the Parliamentarian New Model Army commanded by Sir Thomas Fairfax and Oliver Cromwell. At the conclusion, practically the whole of the King's infantry was either killed or taken, marking the end of any realistic chance of victory for the Royalists in the First Civil War, and assuring the supremacy of Parliament in England (Rayner 2007). The battle boundary lines established by English Heritage in 1995 include over 450 ha of land northwest of the village of Naseby within the county of Northamptonshire. While it is most popular for its recognition as a Registered Battlefield, the area also hosts a number of working farms alongside pond and hedgerow ecosystems with high levels of biodiversity. The interwoven nature of Naseby's natural and military histories contributes to the significance of the terrain itself, as well as the character of the many services it provides.

3.5.1 *Ecosystems & Services Provided*

The character of the terrain at Naseby is particularly important to its historical interpretation, as a number of the topographic features are understood to have affected the decisions of the commanders, thus altering the course of the battle. The most prominent example involves two ridges: Dust Hill in the north and Naseby Ridge in the south. The ridges run approximately east to west, but are not parallel to one another, so the valley between is wider in the west than the east. From the Royalists' position atop Dust Hill, marked today as Rupert's Viewpoint, they were unable to see the New Model Army concealed by Naseby Ridge. However, as the New Model Army headed north out of Naseby Village, crossing the ridge, they gained view of the entire field of battle, including the position of their Royalist enemies (M. Marix Evans 2016, personal communication, 5 August).

While these high ridges are certainly major features in Naseby's historical topography, also prominent today is a succession of low ridges revealed when the woodland was cut down and replaced by the strip allocation system of agriculture dating back before the 1645 battle. The undulations of ridges and furrows impeded the motion of cavalry during the battle, and served to screen the movements of the two armies from one another (English Heritage 1995a). The steepness of the slopes, then covered with scrub, coppice, and gorse, created a terrain intensely hostile to mounted maneuvers (Marix Evans 2014). While the battlefield area may appear superficially to be no more than plots of common agriculture, the military history of the landscape transforms the undulating fields into a setting that, in every detail, illustrates the happenings of the most significant battle of the British Civil Wars. The terrain, in its character and storytelling, provides a unique opportunity for place-based education that expands across topics of military history, English heritage, geology, and landscape ecology, with the potential to inspire a sense of national identity tied to both the battlefield and environmental conservation.

3.5.1.1 Hedges

The iconic Sulby Hedges are the most illustrative example of a biotic landmark that seamlessly weaves ecology and history (Fig. 3.3). Similar to the site's ridges and furrows, the distinctive vegetation is a defining landscape feature, as the battlefield boundary created by the robust foliage is still present today. The hedges were used at that time as a barrier to enclose or exclude animals, and as a protective force against marauders, but they also mark the area in which Cromwell posted his dragoons (English Heritage 1995a). Despite this seemingly uninviting history, the hedges play an important role in the present-day conservation of wildlife diversity. The thick foliage provides food and shelter resources, and also creates stretches of linear corridors across the countryside. Hedgerows function as exemplar corridor structural elements by (1) facilitating species movement, (2) acting as habitat for

Fig. 3.3 View from Fairfax's Viewpoint to Sulby Hedges in Naseby Battlefield. Image provided by Battlefields Trust: https://www.flickr.com/photos/thebattlefieldstrust/. Accessed 31 July 2018

certain species, particularly edge species, and (3) creating a barrier between adjacent fields (Forman and Baudry 1984). For example, hedgerows throughout lowland farming landscapes in Britain comprise one of the most important surviving elements of semi-natural habitat for birds (Hinsley and Bellamy 2000). In addition to providing cover for local and long-distance movement, hedge habitats promote nesting and act as roosting and foraging sites as well (Whittingham and Evans 2004; Davies and Pullin 2007; Wolton et al. 2013). Butterflies are another major beneficiary, with one review study finding that 64% of all British butterfly species have been recorded within hedgerows (Dover and Sparks 2000). Examples of notable species found within the Sulby hedges include the bullfinch (*Pyrrhula pyrrhula*) and the gatekeeper butterfly (*Pyronia tithonus*). Scrambling and climbing plants such as bittersweet (*Solanum dulcamara*) and hedge bingeweed (*Calystegia silvatica*) also rely on the hedge structures for survival (McCollin et al. 2000). These species in turn produce nectar and berries that support higher trophic levels, including the flesh-fly (*Sarcophaga carnaria*) and common garden spider (*Araneus diadematus*). At the ecosystem level, the vegetation supports its surrounding environment by stabilizing the soil and preventing erosion. Hedges serve as guides for contour cultivation, but also slow and disperse surface runoff (Dabney et al. 1999). Therefore, while the iconic hedges are deemed significant primarily for their presence in Naseby's history, they continue to provide measurable ecological benefits.

3.5.1.2 Wetlands

In addition to the Sulby Hedges, the Northamptonshire Natural History Society has recognized and provided interpretive features for two wetland ecosystems within the Naseby Battlefield borders. Tarry's Pond illustrates a focus on native ecology and the ecological potential of well-managed wetland sites. In the summer of 2011, the pond was cleared of decaying leaves and other natural debris in order to create a potential haven for wildlife in need of freshwater habitat. Since the cleaning, a number of species have colonized the area including frogs (*Rana temporaria*), toads (*Bufo bufo*), grass snakes (*Natrix natrix*), and one of the country's most distinctive native species, the peacock butterfly (*Inachis io*). Dragonfly Pond, an intermittent freshwater source, provides a similar service in its ability to host wildlife. Dragonfly nymphs and invertebrate larvae burrow down into the mud when the pond dries, but survive and re-emerge when water returns. The margin of the pond also supports distinctive marsh vegetation including sedges (glaucous sedge (*Carex flacca*)), rushes (jointed rush (*Juncus articulatus*), soft-rush (*J. effuses*), compact rush (*J. conglomeratus*)), and grasses (marsh foxtail grass (*Alopecturus geniculatus*)).

Both Tarry's Pond and Dragonfly Pond are contained within the borders of the registered Naseby Battlefield, yet they provide supporting and cultural ecological services unrelated to the historic label under which they fall. Compared to other freshwater environments such as rivers, streams, and lakes, ponds generally support the highest number of species and have the highest index of species rarity (Williams et al. 2003; Scheffer et al. 2006; Céréghino et al. 2008a; Davies et al. 2008). Although fish biomass is comparatively low, the high abundance of submerged vegetation increases the richness of aquatic birds, plants, amphibians, and invertebrates. This is especially the case in cultivated areas. Ponds within agricultural landscapes, man-made or natural, make a significant contribution to regional aquatic biodiversity (Céréghino et al. 2008a; Davies et al. 2008). In addition to increased species richness, ponds also provide supporting services such as nutrient interception and hydrological regulation. The many environmental benefits present equally valuable educational opportunities. Tarry's Pond and Dragonfly Pond can be examined as powerful model systems for studies in ecology, evolutionary biology, and conservation biology, and can even be used as sentinel systems in monitoring global change (Céréghino et al. 2008b).

3.5.2 Management

3.5.2.1 Ownership

The key to revealing the significance of Naseby's landscape lies in its interpretation. The layered history should be accessible to the modern, casual visitor as well as the scholar, and the responsibility of this great task has been divided among multiple participants over the years. In 1823, a large obelisk commemorating the battle was erected about a mile southeast of the battlefield (English Heritage 1995a). During the 1930s, the Cromwell Association installed a second memorial on the battlefield itself. In 1991, the Northamptonshire Archaeology Unit produced the first interpretive panels for the Battle of Naseby to accompany the two historic monuments (English Heritage 1995a). In 1995, the location was designated a Registered Battlefield by English Heritage with the hopes of introducing viewpoints and visitor access points throughout the privately-owned countryside. However, as news of such plans spread to the resident landowners, so did local panic and rumors of compulsory purchase of the land by the state. This miscommunication, which at the time blunted any sort of visitor access or interpretive development, led to almost 5 years of careful relationship building between the farmers and battlefield historians. While matters remain delicate, the relationship has greatly improved since the initial conflict in 1995, and the importance of landowner respect and consultation has been thoroughly acknowledged by those responsible for site conservation (M. Marix Evans 2016, personal communication, July 24).

In 2001, the Naseby Battlefield Project was established to raise funds for the refurbishment of the two monuments, as well as their integration into the Naseby Battlefield Tour, complete with interpretation panels, guided walks, and parking facilities. Through personal communication, public meetings, and open and honest conversation, lingering tensions with landowners were reduced, and interpretive development and public visitation finally began to flourish. In 2006, Chair of the Project Management Committee, Martin Marix Evans, told BBC news that "after five long years, the Naseby Project [had] truly begun" (BBC 2006). By 2007, the charitable company was established, and since 2008, it has been possible to visit the landscape and follow the events of that day using the Battlefield Trail (Marix Evans 2011). The project was funded in bulk by East Midlands Tourism, Biffaward, Northampton County Council, and the Sealed Knot, England's leading Civil War re-enactment society, but was also made possible by a grant from the Heritage Lottery Fund (Marix Evans 2011). Furthermore, alliances with the Field Studies Council, local primary schools, the Northamptonshire Natural History Society, and even the Royal Society for the Protection of Birds, have all contributed to the active interpretation of Naseby Battlefield for both its political and natural history.

3.5.2.2 Interpretive Features

The viewpoints, trails, and panels available at Naseby are designed to fully immerse visitors into the perspectives of the participating forces during the war. Viewers are encouraged to compare what can and cannot be seen at Fairfax and Rupert's viewpoints, respectively. Moving across the two ridges and through the Broadmoor Valley reinforces the limitations of a static viewing of a battlefield, and emphasizes that one must travel through the landscape to understand the perceptions of the combatants and their leaders (M. Marix Evans 2016, personal communication, 16 August). Independent military historian, and former trustee of the Battlefields Trust, Marix Evans reflects:

> My personal belief is that it is vital to travel, on foot or cycle, through the landscape, rather than attempt to understand it from a static viewpoint. Moving along, it is inevitable that you will pass crops, hedges, trees, and so forth, together with flowers and various sorts of cover. I have seen buzzards and red kites flying over Naseby field, as well as many songbirds.

> Rabbits abound, and one may spot the occasional fox or deer. Insects are numerous and pond life is visible, all because you are following in the footsteps of a 17th century soldier. Immersed in the terrain, you experience far more than what an academic study of a battle would give you. The battle and its landscape are unavoidably intertwined. The more versatile your appreciation of then and now, the richer your experience.

The interpretation boards located at the platforms are intended to help the visitor "make an independent valuation of the scene from the eye-level of [a mounted] officer" (Marix Evans 2014). From the elevated platform, viewers can personally assess the validity of the contested belief that the New Model Army initiated contact. An additional interpretation board at Rupert's Viewpoint provokes further analysis of the landscape, as it juxtaposes a modern map against the 1630 field map, highlighting the differences in terrain and biota. Other helpful interpretive features include three-dimensional diagrams, panoramas, and identifying flagpoles that are displayed when visitors are expected in larger numbers. The Battlefield Trail at Naseby is successful because it not only invites visitors to think thoughtfully and imaginatively about the site's history, but also because it encourages new and creative ideas about the way in which individuals interact with their environment. The installation of such features allows the historic terrain to benefit visitors on a recreational, educational, and emotional level.

Interpretative guides in battlefield parks provide a cultural service to the community by weaving together a national consciousness and helping visitors to develop a sense of place (Ryan 2007). Naseby Battlefield is largely undeveloped, with the key areas and views "unspoilt" (Rayner 2007). The potential of battlefield landscapes is realized when they inspire individuals across generations, rather than a narrow group of enthusiasts. The working members of the Naseby Project have recognized the benefits of broad outreach, and have subsequently devised projects and field days with local colleges and primary schools (Fig. 3.4). As an example of these efforts, an audio package project, devised with the Abbeyfield School, was the recipient of a regional award for Work with Children and Young People in 2010 (Marix Evans 2011). In that instance, primary children were prompted to consider both the natural and historical properties of the landscape under a constructive conservation-type framework by investigating the site's flora and fauna under the aegis of OPAL (Open Air Laboratories) and the Northamptonshire Natural History Society.

3.5.3 Threats

Despite the many successes in management of its historical and environmental resources, the site still faces a number of persisting threats. In the early 1990s, the A14 dual carriageway was built across the southern edge of the battlefield, separating the village of Naseby from the remainder of the battlefield (Rayner 2007). In 1995, when Naseby was added to the Battlefield Register, it was noted that "although [the road is] largely hidden by being set in a cutting, it is as yet uncertain how much

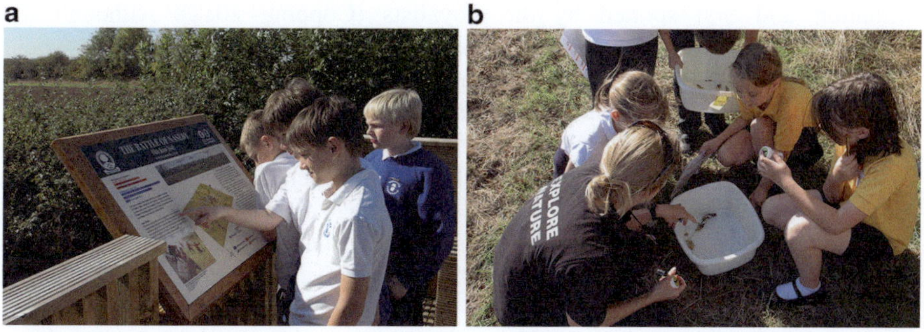

Fig. 3.4 Naseby Battlefield's educational ecosystem services. (**a**) local students reading the information panel at the Sulby Hedges Viewpoint; (**b**) students identifying collected insects during a "Natural History Day" assisted by OPAL (Open Air Laboratories). Images provided by the Naseby Battlefield Project, 2011

traffic noise will be audible in the battlefield area" (English Heritage 1995a). Later sources report consistently negative impacts of the road, describing it as the "most obvious intrusion…[that] cuts a swathe across the landscape," (Partida et al. 2013) and "an irreversible decision…to commit sacrilege across the battlesite of Naseby" (Chandler 1989). As reflected by these critiques, noise and compromised aesthetic have the potential to alter visitor experience and diminish the ecosystems services provided.

The damage is not restricted to the cultural sphere. Habitat loss and deterioration are often directly related to the expansion of roads and urbanization. The pervasiveness of transportation infrastructure in all European countries has fragmented ecosystems on an expansive scale, reducing core habitat area and connectivity (Torres et al. 2016). When road construction and associated human activity interrupt a historically and environmentally significant landscape, the ability of the landscape to provide a robust set of ecosystem services is jeopardized. In this instance, the construction of the A14 provoked a new dedication to the preservation of English battlefields that has led to protect those services. The controversial issue sparked the concern of historians as well as neighboring communities, and eventually resulted in the publication of the Battlefields Register and the establishment of the Battlefields Trust (Rayner 2007; Ryan 2007). While the primary focus of the Register and Trust was never explicitly to preserve ecosystems or biodiversity, that has been a collateral outcome.

Another ongoing threat is the lack of maintenance of the iconic bordering hedges. In their report on the historic environment of Naseby, the Rockingham Forest Trust describe the current state of the hedges as compromising the character of the battlefield and preventing the land from being thoroughly appreciated (Oreszczyn and Lane 2000; Partida et al. 2013). The potential construction of the Kelmarsh wind farm adjacent to the battlefield presents a similar threat to park aesthetics and ecology. The windfarm, if completed, would comprise four 415-ft turbines and two 397-ft turbines built in the area where the Parliamentarians were drawn into battle formation and first spotted by the Royalists (Copping 2012). Although the government-appointed planning inspector acknowledged that the turbines would "introduce another modern element into views" and "detract from the significance of the battlefield," it was ultimately concluded that "the degree of harm would be less than substantial" (Griffiths 2011). Word of the project brought forth heated objections from heritage conservation groups; however, leaders of the project ultimately decided, "in terms of the age of the designated heritage assets affected … [the disturbance] is relatively insignificant" (Griffiths 2011). While the central concern of those contesting the wind farm is historically based, massive wind turbines introduce a variety of threats to the area's ecology as well. The large size and extensive placement of turbines presents potential hazards to birds and bats, while the associated infrastructure including roads and transmission lines can cause further habitat fragmentation and provide avenues for invasion by exotic species (Kuvlesky et al. 2007).

Such conflict between competing uses of the land bring into focus the scope of research and high level of coordination required to preserve and utilize battlefields more effectively as a resource. Academic research about site military and natural history is vital but must be complemented by effective organizational structures and appropriate funding. While the passion for battlefield conservation and interpretation has been fervent since the creation of the Register, financial obstacles often present challenges for the charitable societies dedicated to preserving English heritage. Resolving these challenges can require creative partnerships. For example, in 2013, the Naseby Battlefield Project had hoped to purchase land overlooking the battle site in order to construct a new museum (BBC 2013). However, the project fell short of the £300,000 necessary, and the committee had to adjust planning based on available funds. A new joint project with Naseby Church was then established in order to facilitate the curation of a visitor center museum for the battle within a historic place of worship (BBC 2013).

3.5.4 Naseby Land Cover Assessment

The 2000 land cover data (Fuller et al. 2002) indicate that Naseby Battlefield is composed mainly of agricultural and grassland habitats with small patches of broadleaf woodland (Figure 3.5a). This is a landscape highly representative of the historical British countryside. Many of the small, intermittent man-made lakes and ponds created in the agricultural areas, as well as other natural wetland habitats, are not readily apparent in this coarse-grained dataset. The highly valued hedgerows between fields are also difficult to discern at this resolution. Put together, the wetlands and hedgerows provide a substantial portion of the ecological benefits within this largely agricultural landscape but are not easily identifiable using standard land cover data.

The comparison between the battlefield landscape and the surrounding 3-km buffer demonstrates the representativeness of the site to the greater region. The two areas of analysis are similar in composition. Both the battlefield and its buffer feature large swaths of agricultural land, comprising nearly 50% of both areas

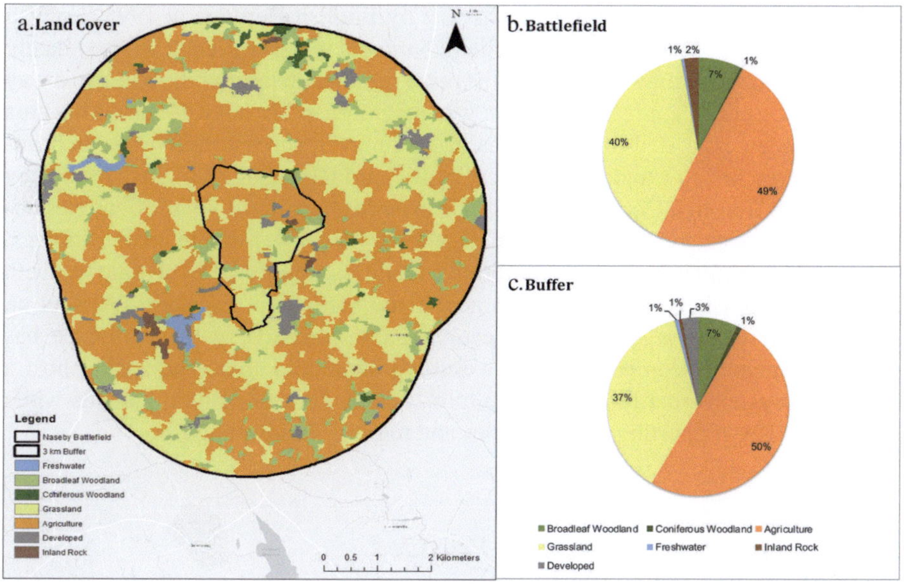

Fig. 3.5 Comparison of land cover in and around Naseby Battlefield. (**a**) Spatial distribution of land cover types has been converted to percentages within (**b**) the Battlefield and (**c**) the surrounding 3-km buffer. Data represent land cover breakdown in 2000 (Fuller et al. 2002)

(Fig. 3.5b, c). Percentages are also similar when assessing the less prominent land cover classes. It is apparent from these data that, at least at present, the land cover within Naseby Battlefield is characteristic of its surrounding and historical area. If these are landscapes that are deemed valuable, then battlefield identification, registration, and management through the tenets of constructive conservation can be a useful tool for their preservation into the future.

3.6 Bosworth Battlefield

The Battle of Bosworth stands alongside Naseby as one of the most iconic battles fought on English soil. The decisive battle on August 22, 1485, saw a dramatic military reversal in which the forces of Henry Tudor defeated a larger royal army led by Richard III, the last king of the House of York (Foard and Curry 2013). At the conclusion of the battle, Henry VII, the most favored alternative candidate for the throne, was crowned victorious, bringing to an end the dynamic struggle known as the Wars of the Roses (English Heritage 1995c; Ingram 2016). Today, Bosworth Battlefield (52°35′07.02" N, 1°25′35.72" W) represents what is "possibly the best-preserved battle site in the country" (Conduit 2004), despite the numerous contradictory theories regarding the location boundary set by English Heritage in 1995 (English Heritage 1995c). The original battlefield borders, which covered approximately 632 ha of land, were drawn to include the outer reasonable limit of the battle in a way that could be easily appreciated on the ground (English Heritage 1995c). This area has since been expanded to 1071.76 ha based on evidence revealed by an archaeological survey in 2009, and the discovery of Richard III's remains in 2012 (Hinckley & Bosworth Borough Council 2014). The Bosworth Battlefield Visitor Centre established by Leicestershire County Council in 1976 has, based on the recent findings, redeveloped its extensive indoor interpretive facilities, adjusted the 6.5-mile footpath trail, and established new viewpoints across the battlefield landscape (AFA et al. 2013; Conduit 2004). Although the evolving interpretation of the landscape has greatly changed its composition, the extensive battlefield area continues to host a mix of ecosystems that provide agricultural, economic, and cultural benefits while simultaneously supporting natural cycles and regional biodiversity.

3.6.1 *Ecosystems & Services Provided*

The only aspect of the 1485 Bosworth Battlefield terrain that can be confirmed with any certainty is the presence of a marsh, mentioned independently by more than one contemporary or near-contemporary writer (English Heritage 1995c). During the time of the battle, the ground is said to have been a flat plain, mainly composed of fenland crossed by streams with an area of peat marsh, known as Fen Hole (Ingram 2016). Since the conclusion of the battle, however, the landscape has seen a considerable amount of activity and development, with land improvement practices occurring as soon as the sixteenth century, and enclosure of the land beginning around 1600 (English Heritage 1995c). The underlying geology of the battlefield area, including mudstone, sandstone, and drift deposits of boulder clay, as well as sand and gravels, results in a varied and agriculturally productive landscape (AFA et al. 2013). In recent years, the battlefield area, farmed by over 12 separate landowners in its southern portion, has been awarded an Agricultural Land Classification of Grade 3: good to moderate quality (AFA et al. 2013). The land use is therefore predominately agricultural, including crops of wheat, barely, and fodder beans (AFA et al. 2013). Although modern farming methods have resulted in a decrease of biodiversity (Jeeves 2010), this tranquil piece of English countryside is still said to provide a "sense of identity and belonging for locals," while "supporting local business" and offering "a range of recreation and leisure interests" (AFA et al. 2013).

3.6.1.1 Grasslands

The Leicester, Leicestershire, and Rutland Biodiversity Action Plan (BAP) recognizes 19 priority habitats in need of conservation and restoration (Jeeves 2010), including field margins and neutral grass fields, two habitats present throughout the

Bosworth Battlefield area. Field margins are defined as buffer zones that mark the boundaries between fields, crops, and roads. These ecosystems provide important wildlife corridors between species-rich areas, reduce soil erosion, enhance crop pollinator populations, and support arable weeds, one of the most threatened categories of plant in the UK (Marshall and Moonen 2002). Seeds from arable weeds, such as fat hen (*Chenopodium album*), are an important resource for many species of farmland bird. Sympathetic management of arable field margins therefore provides support to species such as the grey partridge (*Perdix perdix*), corn bunting (*Emberiza calandra*) and skylark (*Alauda arvensi*).

Species-rich neutral grassland once dominated Leicestershire and Rutland (Fig. 3.6); however, this ecosystem has since been damaged by agricultural practices such as ploughing and reseeding, as well as the use of extensive fertilizer and herbicide treatments (Potts et al. 2009). Characteristic species of this environment, including common knapweed (*Centaurea nigra*), cowslip (*Primula veris*), pepper-saxifrage (*Silaum silaus*), yellow-rattle (*Rhinanthus minor*), adder's-tongue fern (*Ophioglossum vulgatum*), and green-winged orchid (*Orchis morio*), have all been in decline since 97% of the habitat was destroyed between 1930 and 1984. The abundance of flowering plants in these meadows is crucial for pollination as they provide a source of nectar for many invertebrates including the common blue (*Polyommatus icarus*) and meadow brown (*Maniola jurtina*) butterflies as well as

Fig. 3.6 Bosworth Battlefield's meadow and grassland ecosystems. Image provided by the Battlefields Trust: https://www.flickr.com/photos/thebattlefieldstrust/. Accessed 13 Dec 2016

the chimney sweeper moth (*Odezia atrata*) (Öckinger and Smith 2006). Redemore Plain, the likely location of the physical Battle of Bosworth, is described in a Landscape Character Assessment (AFA et al. 2013) as relatively flat, low-lying land featuring linear transport corridors as well as fragmented areas of species-rich meadow. Kendall's Meadow, within Redemore Plain, was designated a Site of Special Scientific Interest (SSSI) as it is valued for its grassland and orchids, and illustrates the biological and scientific benefits of open, grassy environments.

3.6.1.2 Woodlands

The variety of woodland ecosystems within Bosworth Battlefield provides a wide range of ecosystem services. Categorized into four main categories (lowland wood pasture and parkland, mature forest, broadleaved woodland, and wet woodland), these priority habitats regulate local climate and air quality, sequester and store carbon, prevent erosion, and provide habitat for a number of rare and diverse species (Valatin and Starling 2010; Freer-Smith et al. 1997; Milne and Brown 1997; Yan et al. 2003). Lowland wood pasture and parkland is characterized by open grown trees at various densities visible across a matrix of grazed grassland, but the primary interest in these sites, similar to mature forest, is for their ancient trees. English oak (*Quercus robur*) is typically ubiquitous on these sites, but other non-native species may be present as well. Most importantly, these live and rotting trees are incredibly valuable for fungi; the Bosworth habitat specifically hosts five Red Data Book (RDB) species of beetle, two RDB species of spider, and two RBD species of lichen. The mature forest ecosystem is similarly significant for its diverse lichen communities, which include at least 50 species that are known to be locally rare. Additionally, the hollow and decaying wood provides roosting sites for seven species of bat, as well as birds including the redstart (*Phoenicurus phoenicurus*) and barn owl (*Tyto alba*), both BAP species.

In addition to providing habitat for a range of key species, the mature and lowland wood areas are also responsible for the landscape's distinctive scenic quality (AFA et al. 2013). The Shenton Parkland area within the battlefield border is especially notable for its inclusion of these woodlands, as well as its strong recreational and interpretative values (AFA et al. 2013). The broadleaved and wet woodland habitats, although less accessible to the public, are equally significant due to their regulating and supporting services. Approximately 11% of the broadleaf woodland habitat in Leicestershire and Rutland was totally destroyed between 1930 and 1985. The relatively small percentage of woodland that remains includes an ecologically important ancient semi-natural woodland found only on sites that have been continuously wooded since before 1600 A.D. (Jeeves 2010). Nearly 50% of the broadleaf woodland habitat today is fragmented into parcels, each less than 10 ha in size. Wet woodland, although it similarly hosts stands of ancient semi-natural woodland, is differentiated by its particularly high water table and dominant alder and willow species. Together, what is left of the broadleaved and wet woodland habitats has developed into species-rich sites of high conservation value, housing populations of rare species such as the small-leaved lime (*Tilia cordata*) and the sessile oak (*Quercus petraea*).

3.6.1.3 Wetlands

Although the construction of the Ashby Canal between 1768 and 1804 severed the low-lying area of Redemore Plain, the waterway is still a historically and ecologically significant feature within the Bosworth landscape. In fact, the canal was designated a conservation area in 2009 for its historical significance, and included in the 2010 BAP for its importance as an ecological and recreational resource (NWLDC 2001; Hinckley & Bosworth Borough Council 2009; Jeeves 2010). The eutrophic standing water is nutrient-rich and particularly important for the nationally scarce grass-wrack pondweed (*Potamogeton compressus*) and the rare flat-stalked pondweed (*Potamogeton friesii*) (Jeeves 2010). Those who choose to walk the battlefield trail are directed along "an attractive part of the Ashby Canal which supports many species of aquatic wildlife" (Biggs and Biggs 2002).

The canal is not the only significant wetland feature, as the 2010 BAP also includes natural springs and flushes, reedbeds, and floodplains. The natural springs, present on steeper slopes such as Ambion Hill and King Dick's Well, occur where water wells up to the surface from underground aquifers, and are notable for their long history of hydrological continuity. Typical plants of springs include marsh-marigold (*Caltha palustris*), tussock sedge (*Carex paniculata*), great horsetail (*Equisetum telmataei*), soft rush (*Juncus effuses*), and marsh valerian (*Valeriana dioica*). Contrastingly, reedbeds are dominated by stands of common reed (*Phragmites australis*), and are notable in the UK for their support of four species of specialist birds: reed warbler (*Acrocephalus scirpaceus*), bearded tit (*Panurus biarmicus*), marsh harrier (*Circus aeruginosus*), and bittern (*Botaurus stellaris*). Floodplains, which encompass a range of wetland habitats, are not only important for wildlife, but for their role in natural filtration and flood alleviation as well.

3.6.2 Management

3.6.2.1 Ownership

Bosworth Battlefield, located near the town and civil parish of Market Bosworth, is owned in part and managed by the Leicestershire County Council (LLC) (Buhalis et al. 2012). Two other major landowners within the registered battlefield include the Shenton Estate and the Sutton Cheney Estate; the southern portion of the area hosts several private farms as well. The Heritage Centre and County Park are both run by the LLC on land leased from the two major estates, with a small strip along the disused railway owned entirely by the LLC (AFA et al. 2013). Daily maintenance and security of the landscape and associated footpaths is provided by the Ranger Team as a part of the Council's portfolio of county parks (AFA et al. 2013). As the area spans a range of natural, historic, and private land, development and management are often dependent on active cooperation between Natural England, English Heritage, the LLC, the Hinckley & Bosworth Borough Council, the Canal and Rivers Trust, and others (AFA et al. 2013).

Despite the various interests and priorities of stakeholders, the goals set forth in the 2013 Conservation Plan have been widely accepted by the associated parties. Included in this report are not only environmental and historic conservation strategies, but corporate aims as well (Hinckley & Bosworth Borough Council 2014). Specifically, the county looks to maintain the battlefield and surrounding suburbs as a vibrant place to work and live that will "empower communities" and provide "value for money and pro-active services" (Hinckley & Bosworth Borough Council 2014). The Conservation Plan highlights the special nature of the area resulting from "the interplay of many different qualities: physical historical evidence, historical value through continuity and landscape features; the aesthetic value of quintessential, peaceful English landscape, and communal value – the importance of a place giving people a sense of identity, a place for commemoration, as well as being a working agricultural landscape" (AFA et al. 2013). With such a wide scope of benefits, many of the stakeholders have recognized the importance of a holistic and sustainable approach that encourages visitors and supports the local economy, while reducing damage to the fabric of the countryside (AFA et al. 2013).

3.6.2.2 Attractions and Amenities

Two attractions that contribute to this overarching goal of active and holistic conservation are the visitor center and the accompanying battlefield trail. The Bosworth Battlefield Visitor Centre, the first of its kind in England, was established in 1976 in order to provide descriptions of the battle and explain its historical context (Rough Guides 2012). Compared to other on-site, interpretive features such as monuments and viewpoints, the Centre offers a more sophisticated approach that includes extensive visitor amenities such as a bookstore and restaurant, as well as indoor space for conferences, meetings, and educational visits (Buhalis et al. 2012). The Centre helps distinguish Bosworth from other English battlefields as a tourism destination site, with annual visitation rates peaking at 37,000 paying visitors a year (Ryan 2007).

The battlefield trail has a similar allure, as it provides opportunity for both environmental and historical education as well as recreation. The route begins where Richard III's armies were camped, runs over Ambion Hill, from which the whole battlefield can be surveyed, passes through the plains that saw the majority of the fighting, visits the alleged site of Richard's death, and passes close to where Henry Tudor's troops camped (Conduit 2004). The majority of the trail crosses fields and follows quiet footpaths. The grassy track follows a footbridge over a stream, passes through areas of woodland, and concludes along the Ashby Canal (Biggs and Biggs 2002). In an effort to diversify and increase visitation, the LLC has begun to reach out to older visitors in an attempt to facilitate multi-generational experiences between children and their grandparents (Buhalis et al. 2012). The renovated interpretative facilities, supported by the Heritage Lottery Fund in 2003, have developed an environment in which children and adults alike are able to discover how the Battle of Bosworth finally ended the feuding between the great houses of York and Lancaster (Biggs and Biggs 2002).

3.6.3 Threats

As noted, numerous landowners manage the area; while the 2013 Conservation Plan was seemingly well-received, separate entities have expressed different views and aspirations regarding the future of their land. These interests cover a wide scope, including farming, conservation, and educational, economic and environmental ambitions; however, if the area is to continue to provide its many, diverse services, it may be necessary to emphasize conservation more than it has been in the past. The land has been impacted by agricultural practice since the late Medieval and post Medieval periods, which each brought significant change to the landscape through field reorganization and the planting of hedgerows and trees (AFA et al. 2013). By 1807, much of the area had been enclosed, low-lying land was drained, and fences were grown up to establish property divides (English Heritage 1995c). As communities developed around the successful agriculture, the need for improved transportation infrastructure grew, resulting in the construction of the canal and subsequent railway. Although the area's rate of industrialization has been significantly reduced as preservation concerns have increased, threats to key habitats and species continue into the present.

Modern agricultural practices, which include fertilizer, pesticide and herbicide use, and mowing, all contribute to the decline of grassland and woodland species. Spray drift and eutrophication in particular have contaminated a number of local wildlife sites, causing impoverishment of woodland flora and a loss of native species (Jeeves 2010). Over-management is particularly damaging to lowland wood pasture and mature trees, as removal of deadwood through perceptions of tidiness is occurring at sites with high public use (Jeeves 2010). Water-based ecosystems face similar threats, as pollution by toxic chemicals causes a loss of fish and amphibians and the accumulation of poisons in the food chain. River straightening, dredging, and other flood defense works employed in an attempt to regulate natural systems have also resulted in a decline of priority habitats. Drainage of springs and groundwater extraction for agricultural irrigation also indicate a potential mismanagement of ecosystem resources. Constructive conservation of Bosworth Battlefield has resulted in a flourishing historical landscape commended for its perceptive and welcoming facilities; however, the ability of the land to provide its multitude of services relies on thorough and balanced management of its coupled human and natural systems (Smith et al. 2012). If the aforementioned threats continue to reduce priority habitats, the historic, cultural, and economic aspects of the interconnected landscape will likely become diminished.

In September 2018 Hinckley and Bosworth Council planning committee agreed to permit Horiba Mira to build a test track for electric vehicles on land which it owns on the western edge of the Bosworth battlefield. In the Battlefields Trust's view the development will cause substantial harm to the position where Henry Tudor first saw Richard III's army. This determined Henry's subsequent tactical decisions and approach to engage in battle. Despite identifying this approach as a key element of the internationally important registered battlefield, Historic England was unwilling to characterise the harm as substantial and this judgement appears to have been central to the planning committee's decision to pass the application (Battlefields Trust 2018). At the time of writing, this work is at a very early stage and it is not yet possible to form a clear view of the harm that will be caused. However, it serves as a reminder of the continuing pressure on battlefield sites from the threat of development.

3.6.4 Land Cover Assessment

In order to assess larger patterns in land cover usage and change, the aggregate classes shown in Table 3.2 were further simplified into three broad categories: agricultural, developed, and natural. The first category includes any land set aside for agricultural purposes, thus incorporating subclasses such as horticultural land, tilled land, and orchards. Developed land includes areas described as urban, suburban, or rural, while natural land includes areas of woodland, grassland, fresh and saltwater, and undeveloped coastlines. The purpose of this broader aggregation was to compare the expansion or reduction of each category's area over time.

Between 1990 and 2007, the overall changes in land cover are strikingly similar between Bosworth Battlefield and its surrounding 3-km buffer (Fig. 3.7). In both areas, agricultural land cover decreased and natural areas increased. This indicates a general trend towards the restoration of natural areas, perhaps through the conversion of abandoned agricultural plots to grassland and woodland areas. It is noteworthy that the increase in natural areas occurred concurrent with the site's addition to the Battlefield Register in 1995. Developed areas simultaneously decreased within both the battlefield and its boundary over the 17-year period; the reduction of developed area within the battlefield is slightly greater (13.5%) than that within the buffer (10.2%). This discrepancy, although slight, would be consistent with the acknowledgement of the historical significance of the battlefield landscape through the Battlefield Register. Nevertheless, the primary conclusion drawn from the land cover analysis is that the changes within the battlefield closely resembled those in the surrounding area, both following trajectories towards more natural landscapes.

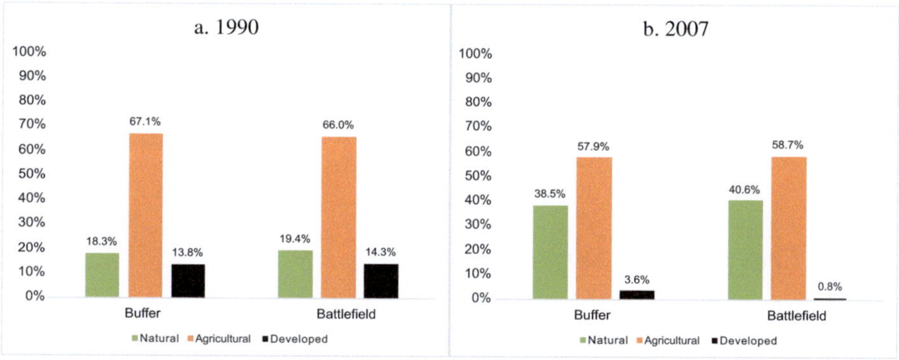

Fig. 3.7 Comparison of simplified land cover in and around Bosworth Battlefield in the years (**a**) 1990 and (**b**) 2007. Despite an overall increase in natural areas in both the battlefield and its buffer, the similar pattern of land cover change between the two years indicates a need for statutory protection of the battlefield area

3.7 Maldon Battlefield

The area designated by English Heritage as Maldon Battlefield (51°43′02.74" N, 0°42′05.58″ E) is often acknowledged more for its status as a Local Wildlife Site than for its historical significance. The Registered Battlefield borders officially include approximately 42 ha of mixed coastal farm and salt marsh along the western bank of the River Blackwater, but the neighboring Northey Island (51°43′19.74" N, 0°43′18.47″ E) is also understood to represent key components of the battle's landscape.

The Battle of Maldon occurred on August 10, 991 A.D. between the Anglo-Saxons and a Danish Viking Army. It is the oldest site included on the list of Registered Battlefields, and marks the first occasion on which the Vikings met resistance from an English army (English Heritage 1995b). The Viking's boats, commanded probably by Swein "ForkBeard" or possibly by Olaf Tryggvason, arrived in the Blackwater estuary and landed on Northey Island, hoping to raid the burh and mint at Maldon (Rayner 2007). Initially hindered by the high tide, the Vikings attempted negotiation; however, Brihtnoth, the Anglo-Saxon commander, refused to pay the invaders to depart and instead challenged them to battle. When the water retreated, the opposing forces met in savage hand-to-hand combat. At the conclusion of the battle, Saxon Ealdorman Brihtnoth had been slain, and the Viking Army was victorious; however, the loss in numbers had been so great the invaders returned to their ships and departed, rather than pursuing Maldon.

Despite the topographical indicators present in the Old English poem "The Battle of Maldon" that support the boundaries determined by English Heritage (Laborde 1925; Petty and Petty 1976), there is still considerable debate over the precision with which this historical event has been located (Nunn 2013). This controversy does not inhibit the services provided by this site, however, as both Maldon Battlefield and Northey Island are owned by the National Trust and managed mainly as wildlife sanctuaries. The coastal ecosystems present at these protected sites are therefore able to provide a combination of provisioning, supporting, and cultural services on a local and national scale.

3.7.1 Ecosystems & Services Provided

Maldon lies on the estuary of the River Blackwater, which flows into the North Sea. Northey Island (121.5 ha) and the Registered Battlefield (42 ha) compromise approximately 163 ha of mixed coastal farm and salt marsh that support a plethora of plant and animal life. Northey Island is accessible from the mainland only during low tides when the narrow causeway is exposed. Although coastal changes and the building of sea walls have altered some features of the original battlefield, it is still a rewarding site for visitors (Rayner 2007).

3.7.1.1 Mudflats

Mudflats are fine-grained habitats characteristic of estuaries and other protected, low-energy environments (Larsen and Doggett 1991). Such areas, where freshwater seepages provide transition from fresh to brackish conditions, support specialist invertebrate species that rely on particular habitat characteristics such as thin films of water or oxygenated surface layers of mud (EBP 2012). Mudflats and seepages also provide habitat for generalist invertebrates, especially those that have an aquatic stage in their life cycle. Due to the availability of such particular ecosystems, a number of Priority List invertebrate species can be found in the Maldon District. Notable species include the small heath butterfly (*Coenonympha pamphilus*), the white-letter hairstreak butterfly (*Satyrium w-album*), the starlet sea anemone (*Nematostella vectensis*), a saline lagoon specialist, and *Anisodactylus poeciloides*, a seed-eating ground beetle found at the margins of salt marshes and seawall seepages (EECOS 2007).

The expansive communities of invertebrates present throughout the intertidal mudflats are of vital importance to the thousands of migratory birds that pass through and overwinter in the area (EECOS 2007). International migrant birds, including the brent goose (*Branta bernicla*), dunlin (*Calidris alpina*), redshank (*Tringa totanus*), godwits (*Limosa lapponica*) and wigeon (*Anas penelope*), are all supported by the enormous biomass of invertebrates available in the mudflat ecosystems that exist on Northey Island and at Maldon Battlefield (Masero and Pérez-Hurtado 2001; EECOS 2007). These environments are crucial not only for their benefit to wildlife, but for their role in ecological succession as well. The development of saltmarsh vegetation is dependent on the presence of intertidal mudflats. It is therefore no wonder that the habitat types in Essex County with the most international designations are coastal mudflats and marshes (EBP 2012; EECOS 2007).

3.7.1.2 Salt Marsh

In the UK, the upper, vegetated portions of intertidal mudflats, particularly those that lie between mean high-water neap tides and mean high-water spring tides, are considered coastal salt marshes (EBP 2012). In total, the Essex County estuaries support approximately 2878 ha of salt marsh extending from the River Stour through to north Thames (EBP 2012). On its own, Northey Island provides 90 ha of this threatened habitat, as three-fourths of the island's area consists of salt marsh and muddy creeks (National Trust 2014). Present on the island are large assemblages of saltmarsh plants, from pioneer to upper marsh communities, and scarce species such as shrubby sea-blight (*Suaeda spp.*) and golden samphire (*Limbarda crithmoides*) (National Trust 2014). These coastal habitats provide important high-tide refuges and breeding sites to wading birds and wildfowl, and they act as food sources for passerine birds, especially in autumn and winter (EBP 2012). They also provide habitats for fishes, act as nutrient and sediment sinks, and establish coastline protection, and the restoration of salt marshes has been given particular attention (Colclough et al. 2005; Moller and Spencer 2002; Shepherd et al. 2005; Garbutt and Wolters 2008).

Unlike Northey Island, the Maldon Battlefield area has had the structure and composition of its salt marsh altered by grazing, which reduces the height of vegetation and the diversity of plant and invertebrate species (EBP 2012). Such changes have, in turn, favored species associated principally with agricultural systems, such as the corn bunting (*Emberiza calandra*), turtle dove (*Streptopelia turtur*), gray partridge (*Perdix perdix*), and skylark (*Alauda arvensis*), all of which are included under Essex County BAPs (EECOS 2007). The diversity of habitats in the coastal strip surrounding Maldon, including the pristine salt marsh and coastal grazing salt marsh alongside arable land, has allowed these species to flourish. Due to their immense value to bird, plant, and invertebrate communities, the Essex marshes as a whole are among the top five coastal wetlands in the country (National Trust 2014).

3.7.1.3 Estuary

Although Northey Island and the Registered Maldon Battlefield are important conservation sites in their own right, they both contribute to the significance of the internationally important Blackwater Estuary (Fig. 3.8). The Blackwater is not only a Site of Special Scientific Interest (with Local Wildlife Sites spread throughout), but also a Special Protection Area and a Special Area of Conservation because it has one of the largest and least disturbed areas of salt marsh in East Anglia (National Trust 2014). As mentioned previously, the estuary supports large populations of wintering birds, and Northey Island in particular becomes a highly valuable

Fig. 3.8 Maldon Battlefield's estuary and mudflat ecosystems. Julian Humphrys leads the Battlefields Trust members on a tour of Maldon's Battlefields Trust Walk. Image provided by the Battlefields Trust 2012. https://www.flickr.com/photos/thebattlefieldstrust/. Accessed 13 Dec 2016

high-tide roost when the floodwaters cover the inner Blackwater (National Trust 2014). The estuary has also been recognized for its role in preserving marine species; offshore the Blackwater has been designated as a Marine Conservation Zone, mainly due to its role as the national stronghold for a threatened species of native oyster (National Trust 2014).

The Blackwater also plays an important role in the local and national movement of wildlife populations. The rushing water of the estuary, which bisects much of the district with a broad, and to many species inhospitable, environment, can act as a significant physical barrier. But for others, the Blackwater serves as a corridor, bringing some of the most internationally important species into the heart of the district (EECOS 2007).

The estuary also provides a number of provisioning services. In addition to marketable seafood, the estuary provides a source of salt. The Maldon Crystal Salt Company, established in 1882, has been producing salt for more than 200 years and represents the only place in the country where sea salt is still manufactured (Christy 1906). Beyond providing a useful and plentiful product, Maldon salt has also contributed to the reputation of the area, as the town was at one point recognized as a center of the salt-making trade (Christy 1906).

3.7.1.4 Intertidal Zone

Isostatic adjustment since the glacial withdrawal at the end of the last Ice Age has caused the South East of England to slowly sink (English Heritage 1995b). This isostatic rebound, combined with contemporary global warming and sea level rise, has resulted in an appreciable increase of relative sea level since the Battle of Maldon occurred in 991 A.D., which has subsequently altered a number of the coastal landscapes. While the rise has led to the establishment of extensive intertidal salt marsh and flood plains, the trend may ultimately result in a loss of these crucial habitats. In response to this concern, Northey Island became the subject of the country's first monitored coastal realignment project in 1991 (National Trust 2014). As an experimental study, an 0.70-ha area was enclosed inside the existing sea wall and breaches were constructed to allow one in three tides to cover the new area. The intricately planned project provided critical insights that may inform future management (National Trust 2014). By adjusting to a changing environment and remaining mindful of the location's historic character, the Northey Island project exemplifies the successful application of constructive conservation. As the practice of managed realignment becomes more widespread, so will an increase in the amount of salt marsh and mudflat habitats across the district of Maldon (EECOS 2007).

3.7.2 Management

3.7.2.1 Ownership

The National Trust, short for the National Trust for Places of Historic Interest or Natural Beauty, is an organization in England, Wales, and Northern Ireland dedicated to the conservation of heritage. Given statutory powers through the National Trust Act of 1907, the independent charity aims to permanently preserve and protect the character of natural and historic sites for the benefit of the nation (National Trust 2013). The Maldon Registered Battlefield and Northey Island are both owned entirely by the National Trust. Other invested stakeholder organizations, including the Essex Wildlife Trust, the Maldon District Council, and Historic England, work in collaboration to help describe and preserve the biodiversity present. Protection of the land, however, is ultimately dictated by its ownership. The National Trust is actively pursuing eastward expansion of the Maldon footprint along the Blackwater Estuary so Northey Island and the adjacent battlefield grounds can continue to flourish as an ecological hub.

3.7.2.2 Visitation

Maldon Battlefield and Northey Island are advertised as wild and undeveloped. The sites are depicted as an area where one comes face-to-face with nature in its most dynamic forces: strong winds and fast-moving tides (National Trust 2014). At low tide, a rough causeway is revealed, surrounded on all sides by extensive marshland and mudflats. At high tide, the causeway is fully submerged, isolating visitors on the island to fully explore the coastal environment and its many unique inhabitants. Described as a birdwatcher's paradise, the island invites guests to observe the wildlife from special hides in order to pursue rare species without disturbing them (National Trust 2014). The seemingly inaccessible nature of Northey Island has allowed it to maintain an intriguing air of mystery. An annual "Castaway" camping event, during which friends and family are encouraged to partake in guided walks and geocaching, draws visitors who wish to experience Northey's peaceful solitude (National Trust 2014). Such events occur not only to entertain guests, but also to illustrate the local character of the land in terms of its assemblages of animals and plants (EECOS 2007). The ecological oddities and unique landscapes, including the battlefield, incite a desire to understand, conserve, and maintain the intrinsic character of the land (EECOS 2007). The attitudes produced by this understanding facilitate the creation of local biodiversity action plans, which allow for the further recognition and conservation of local distinctiveness, whilst simultaneously contributing to the goals of regional and national conservation objectives (EECOS 2007).

3.7.3 Threats

Although the National Trust prevents the direct influence of threats such as development within the Maldon landscape, there are persistent anthropogenic and environmental pressures that contribute to the degradation of its habitats at the boundaries. Centuries of over-exploitation, habitat modification, and pollution have led to a loss of estuarine and coastal habitats at the international level in terms of reduced area, biodiversity loss, and loss of ecological resilience (Lotze et al. 2006; Garbutt and Wolters 2008). In Essex County, the overall net saltmarsh area lost between 1973 and 1998 is 25%, which represents a loss of nearly 1000 ha at an average rate of approximately 40 ha per year (EBP 2012). In the past, a portion of this loss was attributed to land claim for industry and port facilities; however, today large-scale land claim schemes for agriculture are considered rare (EBP 2012). Comparatively, waste disposal is still relatively common alongside marina development on saltmarsh sites. Similar developments, such as coast protection works or the dredging of shipping channels, may also impact sediment dynamics, altering flow and weakening flood defenses (EBP 2012).

Other human influences that typically contribute to the decay of coastal environments include turf cutting, oil pollution, recreational pressures, agricultural improvement including re-seeding and draining, and eutrophication (EBP 2012). Erosion and "coastal squeeze" are additional stressors that are particularly pronounced in southeast England. Erosion of the seaward edge of salt marshes occurs widely in high-wave energy locations as a result of coastal processes (EBP 2012). Additionally, many salt marshes are quickly dissolved when "squeezed" between an eroding seaward edge and fixed flood defense walls. As the area of coastal environments continues to shrink, the risk of rising sea levels only increases. Because the mudflats that surround Northey Island's grass field are lower than those on the mainland, the island faces frequent overtopping during spring tides and storm surges (EECOS 2007). Increasing rates of sea level rise exacerbate the situation, resulting in the potential loss of freshwater habitats (National Trust 2014). Recognition of these pressures has led to deliberate coastal realignment at Northey Island, which is likely to be practiced more widely in the future and will hopefully contribute to the security of coastal habitats (EECOS 2007).

While the grassy grounds within the Registered Battlefield borders are less impacted by coastal degradation, these habitats are vulnerable to other pressures, such as human disturbance and agriculture encroachment along the borders of National Trust owned land. In the winter months, the fields on both the island and the battlefield are important feeding grounds and refuge for up to 5000 brent geese. The birds rely on a plentiful supply of short grass for their survival. The depletion of fine-scale mosaics of arable, pastoral, and semi-natural habitats, in tangent with an increase in agrochemicals, has resulted in a national decline of similarly important agricultural species.

3.7.4 Land Cover Assessment

Of the three case studies presented here, Maldon is unique in having a single land owner. This simplifies its management in some ways, and a comparative analysis of land cover change between Maldon Battlefield and its surrounding 3-km buffer clearly indicate the advantage of statutory protection in preserving and restoring natural ecosystems and their services. Owned and managed by the National Trust, the area contained within Maldon Battlefield is entirely protected from expansion of developed areas and agriculture. The surrounding area, however, has seen an increase in both its agricultural and developed areas from 1990 to 2007 (Fig. 3.9). While the battlefield has experienced an increase in natural areas of about 11% during the 17-year period, natural areas in the surrounding buffer have decreased by close to 13%. The diverging trends between the battlefield and its buffer were more apparent for Maldon than any of the other battlefield sites examined. The results

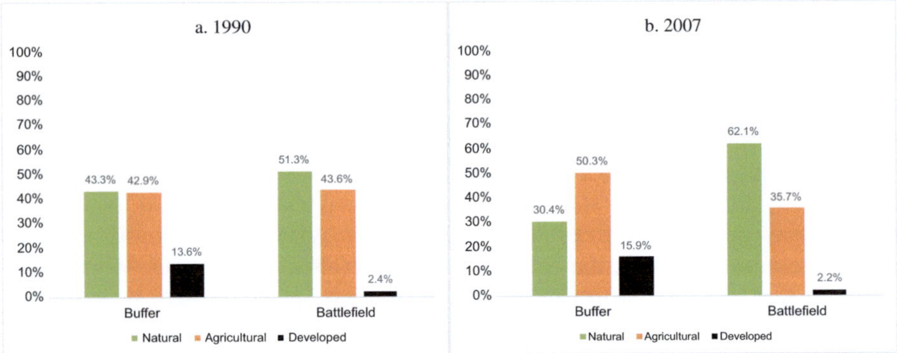

Fig. 3.9 Comparison of simplified land cover in and around Maldon Battlefield between (**a**) 1990 and (**b**) 2007. Natural areas within the battlefield borders increased while agriculture and developed areas decreased within the 17-year period. The opposite occurred in the buffer area, as natural areas decreased and agriculture and developed areas expanded. The pattern of divergence evident at Maldon demonstrates the importance of statutory protection in preserving natural areas and their ecosystem services

indicate that while the recognition conveyed by listing sites on the Battlefield Register alone is an important first step in conserving these ecosystems and their associated services, it seems that a clear system of ownership and management, similar to the U.S. National Battlefield Park system described in Madron and Tilton (Chap. 2 of this book), for example, still conveys the highest likelihood of long-term conservation success.

3.8 Conclusions

England's extensive military history has resulted in over 500 recognized battlefields spread throughout the country. In 1995, the Register of Historic Battlefields was established to recognize the importance of these landscapes. To date, Historic England has registered 46 sites. Although the registered sites are not awarded statutory protection, local communities and scholars have begun to acknowledge the many diverse values of these multi-layered landscapes. The opportunity to experience and understand nature and history simultaneously has resulted in increased communal worth of the lands, which in turn has inspired passionate movements for their more comprehensive management. From this movement, the concept of constructive conservation has emerged as a flexible and adaptive management style that remains faithful to history while considering the ecological value of the land. The approach has allowed a wide range of ecosystem services to flourish across sites like Naseby, Bosworth, and Maldon Battlefields.

Through its personable interpretive features, Naseby Battlefield connects viewers to their landscape in a manner that provokes not only a sense of pride and identity, but curiosity and creativity as well. The hedgerow and small wetland ponds increase landscape biodiversity, provide opportunities for research and education, and enhance social and cultural identity. A comparison of the land cover of the site to its surroundings indicates that the landscape is broadly representative of the historical British countryside, with an intermixture of grassland and agricultural patches. Notably, many of the small landscape features that provide a large number of the ecosystem services at the site are not resolvable on standard land cover products.

Bosworth Battlefield's Conservation Plan, developed in 2013, illustrates the results of purposeful and organized constructive conservation. Bosworth's ability to effectively manage overlapping historical and environmental themes is attributable to the funding provided by the Heritage Lottery Fund, as well as cooperation with private landowners and environmentally focused organizations. These partnerships acknowledge the need for a flexible management strategy that continuously facilitates intergenerational and cultural identity alongside the preservation of priority habitats. Perhaps in response to this collaborative effort among multiple stakeholders, land cover change over the past couple decades has followed similar trajectories inside the formally recognized battlefield boundary and within a 3-km buffer surrounding the battlefield. For both analysis areas, the amount of land in natural cover increased by about 20%.

In contrast, Maldon Battlefield represents a more traditional protected area model, where the entire landscape is owned and managed by a single entity, the National Trust. This coastal landscape commemorates the oldest battle recognized on the Register of Historic Battlefields. The site's value is clearly recognized as both historical and ecological, and its significance as an ecological hub and wildlife sanctuary has played a leading role in its acquisition and conservation. Here we see the effectiveness of control by a sole owner in protecting and even restoring natural areas and their historic character when surrounded by wide-scale habitat degradation and land cover change.

These examples illustrate the considerable range in conditions under which constructive conservation can be successfully applied. The approach is not a prescriptive mandate of management rules, but instead it is meant to inspire a creative and careful consideration of the multitude of potential regulating, supporting, provisioning, and cultural services delivered by the unique ecosystems at each particular site. The ecosystem services provided at each site, whether they are agriculturally influenced, highly heterogeneous, or dominantly coastal, should be inventoried, weighted, and properly accounted for in management decisions if true constructive conservation is to be achieved. This holistic approach provides a guide for cooperation between representative parties of scientists, naturalists, historians, and community members in order to prioritize the many ecological and cultural benefits of battlefield landscapes.

Acknowledgements We would like to offer special thanks to the Battlefields Trust for their generous support and overwhelming enthusiasm for this project and its goals. We also wish to acknowledge the invaluable contributions provided by Martin Marix-Evans, Glenn Foard, and Stephen P. Nunn, alongside a number of experts from the National Trust and Historic England. Assistance provided by staff at the University of Richmond was greatly appreciated, including technical guidance by Kim Browne and Taylor Holden in the Department of Geography's Spatial Analysis Lab, and library support by Samantha Guss at the University's Boatwright Memorial Library.

References

Allison Farmer Associates (AFA), Leicester County Council, English Heritage. (2013). *Bosworth Battlefield: The way forward.* http://www.hwa.uk.com/site/wp-content/uploads/2015/03/EB-23.pdf. Accessed 19 Sept 2016.

Battlefields Trust. (2012). *Battlefields Trust statement on planning* (Unpublished internal document).

Battlefields Trust. (2018). *Hinckley and Bosworth Borough Council agree plans to build on Bosworth Battlefield.* http://www.battlefieldstrust.com/news.asp?NewsArticleID=233&Refresh=21%2F5%2F2019+3%3A00%3A36+PM. Accessed 21 May 2019.

Bessinger, J. B. (1963). Maldon and the olafsdrapa: An historical caveat. In S. B. Greenfield (Ed.), *Studies in old English literature* (pp. 23–35). Oxford: Clarendon Press.

Biggs, P. A., & Biggs, S. (2002). *Leicestershire & Rutland walks with children.* Ammanford: Sigma Leisure.

British Broadcasting Corporation (BBC) News. (2006). *Cash boost for Naseby battlefield: A project to boost tourism at Northamptonshire's historic Naseby battlefield site has secured funding of more than £200,000.* http://news.bbc.co.uk/2/hi/uk_news/england/northamptonshire/4772046.stm. Accessed 2 Sept 2016.

British Broadcasting Corporation (BBC) News. (2013). *Naseby battlefield visitor centre to open in church.* http://www.bbc.com/news/uk-england-northamptonshire-22690383. Accessed 9 Sept 2016.

Brooks, R. (2005). *Cassell's battlefields of Britain and Ireland.* London: Weidenfeld & Nicolson.

Brown, G. B. (1905). *the care of ancient monuments: An account of the legislative and other measures adopted in European countries for protecting ancient monuments and objects and scenes of natural beauty, and for preserving the aspect of historical cities.* Cambridge: University Press.

Bruce-Lockhard, L. (2008). *Conservation principles, policies and guidance for the sustainable management of the historic environment.* https://historicengland.org.uk/advice/constructive-conservation/. Accessed 3 Apr 2017.

Buhalis, D., Darcy, S., & Ambrose, I. (Eds.). (2012). *Best practice in accessible tourism: Inclusion, disability, ageing population and tourism.* Bristol: Channel View Publications.

Catling, C. (2013). *Constructive conservation: sustainable growth for historic places.* Pureprint Group Ltd. https://content.historicengland.org.uk/images-books/publications/constructive-conservation-sustainable-growth-historic-places/Acc_ConConservation.pdf/. Accessed 15 June 2016.

Céréghino, R., Ruggiero, A., Marty, P., & Angélibert, S. (2008a). Biodiversity and distribution patterns of freshwater invertebrates in farm ponds of south-western French agricultural landscapes. *Hydrobiologia, 597*(1), 43–51.

Céréghino, R., Biggs, J., Oertli, B., & Declerck, S. (2008b). The ecology of European ponds: defining the characteristics of a neglected freshwater habitat. *Hydrobiologia, 597*(1), 1–6.

Chandler, D. G. (1989). Introduction: Regular and irregular warfare. *The International History Review, 11*(1), 2–13.

Christy, M. (1906). A history of salt-making in Essex. *Essex Naturalist: Being the Journal of the Essex Field Club, 14–15*, 193–204.

Colclough, S., Fonseca, L., Astley, T., Thomas, K., & Watts, W. (2005). Fish utilization of managed realignments. *Fisheries Management & Ecology, 12*, 351–360.

Conduit, B. (2004). *Battlefield walks in the Midlands*. Ammanford: Sigma Leisure.

Copping, J. (2012). Wind farm to be built at site of decisive Civil War battle. *The Telegraph*. http://www.telegraph.co.uk/news/earth/energy/windpower/8999705/Wind-farm-to-be-built-at-site-of-decisive-Civil-War-battle.html. Accessed 16 Aug 2016.

Coskun, B. S. (2015). *Constructive conservation: a British approach to conservation*. Paper presented at the 3rd international conference on documentation, conservation, and reuse of architectural heritage and landscape, Mimar Sinan Fine Arts University, Turkey, 26 Oct 2015.

Dabney, S. M., Liu, Z., Lane, M., Douglas, J., Zhu, J., & Flanagan, D. C. (1999). Landscape benching from tillage erosion between grass hedges. *Soil and Tillage Research, 51*(3–4), 219–231.

Davies, Z. G., & Pullin, A. S. (2007). Are hedgerows effective corridors between fragments of woodland habitat? An evidence-based approach. *Landscape Ecology, 22*, 333–351.

Davies, B. R., Biggs, J., Williams, P. J., Lee, J. T., & Thompson, S. (2008). A comparison of the catchment sizes of rivers, streams, ponds, ditches and lakes: implications for protecting aquatic biodiversity in an agricultural landscape. *Hydrobiologia, 597*(1), 7–17.

Department for Communities and Local Government. (2012). *National planning policy framework*. London: Crown.

Dover, J., & Sparks, T. (2000). A review of the ecology of butterflies in British hedgerows. *Journal of Environmental Management, 60*(1), 51–63.

Drury, P., & McPherson, A. (2008). *Conservation principles, policies and guidance*. English Heritage. https://historicengland.org.uk/advice/constructive-conservation/conservation-principles/. Accessed 28 June 2017.

English Heritage. (1995a). *English Heritage battlefield report: Naseby 1645*. https://content.historicengland.org.uk/content/docs/battlefields/naseby.pdf. Accessed 8 Sept 2016.

English Heritage. (1995b). *English Heritage battlefield report: Maldon 991*. https://content.historicengland.org.uk/content/docs/battlefields/maldon.pdf. Accessed 10 Sept 2016.

English Heritage. (1995c). *English Heritage Battlefield report: Bosworth 1485*. http://politics.leics.gov.uk/documents/s86833/12.%20Bosworth%20Battlefield%20Conservation%20Plan%20-%20Appendix%20A%20-%20Annex%202%20part%202.pdf. Accessed 15 Sept 2016.

English Heritage. (2016). *About us: Our history*. English Heritage. http://www.english-heritage.org.uk/about-us/our-history/. Accessed 6 June 2016.

Essex Ecology Services (EECOS). (2007). *EB070 Maldon district nature conservation study*. https://www.maldon.gov.uk/publications/LDP/pre-submission/5%20Natural%20Environment%20and%20Green%20Infrastructure/EB070%20Maldon%20District%20Nature%20Conservation%20Study.pdf. Accessed 21 Sept 2016.

Essex Biodiversity Project (EBP). (2012). *Essex Biodiversity Action Plan 2010–2020: Vision for Essex Coastal Saltmarsh.*. http://www.essexbiodiversity.org.uk/app/webroot/files/ebap2010/13%20COASTAL%20SALTMARSH%20BAP.pdf. Accessed 27 July 2016.

Ferguson, N. (2013). Biting the bullet: The role of hobbyist metal detecting within battlefield archaeology. *Internet Archaeology, 33*. https://doi.org/10.11141/ia.33.3.

Foard, G. (2003). *Maldon Battlefield and Campaign*. The UK Battlefields Resource Centre. http://www.battlefieldstrust.com/media/149.pdf. Accessed 14 Jan 2019.

Foard, G. (2008). *Conflict in the pre-industrial landscape of England: A resource assessment*. University of Leeds. http://www.battlefieldstrust.com/resource-centre/battlefieldsuk/periodpageview.asp?pageid=831. Accessed 16 July 2016.

Foard, G., & Curry, A. (2013). *Bosworth 1485: A battlefield rediscovered*. Oxford: Oxbow Books.

Foard, G., & Morris, R. (2012). The archaeology of English battlefields: conflict in the pre-industrial landscape (CBA research report 168). *York: Council for British Archaeology, 86*(333), 943–944.
Forman, R. T., & Baudry, J. (1984). Hedgerows and hedgerow networks in landscape ecology. *Environmental Management, 8*(6), 495–510.
Freer-Smith, P. H., Holloway, S., & Goodman, A. (1997). The uptake of particulates by an urban woodland: Site description and particulate composition. *Environmental Pollution, 95*(1), 27–35.
Fuller, R. M., Groom, G. B., & Jones, A. R. (1994). The Land Cover Map of Great Britain: An automated classification of Landsat Thematic Mapper data. *Photogrammetric Engineering and Remote Sensing, 60*(50), 553–562.
Fuller, R. M., Smith, G. M., Sanderson, J. M., Hill, R. A., & Thomson, A. G. (2002). The UK Land Cover Map 2000: Construction of a parcel-based vector map from satellite images. *Cartographic Journal, 39*, 15–25.
Garbutt, A., & Wolters, M. (2008). The natural regeneration of salt marsh on formerly reclaimed land. *Applied Vegetation Science, 11*(3), 335–344.
Griffiths, P. (2011). Appeal decision APP/Y2810/A/11/2154375. The Planning Inspectorate. http://www.battlefieldstrust.com/cms/upload/docs/274/windfarm_decision_document.pdf. Accessed 21 Aug 2016.
Hinckley & Bosworth Borough Council. (2009). *Ashby Canal conservation area*. http://www.hinckley-bosworth.gov.uk/download/downloads/id/1494/appraisal.pdf. Accessed 14 Sept 2016.
Hinckley & Bosworth Borough Council. (2014). *Bosworth Battlefield: The way forward* (Report of deputy chief executive). http://moderngov.hinckley-bosworth.gov.uk/documents/s3690/BosworthBattlefieldCA.pdf. Accessed 18 Sept 2016.
Hinsley, S. A., & Bellamy, P. E. (2000). The influence of hedge structure, management and landscape context on the value of hedgerows to birds: A review. *Environmental Management, 60*(1), 33–49.
Historic England. (2016). *Registered battlefields*. Historic England. https://historicengland.org.uk/listing/what-is-designation/registered-battlefields/. Accessed July 8 2016.
Historic England. (2017a). Download listing data. *Historic England*. https://historicengland.org.uk/listing/the-list/data-downloads/. Accessed 5 Sept 2017.
Historic England. (2017b). *Battlefields registration selection guide*. https://historicengland.org.uk/images-books/publications/dsg-battlefields/heag072-battlefields-rsg/. Accessed 14 May 2018.
Hourdequin, M., & Havlick, D. G. (2016). *Restoring layered landscapes: History, ecology, and culture*. Oxford: Oxford University Press.
Ingram, M. (2016). *Bosworth 1485*. Dundurn: Battle story.
Jeeves, M. (2010). *Space for wildlife Leicester, Leicestershire and Rutland Biodiversity Action Plan 2010–2015*. Leicester & Rutland Wildlife Trust. http://www.lrwt.org.uk/media/uploads/miscellaneous/space_for_wildife_2010-2015__2011_revision_.pdf. Accessed 19 Sept 2016.
Kuvlesky, W. P., Jr., Brennan, L. A., Morrison, M. L., Boydston, K. K., Ballard, B. M., & Bryant, F. C. (2007). Wind energy development and wildlife conservation: Challenges and opportunities. *Wildlife Management, 71*(8), 2487–2498.
Laborde, E. D. (1925). The site of the Battle of Maldon. *The English Historical Review, 40*(158), 161–173.
Larsen, P. F., & Doggett, L. F. (1991). The macroinvertebrate fauna associated with the mud flats of the Gulf of Maine. *Journal of Coastal Research, 7*(2), 365–375.
Lotze, H. K., Lenihan, H. S., Bourque, B. J., Bradbury, R. H., Cooke, R. G., Kay, M. C., Kidwell, S. M., Kirby, M. X., Peterson, C. H., & Jackson, J. B. C. (2006). Depletion, degradation, and recovery potential of estuaries and coastal seas. *Science, 312*, 1806–1809.
Magnus, L. (2015). *Valuing our past enriching our future: Historic England corporate plan 2015 to 2018*. Historic England. https://content.historicengland.org.uk/images-books/publications/he-corp-plan-2015-18/he-corp-plan-2015-18.pdf/. Accessed 15 June 2016.

Marix Evans, M. (2011). Naseby Field, the site of the battle on 14 June 1645. *Battlefield Trust, 16*(3), 10–13.
Marix Evans, M. (2014). Presenting naseby. *Arms & Armour, 11*(1), 16–31.
Marsh, S. (2016). *Battlefield threats: A policy approach*. Battlefields Trust. http://www.battlefieldstrust.com/cms/upload/docs/343/20160409managing_battlefield_threats_policy_paper_v1_1.pdf. Accessed 7 July 2016.
Marshall, E. J. P., & Moonen, A. C. (2002). Field margins in northern Europe: their functions and interactions with agriculture. *Agriculture, Ecosystems and Environment, 89*, 5–21.
Masero, J. A., & Pérez-Hurtado, A. (2001). Importance of supratidal habitats for maintaining overwintering shorebird populations: how redshanks use tidal mudflats and adjacent saltworks in southern Europe. *The Condor, 103*(1), 21–30.
McCollin, D., Jackson, J. I., Bunce, R. G. H., Barr, C. J., & Stuart, R. (2000). Hedgerows as habitats for woodland plants. *Environmental Management, 60*, 77–90.
Milne, R., & Brown, T. A. (1997). Carbon in the vegetation and soils of Great Britain. *Environmental Management, 49*(4), 413–433.
Moller, I., & Spencer, T. (2002). Wave dissipation over macrotidal saltmarshes: Effects of saltmarsh edge typology and vegetation change. *Journal of Coastal Research, 36*, 506–521.
Morton, D., Rowland, C., Meek, L., Marston, C., Smith, G., & Simpson, I. C. (2011). Final report for LCM2007 – the new UK land cover map. CS Technical Report, No 11/07. NERC/Centre for Ecology & Hydrology. 108 pp. (CEH project number: C03259).
Murray, D. (2015). *An archaeological survey of the United Kingdom, 1896*. London: Forgotten Books.
National Trust. (2013). *Governance handbook fourth edition July 2013*. https://www.nationaltrust.org.uk/documents/governance-handbook-4th-edition-2013.pdf. Accessed 21 July 2016.
National Trust. (2014). *Northey Island and South House Farm statement of significance* (Unpublished internal document).
North West Leicestershire District Council (NWLDC). (2001). *Ashby Canal conservation area appraisal and study*. https://www.nwleics.gov.uk/files/documents/ashby_de_la_zouch_canal_conservation_area_appraisal_and_study/Ashby%20de%20la%20Zouch%20Canal%20Conservation%20Area%20Appraisal%20and%20Study.pdf .Accessed 12 Sept 2016.
Nunn, S. P. (2013). 991- was it really the Battle of Heybridge? *The Journal of the Maldon Archaeological and Historical Group, 26*, 18–25.
Öckinger, E., & Smith, H. G. (2006). Semi-natural grasslands as population sources for pollinating insects in agricultural landscapes. *Applied Ecology, 44*(1), 50–59.
Oreszczyn, S., & Lane, A. (2000). The meaning of hedgerows in the English landscape: different stakeholder perspectives and the implications for future hedge management. *Environmental Management, 60*, 101–118.
Partida, T., Hall, D., & Foard, G. (2013). *A Battlefield landscape: Naseby and the upper Welland*. Historic Environment Report: Zone 1 preparatory to an atlas of Northamptonshire: The Medieval and Early-Modern Landscape. Oxford: Oxbow Books.
Petty, G. R., Jr., & Petty, S. (1976). Geology and the Battle of Maldon. *Speculum, 51*(3), 435–446.
Pollard, T. (2003). The value of enmity: Remaking and revisiting historic battlefields in the United States and Britain. *Landscapes, 4*(2), 25–34.
Potts, S. G., Woodcock, B. A., Roberts, S. P. M., Tscheulin, T., Pilgram, E. S., Brown, V. K., & Tallowin, J. R. (2009). Enhancing pollinator biodiversity in intensive grasslands. *Applied Ecology, 46*(2), 369–379.
Rayner, M. (2007). *English battlefields: 500 battlefields that shaped English history*. Gloucestershire: Tempus Publishing Limited.
Rough Guides. (2012). *The Rough Guide to Britain*. London: Penguin.
Ryan, C. (Ed.). (2007). *Battlefield tourism: History, place, and interpretation* (pp. 30–43). Oxford: Elsevier Ltd.
Sargent, A. (2001). "RCHME" 1908–1998: a history of the Royal Commission on the Historical Monuments of England. *Transactions of the Ancient Monuments Society, 45*, 57–80.

Scheffer, M., Van Geest, G. J., Zimmer, K., Jeppesen, E., Søndergaard, M., Butler, M. G., Hanson, M. A., Declerck, S., & De Meester, L. (2006). Small habitat size and isolation can promote species richness: second-order effects on biodiversity in shallow lakes and ponds. *Organismal Biology, 112*(1), 227–231.

Shepherd, D., Jickles, T., Andrews, J., Cave, R. R., Ledoux, L., Turner, K., Watkinson, A., Aldridge, J., Malcolm, S., Parker, R., & Young, E. (2005). *Integrated modeling of an estuarine environment: an assessment of managed realignment options* (Tyndall Centre for Climate Change Research, Technical Report 21). Norwich: University of East Anglia.

Smith, P., Ashmore, M. R., Black, H. I. J., Burgess, P. J., Evans, C. D., Quine, T. A., Thomson, A. M., Hicks, K., & Orr, H. G. (2012). REVIEW: The role of ecosystems and their management in regulating climate, and soil, water and air quality. *Applied Ecology, 50*(4), 812–829.

Technische Universität Darmstadt. (2013). Is a constructive conservation the last chance for biodiversity? Pragmatic approach to saving what can be saved. *ScienceDaily*. https://www.sciencedaily.com/releases/2013/10/131010104925.htm. Accessed 8 Sept 2016.

Torres, A., Jaeger, J. A. G., & Alonso, J. C. (2016). Assessing large-scale wildlife responses to human infrastructure development. *PNAS, 113*(30), 8472–8477.

Valatin, G., & Starling, J. (2010). *UK NEA economic analysis report: 8-Valutation of ecosystem services provided by UK woodlands*. http://uknea.unep-wcmc.org/LinkClick.aspx?fileticket=TxLTiDHKooI%3D&tabid=82. Accessed 17 Sept 2016.

Whittingham, M. J., & Evans, K. L. (2004). The effects of habitat structure on predation risk of birds in agricultural landscapes. *Ibis, 146*, 210–220.

Williams, P., Whitfield, M., Biggs, J., Bray, S., Fox, G., Nicolet, P., & Sear, D. (2003). Comparative biodiversity of rivers, streams, ditches and ponds in an agricultural landscape in Southern England. *Biological Conservation, 115*, 329–341.

Wolton, R. J., Morris, R. K. A., Pollard, K. A., & Dover, J. W. (2013). *Understanding the combined biodiversity benefits of the component features of hedges* (Report of Defra project BD5214).

Yan, Z., Baoyuan, L., Qingchun, Z., & Yun, X. (2003). Effect of different vegetation types on soil erosion by water. *Acta Botanica Sinica, 45*(10), 1204–1209.

Chapter 4
Old Forts and New Amenities in the Southern Plains

Jason P. Julian

Abstract The Southern Plains of the United States (U.S.), specifically the states of Texas and Oklahoma, is a region of transition. Physically, it represents the transition from the humid, forested eastern U.S. with mostly perennial water resources to the dry western U.S. with grasslands, deserts, and mostly ephemeral water resources. Socioeconomically, it represents the transition from the densely populated eastern U.S. to the wild open spaces of the western U.S. Historically and culturally, it represents the transition from the French/English colonies of the eastern half of the U.S. to the Spanish territory of the Southwest. Later, it would represent the transition from the eastern pre-Civil War states to the western post-Civil War states. The Southern Plains also represent a transition in time when U.S. settlers were moving into western Native American lands. This occupation led to many intense battles between the European/American settlers and various Tribal Nations. Between 1821 and 1890, many forts were built in response to these conflicts and also to promote new settlements. Of these, 33 have been protected as publicly accessible places, including museums, state parks, national historic sites, city parks, resorts, and even a U.S. Department of Agriculture research facility. This chapter inventories and discusses the historical, cultural, and natural values of these 'protected forts' within the context of ecosystem services that have evolved from these sites.

4.1 Introduction

4.1.1 Forts Are Valuable Protected Places

Places have been protected because of their recognized historical, cultural, or natural values. Preserving these values is important for our well-being, whether it is reminiscing over our past, interacting with others who have similar values, or enjoying the benefits of the natural environment. The vast majority of protected places

J. P. Julian (✉)
Department of Geography, Texas State University, San Marcos, TX, USA
e-mail: jason.julian@txstate.edu

T. R. Lookingbill, P. D. Smallwood (eds.), *Collateral Values*, Landscape Series 25,
https://doi.org/10.1007/978-3-030-18991-4_4

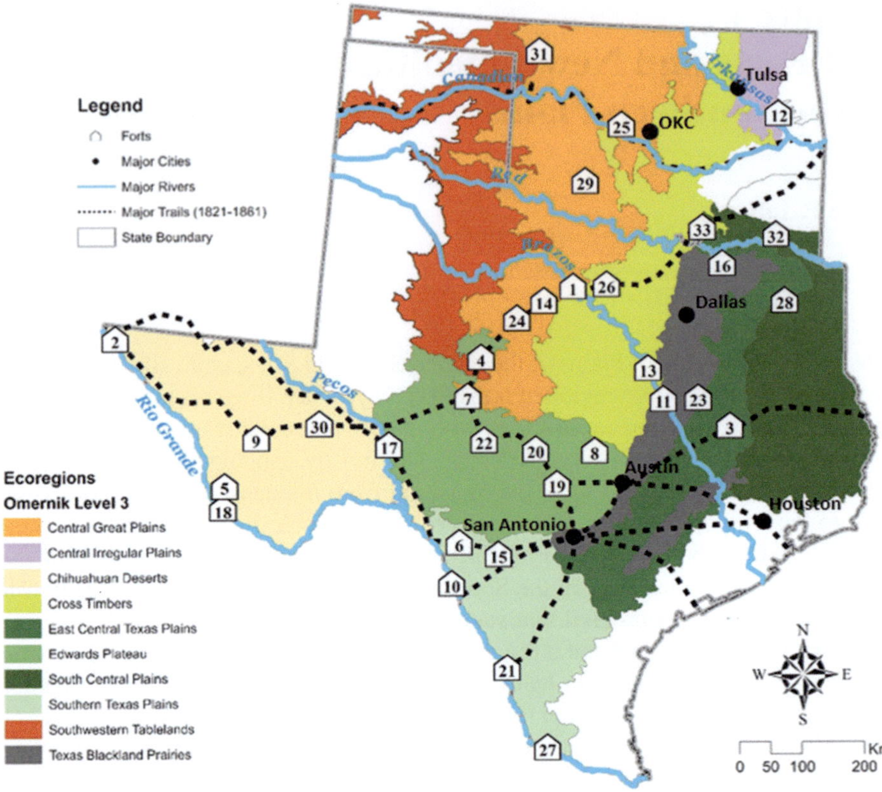

Fig. 4.1 Forts from the Southern Plains Indian Wars that have been commemorated, protected, and open to the public in present-day Oklahoma and Texas. The number on the fort corresponds to its alphabetical Fort ID in Table 4.3

embody one or two of these values. Rarely does a protected place encompass all three. In this chapter, I show how protected forts from the Southern Plains Indian Wars have historical, cultural, and natural values (Fig. 4.1).

Historically, these forts represent the beginnings of statehood for both Texas (TX) and Oklahoma (OK). They promoted settlement in a hostile environment, while also serving transportation and trade routes (Field 2006; Frantz 1970). Culturally, forts were and are the meeting place of diverse societies, including settlers from across North America and Europe, the military, and multiple Tribal Nations (Gwynne 2010; Michno 2011; Wooster 1987). Not only are they popular cultural heritage tourism destinations, but they also serve as sites for a wide range of modern cultural activities. Naturally, these protected places provide precious habitat and ecological functions in two states with relatively little protected land compared to their size and the rest of the U.S. They also provide the public with unique nature experiences and recreational opportunities across a spectrum of ecosystems, from deserts to temperate forests.

In my assessment of historical, cultural, and natural values, I investigated all forts in present-day TX and OK that were established 1821–1890 and used for war activities and support during the Southern Plains Indian Wars, which have also been referred to as the Texas-Indian Wars or the Comanche Wars. Accordingly, these forts include U.S. government installations, Texas Rangers posts, trading posts, lodging facilities, and settlers' forts, which were established by families, rancher groups, or local militias to protect settlers and livestock from raids by small parties of roaming Indians (Hannings 2006). Using the above criteria, I found a total of 73 forts. Of these, 33 have been commemorated, protected, and open to the public (Fig. 4.1). I collected information on these 33 forts related to their geography, history, and current amenities with a focus on ecosystem services, the benefits societies obtain from ecosystems (MEA 2005). Data was collected from a variety of online and literature sources. The online encyclopedias hosted by the Texas State Historical Association and Oklahoma Historical Society provided comprehensive overviews of all forts located within each state. Additionally, the Texas Historic Sites atlas provided locations for every historic marker associated with a recognized fort. Information about forts that were converted to local, county, state, and national parks was obtained from the parks' websites. Data on Forts Bliss and Sill, the only active forts in this list, was requested from the U.S. Army or obtained from the base website. Online sources were supplemented with Hannings's (2006) encyclopedia of U.S. forts, Pierce (1969) and Alexander and Utley's (2012) reviews of historical military sites in Texas, and Awbrey and Dooley's (1992) list of roadside historical markers in Texas. Finally, I visited all but three of the forts during 2016 and interviewed all the directors/managers to collect additional information on ecosystem services and societal benefits not provided in texts or on websites.

4.1.2 Background on Indian Wars and Forts

The American Indian Wars occurred over hundreds of years from colonial times till the early 1900s. Some of the most widespread fighting and deadliest battles occurred in the Southern Plains (Gwynne 2010; Michno 2011), what is now TX and OK. While there were forts in eastern TX used to battle American Indian tribes such as the Karankawas from the late 1600s to the early 1800s, the recognized start of the Southern Plains Indian Wars was Mexico's independence in 1821. The new Mexican Republic welcomed and incentivized American traders and settlers, largely to establish an official political and economic presence, and push out the Native Americans (Field 2006). Some even claim that the impetus for this policy was to create a buffer between the core of Mexico and its nemesis the Comanches (Gwynne 2010). During this period, several important trade and settlement routes were established and fortified. Around the same time, numerous tribes were relocated to the Indian Territory of what is now OK, with a surge following the Indian Removal Act of 1830 (Field

2006). Several military forts were established around this new 'Permanent Indian Frontier' to protect the boundary with white settlements, protect the 'Civilized Tribes' from the 'hostile' Plains Indians, police the territory, and enforce U.S. laws and policies (e.g., Forts Gibson, Towson, and Washita).

The movement of both whites and new Indian tribes into the Southern Plains was an invasion into Comancheria, the empire established by the Comanches during the eighteenth century. This act was met with widespread and brutal attacks on white settlers and travelers, and of course retaliatory attacks on the Comanches (Gwynne 2010; Michno 2011). This war with the Comanches, their main allies the Kiowas, and other engaged tribes continued until 1875 when Quanah Parker and the Comanches surrendered. Other tribes and renegades continued to resist until about 1890, coincidently the same year that the Census Bureau declared the frontier was officially closed (Michno 2011). Thus, this Southern Plains Indian War forts analysis covers the period 1821–1890. Over these 70 years, forts were established for multiple purposes: new settlements; preservation of law and order in the new frontier; trading posts for military, settlers, and Indians; enforcement of Indian displacement; soldier lodging; and various military activities (Field 2006).

The Southern Plains forts were important not only for regional development, but nationally as well. In 1848, the acquisition of the southwestern U.S. from Spain and the subsequent discovery of gold in California created the need for increased federal defense for westward emigrants and settlements, particularly across TX and present-day OK (1907 statehood) because this area provided for year-round trails not impacted by cold weather (Fig. 4.1). Indeed, most of the military forts in TX were built 1848–1855. During the Civil War (1861–1865), several of these forts were used by the Confederacy. Following the Civil War, the U.S. government reoccupied some of the old forts, created several new forts across TX and present-day OK, and invested a lot of money and resources in these forts (Smith 1999). Many of the Southern Plains forts, post-Civil War, became posts for the 'Buffalo Soldiers,' who played a key role in the Indian Wars from 1866–1890.[1]

Following the Southern Plains Indian Wars, a few of the forts continued to be used for military purposes periodically (e.g. Fort Reno as a German POW camp during WWII). But with the expansion of settlements and the Southern Pacific Railroad, along with the nation's shift in focus to global conflicts, the need for forts in this region diminished and almost all of the forts were abandoned by the beginning of the twentieth century (Wooster 1987; Hannings 2006). Only two of the forts became permanent military installations, Fort Bliss in TX and Fort Sill in OK.

[1] Buffalo Soldiers, so named for 'their dark skin, curly hair, and fierce fighting spirit,' were all-black U.S. Army cavalry and infantry regiments formed after the Civil War. Buffalo Soldiers played prominent roles in many military campaigns during the Indian Wars (more than 125 engagements), as well as the Spanish-American War in 1898. They also served as some of the first national park rangers (NPS 1993).

4.1.3 How the Forts Became Protected Places

After the forts were abandoned, most were stripped or destroyed for building materials (Alexander and Utley 2012; Alexander and Utley 2015; Field 2006). Some were saved by local citizens moving in to the fort buildings and using them for residence or business (Concho, Davis, McKavett). The Texas settlers' forts (Boggy, Cibolo, Inglish, Leaton, Parker) remained with the original landowner, but over time fell into disuse and the structures deteriorated. It is important to note that none of the Texas forts were on properties owned by the federal government, but instead were leased from private landowners or the state (Smith 1999). Even the land for Fort Bliss was not purchased by the federal government until after 1890. The Oklahoma forts, on the other hand, were on federal property. Following their abandonment, most were transferred to their respective Tribal Nations: Fort Towson (Choctaw), Fort Washita (Chickasaw), Fort Gibson (Cherokee). Fort Supply was transferred to the Department of Interior, and later to the state, where in 1908 it became Oklahoma's first state-operated mental institution. Fort Reno was eventually transferred to the Department of Agriculture (USDA) and became a research center.

Of the approximately 73 forts constructed 1821–1890 and used for the Southern Plains Indian Wars in present-day OK and TX, only 33 ended up being commemorated, protected, and open to the public (Fig. 4.1). These forts became protected places (and open to the public) by various means over a long timeline (Fig. 4.2).

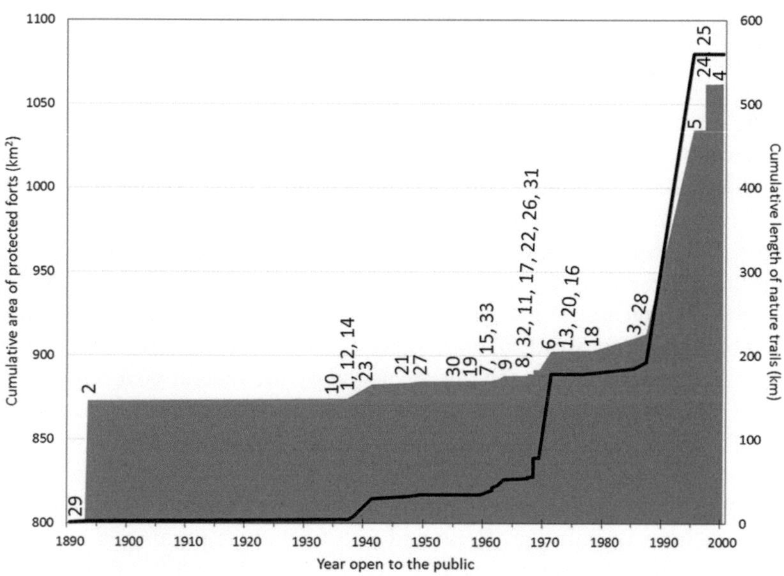

Fig. 4.2 Cumulative area of protected forts (grey area) and length of nature trails (solid black line) over time. Chart labels correspond to the alphabetical Fort ID in Table 4.3. The first fort open to the public, Fort Sill [29] in 1890, has an area of 379 km^2

Except the two U.S. Army bases, the first fort open to the public was Fort Duncan in 1935 when it became a city park. During the latter half of the 1930s, the Works Progress Administration (WPA) and Civilian Conservation Corps (CCC) were responsible for restoring several of the forts, building public parks, and creating amenities that included lakes, picnic areas, camping sites, and hiking trails. Fort Parker Lake, for example, was created by a dam built by the CCC in 1939. Fort structures restored by the CCC and WPA included Belknap, Gibson, Griffin, and McIntosh. Shortly after WWII, two forts became schools, as well as historical sites: McIntosh in 1947 and Ringgold in 1949. It was not until 1955 that other forts became restored and officially open to the public, beginning with Fort Stockton and followed by Martin Scott (1959), Inge (1961), and Concho (1961). All four of these forts were protected through the efforts of local activists and heritage associations in coordination with the city or county. Heritage associations and historical societies were responsible for protecting and restoring three other forts in following decades: Croghan (1967), Mason (1975), Inglish (1976). Most of the forts became protected 1962–1987 through purchases by or donations to the state, and they became state parks or official historical sites. Two forts were purchased privately and became resorts, Clark in 1971 and Cibolo in 1995 [note: Forts Cibolo, Cienega, and Morita are on the same property, but grouped for analysis purposes and collectively referred to as Cibolo]. Finally, two forts were protected by foundations set up by the landowner and became open to the public, Phantom Hill in 1997 and Chadbourne in 1999.

In order to be protected for public use, a few of the forts had to be moved. Fort Inglish was moved 2.0 km (1.2 mi) SW because a hospital was built on the original site. Fort Graham (now part of Old Fort Park) was moved 1.6 km (1.0 mi) E due to flooding from the impoundment of Lake Whitney. Fort Bliss was moved several times, once due to Rio Grande flooding and two other times due to municipal planning. Fort Fisher was moved approximately 1.0 km (0.6 mi) downstream along the Brazos River to be adjacent to I-35 and attract more tourism. The actual fort sites of Fisher, Boggy, and Sherman were lost, but all three were commemorated with a park at their estimated location (or at least a campground and cemetery in the case of Sherman). Consequently, these are the only three forts without ruins or restored fort buildings. The other 30 forts were all restored to varying standards. At three of the forts (Inge, Lancaster, Phantom Hill), only ruins were restored. The rest had at least one building reconstructed, with many almost completely restored.

Currently, the forts are owned and managed by the federal government (n = 5), the state (12), county (4), city (7), non-profit organizations (NPO; 3), and private entities (2). Of the federal properties, two are active military installations (Sill and Bliss), one is owned by the National Park Service (Davis), one is owned by the U.S. Army Corps of Engineers (Graham), and one is owned by the USDA (Reno). The fort structures of both Fort Reno and Fort Graham are managed by NPOs. Of the state properties, five are managed by the Texas Parks and Wildlife Department (TPWD; Boggy, Leaton, Parker, Richardson, and Sherman), three are managed by the Texas Historical Commission (THC; Griffin, Lancaster, and McKavett), and four are managed by the Oklahoma Historical Society (OHS; Gibson, Supply, Towson, and Washita). The three managed by THC were acquired from TPWD in 2008. In 1992, TPWD released to the City of Groesbeck the historical site of *Old Fort Parker*, which is now owned and managed by an NPO. Forts Belknap, Inge, Inglish, and Mason are owned by their respective counties. And Forts Concho, Duncan, Fisher, Martin Scott, McIntosh, Ringgold, and Stockton are owned by their respective cities. While only three forts are owned by NPOs (Croghan, Chadbourne, and Phantom Hill, with the latter two being foundations established by the landowner), NPOs and volunteer organizations play a key role in the maintenance, operation, and events of most of the forts. The two forts owned by private entities are resorts whose amenities can be enjoyed by anyone willing to pay for accommodations. Fort Cibolo is a high-priced luxury resort, while Fort Clark Springs is much more affordable and has campgrounds and low-priced motel rooms located in the original cavalry barracks.

As is summarized later, these 33 forts are used for multiple purposes and activities, but the primary mission of the forts is as follows. Just over half are historical sites or museums: Chadbourne, Concho, Croghan, Davis, Gibson, Griffin, Inglish, Lancaster, Leaton, Martin Scott, Mason, McKavett, Phantom Hill, Stockton, Supply, Towson, and Washita. Of these, Fort Davis is the only National Historic Site. Fort Leaton also serves as the western visitor center for Big Bend Ranch State Park. Fort Supply is also part of a correctional facility and a behavioral/mental health center. Hamilton Creek Park is located on the original site of Fort Croghan, and is thus included in ecosystem service analyses. Likewise, Rio Concho Park is located on the original site of Fort Concho, and is also included in analyses. Five are city or county parks primarily used for local community activities and recreation: Belknap, Duncan, Fisher, Graham, and Inge. Fort Duncan is an agglomeration of multiple parks and city facilities, including recreational fields, golf course, library, museum, multi-purpose center, and an elementary school. Fort Fisher Park also contains the Waco tourist information center and the Texas Rangers Hall of Fame and Museum, which serves as the official museum and principal repository for artifacts and archives related to the Texas Rangers whose unorthodox battle techniques played a pivotal role in the Southern Plains Indian Wars. Four forts are state parks used for camping and outdoor recreation: Boggy, Parker, Richardson, and Sherman (the last named Lake Bob Sandlin State Park). The historic site of *Old Fort Parker* is now separate from the state park, but both are included in ecosystem service analyses. Two of the forts are now schools. Fort McIntosh is the site for Laredo Community College and its Lamar Bruni Vergara Environmental Science Center (LBVESC). Fort Ringgold is now the administrative complex for Rio Grande City Consolidated Independent School District and also contains Ringgold Elementary and Academy for Academic Enhancement Middle School. Two of the fort properties are resorts (Cibolo, Clark); however, both offer public programs and educational opportunities. One is a USDA research facility, but also houses the *Historic Fort Reno* historical site and museum, as well as the U.S. Cavalry Association headquarters and library. Of the 33 forts, only two remain as active military installations, Fort Sill in OK and Fort Bliss in TX.

4.2 Physical Geography, Land Use, and Ecoregions of the Forts

Before discussing the ecosystem services provided by the forts, it is helpful to understand the environment in which the forts were established and their current ecosystems. The Southern Plains of OK and TX have a semi-arid climate with extreme variations in precipitation and temperature, both annually and inter-annually. Consequently, floods and droughts are common, with the latter occurring in approximate decadal cycles that coincide with the La Niña phase of the El Niño-Southern Oscillation. Summers are typically very hot and dry. There is a strong precipitation gradient decreasing from east to west, which has largely dictated the natural vegetation pattern of forests in the east, tallgrass/mixed prairie in the mid-section, shortgrass prairie in the northwest, and a mix of desert grassland and shrubland (depending on soil type and elevation) in the southwest (Griffith et al. 2007; Woods et al. 2005).

Following the Indian Wars, TX and OK were converted to an agriculture-dominated landscape with croplands in ecoregions with productive soils (Central Great Plains, Central Irregular Plains, Texas Blackland Prairies, and river valleys of most ecoregions), pasture/rangeland in less fertile ecoregions (Southwestern Tablelands, East Central Texas Plains, Southern Texas Plains, Cross Timbers, Edwards Plateau, Chihuahuan Deserts), and plantation forestry in the forested ecoregion (South Central Plains). Extensive livestock grazing over the past 150 years suppressed the prairie fire regime and led to the spread of several invasive tree and shrub species, most notably eastern redcedar (*Juniperus virginiana*), Ashe juniper (*Juniperus ashei*), honey mesquite (*Prosopis glandulosa*), creosote bush (*Larrea tridentata*), and catclaw acacia (*Acacia greggii*) (Archer 1994). Urban development spread rapidly beginning in the mid-twentieth century, particularly along the I-35 corridor. The twentieth century was also a period of extensive resource mining, oil/gas drilling, and wind power development. The net result of these intensive and extensive land uses is that the Southern Plains is now a mosaicked landscape with only small areas of native prairie remaining. Although almost all of the forts were originally located on prairie, most of these grasslands have changed into shrublands or woodlands (Table 4.1). And now more than half of the forts are located in urban or suburban environments.

Table 4.1 Current land use and land cover surrounding the protected forts

	Land Use			
Land Cover	Rural/Agricultural	Suburban	Urban	Total
Grassland	1	2	2	5
Shrubland	7	1	7	15
Woodland	5	4	1	10
Forest	3	0	0	3
Total	16	7	10	33

4.3 Current Ecosystem Services of the Southern Plains Forts

4.3.1 Classification of Ecosystem Services

There is not an agreed upon or standard ecosystem services (ES) classification. The most thorough classification is the Common International Classification of Ecosystem Services (CICES); however, it has been found to be overly complex, has unclear terminology, and does not properly frame cultural services (Haines-Young 2016), a focus of this chapter. Thus, I instead used a more simplified classification that better accounts for the cultural ES provided by forts: the Economics of Ecosystems & Biodiversity (TEEB) framework (TEEB 2010). I adapted this framework to create a simplified classification of the ES provided by forts (Table 4.2). For each fort/park, I rated each ES as purposeful (P), incidental (I), or nonexistent (N). These ratings provide context for whether the fort is using resources to 'purposefully' manage for the ES as opposed to the ES taking place on the property 'incidentally.' In some cases, the ES may be there incidentally (e.g. Aesthetics and Inspiration from dark sky), but if the fort has designed and allocated resources for stargazing programs, it would receive a 'purposeful' rating in this category. These types of distinctions are made in each category below.

In addition to ecosystem services provided by the forts, I also identified other key variables including size of the protected area, its connectivity to other protected areas, ecoregion, surrounding land cover/use, length of nature trails, and annual average visitation (over recent years).

4.3.2 Provisioning Services

4.3.2.1 Habitat

By virtue of containing soil and vegetation (and in most cases water), all of the forts provide at least incidental habitat to a number of species. The total protected area of this potential habitat of the 33 forts is 1061 km^2, ranging from 0.6 ha (1.5 acres; Inglish) to 49,396 ha (122,061 acres; Bliss). This area has been increasing since 1935 (Fig. 4.2) and is still increasing as some forts acquire adjacent lands (e.g. Fort Davis NHS adding 17 ha in 2011). There is also potential for this area to increase with restoration of other Indian War forts such as Fort Bird, which is located in Arlington's River Legacy Park but has yet to be commemorated by more than a historical marker. These areas represent the quantity of potential habitat. A measure of habitat quality is the amount of 'connected and protected' habitat, particularly in highly modified landscapes such as the Southern Plains (Fischer and Lindenmayer 2007). The largest 'connected and protected' fort within the boundaries of TX and OK is Fort Cibolo, which connects Chinati Mountains State Natural Area and easement (23,672 ha) to its west and Big Bend Ranch State Park (136,588 ha) to its east.

Table 4.2 Ecosystem services classification, adapted from the TEEB framework (TEEB 2010)

Category	Ecosystem Service	Definition
Provisioning	Habitat	The presence of food, water, and shelter needed by plants and animals to survive. Biodiversity and biological/pest control are byproducts of this service.
	Food	The presence of gardens/farms, ranching, or fishing that directly provide for human consumption.
	Raw materials and energy	The presence of materials used directly for construction, fuel, and energy production. Includes underground resources currently being mined, hydroelectric dams, and wind turbines.
	Fresh water	The presence of surface or ground water currently being used for any purpose.
Regulating	Climate and air quality	The presence of trees or other vegetation that remove pollutants from atmosphere, store/sequester carbon, contribute to the water cycle, and regulate air temperature.
	Water quality and flood control	The presence of wetlands or riparian habitats that mitigate downstream flooding and filter pollutants/pathogens.
	Soil protection/ fertility	The presence of vegetation, structures, or activities specifically designed to prevent soil erosion or promote soil fertility, such as composting.
	Pollination	The presence of habitat and vegetation on fort property, and a farm within 3 km radius of fort (typical forage distance of honey bees).
Cultural	Recreation for mental and physical health	The presence of features (e.g., nature trails, greenspace, water bodies) and activities (e.g., hiking, sports, hunting) that promote physical and mental health.
	Aesthetics and inspiration	The presence of unique, preserved, or scenic landscapes and activities that promote appreciation of arts, culture, or science.
	Identity and spirituality	The presence of natural features that are culturally significant, have sacred/religious meaning, or create a sense of place/pride.
	Tourism	The presence of programs designed to attract visitors for regional economic benefit.

For each ecosystem service for each fort, I assigned a rating of purposeful (P), incidental (I), or nonexistent (N)

The Fort Cibolo property is thus a centerpiece to a 1724-km^2 connected protected area that provides a potential migration corridor over 130 km in length. Beyond state boundaries, Fort Bliss is connected to approximately 12,000 km^2 of protected lands, including its property in New Mexico (and associated DoD lands) along with Franklin Mountains State Park, Lincoln National Forest, and White Sands National Park. If all adjacent federal and state lands are included, this connected and protected habitat extends across most of the southwestern U.S. Other notable protected and connected fort lands are the 1104-ha Davis Mountains State Park (Davis), the 23,885-ha Wichita Mountains National Wildlife Refuge (Sill), and 86,043 ha of Lake Texoma USACE lands and other parks/refuges adjacent to Fort Washita. In all, there are 3978 km^2 of protected and connected fort lands within TX and OK. Not included in this figure are adjacent private lands with active environmental management such as the 566 ha property surrounding Fort Phantom Hill.

Another measure of habitat quality is coverage of native vegetation (Fischer and Lindenmayer 2007). As mentioned in Sect. 4.2, only a few small areas of native prairie remain in the Southern Plains. The protected forts have added to this inventory considerably through native plantings, prescribed fires, exotic/invasive vegetation removal, or grazer control. Fort Cibolo again tops the list in this measure of habitat quality, with almost 3000 ha of restored native prairie, which took years and millions of dollars to accomplish. Fort Reno protects more than 1200 ha of native tallgrass prairie that has never been plowed. Fort Sill maintains several large areas of native prairie, including a 970 ha tallgrass prairie preserve. Other forts that have actively maintained or restored native prairie or are in the process include Boggy, Bliss, Clark, Concho, Davis, Lancaster, Martin Scott, McIntosh, Parker, and Sherman. Fort McIntosh also has a native plant nursery with over 10,000 plants raised since 2003.

Water features on the fort properties also provide valuable habitat. Only five forts do not have water features on or adjacent to their property. Some have a variety of aquatic habitats. Fort Sill, for example, has approximately 220 lakes/ponds (142 managed as fisheries; rest designated as wildlife use), 452 wetlands (55 ha of palustrine; 10.5 ha of riverine; 308 ha of lacustrine), and approximately 800 km of rivers/streams. Aquatic and riparian habitats have been purposefully restored on a few of the forts, including a wetland on McIntosh and ~20 km of riparian buffers on Cibolo. Given that all of these Southern Plains forts are within the Central Flyway, these aquatic and riparian habitats likely serve millions of migratory birds every year. Six of the forts are located along or connected to the Rio Grande, which is a popular stopover for hundreds of bird species (Johnsgard 2012).

Overall, the large areas, high quality, and diversity of environments among the protected forts (Table 4.1; Fig. 4.1) provide a wealth of habitat for fish and wildlife. There is also high biological diversity within the protected lands of some of the larger forts. On Fort Bliss for example, approximately 335 species of birds, 58 species of mammals, 39 species of reptiles, and 8 species of amphibians have been documented in this desert shrub-grassland ecosystem (DPW 2015). Even on the much smaller 212-ha Fort Davis National Historic Site, there were 15 species of mammals, 125 species of birds, 39 species of reptiles, 10 species of amphibians, and 368 species of vascular plants (NPS 2016). Some forts have reintroduced native animals such as bison (Chadbourne, Cibolo) and wild turkey (Cibolo). Texas longhorn cattle, though not native, is a historically and culturally important species that is managed for on several fort properties (Chadbourne, Cibolo, Griffin). Fort Griffin is home to the Official State of Texas Longhorn Herd and its breeding program. During extreme droughts, Fort Sill allows longhorn cattle from the Wichita Mountains National Wildlife Refuge to graze on its grasslands. Other notable species managed for among the forts are pallid bats (*Antrozous pallidus*) at Fort Leaton's bat house (Fig. 4.3) and the threatened Texas horned lizard (*Phrysonsoma cornutum*) at Fort Lancaster by not using pesticides on harvester ants, the lizard's

Fig. 4.3 Habitat features in the desert shrubland ecosystem at Fort Leaton (top). The fort provides incidental habitat for nesting birds (bottom-left) and purposeful habitat for pallid bats (bottom-right)

primary food source. The only endangered animal species I found managed for was the black-capped vireo (*Vireo atricapillus*) at Fort Sill. Fort Bliss manages for several threatened/endangered plant species.

Overall and consistently, the forts that purposefully manage for habitat (Table 4.3) take an ecosystem-based approach rather than managing for specific species. This approach is documented in the Integrated Natural Resources Management Plans (INRMP) developed every five years at Forts Bliss and Sill (DPW 2015; Stout 2014), but was also evident in my discussions with other fort managers. An important component of this ecosystem-based approach at several forts was control of species (i.e. feral hogs, deer) that damage the habitat for other species, or degrade other ecosystem services. Feral hogs have become quite a nuisance in TX and OK, impacting soil, native vegetation, and agricultural productivity, resulting in at least $3 billion in damages and costs for control (Adams et al. 2005). While not having as intense of an impact as feral hogs, deer overpopulations reduce the abundance and health of native plant communities, which impacts other species. Feral hogs and/or deer are controlled (via hunting or trapping) at Boggy, Bliss, Cibolo, Clark, and Sill.

4.3.2.2 Food

Many ecoregions of the Southern Plains are known for their rich, productive soils, particularly the Central Great Plains, Central Irregular Plains, and Texas Blackland Prairies. Many forts are still taking advantage of these fertile soils with small gardens or orchards, including Belknap, Cibolo, Clark, Concho, Gibson, Inglish,

Table 4.3 Inventory of ecosystem services currently being provided by the protected forts

			PROVISIONING				REGULATING			
Fort ID	Fort Name	Land Use-Cover	Habitat	Food	Raw Materials and Energy	Fresh Water	Climate/Air Quality	Water Quality and Flood Control	Soil Protection/ Fertility	Pollination
1	Fort Belknap	R-W	I	P	N	N	P	N	N	P
2	Fort bliss	U-S	P	P	P	P	P	P	P	P
3	Fort boggy	R-F	P	I	N	P	P	I	I	N
4	Fort Chadbourne	R-S	I	P	N	N	I	N	N	N
5	Fort Cibolo	R-S	P	P	P	P	P	P	I	P
6	Fort Clark	U-S	P	P	N	P	I	P	I	P
7	Fort Concho	U-S	P	P	N	P	P	P	P	P
8	Fort Croghan	U-W	I	P	N	P	I	I	N	I
9	Fort Davis	S-S	P	N	N	P	P	P	I	N
10	Fort Duncan	U-S	I	I	N	I	P	I	I	N
11	Fort fisher	U-G	I	I	N	I	I	I	N	I
12	Fort Gibson	S-W	I	P	N	I	I	I	N	I
13	Fort Graham	S-W	I	P	N	I	I	I	N	I
14	Fort griffin	R-S	P	P	N	P	I	I	N	I
15	Fort Inge	R-S	I	P	N	P	I	I	N	I
16	Fort Inglish	U-G	I	P	N	N	I	N	N	I
17	Fort Lancaster	R-S	P	N	N	P	P	I	N	N
18	Fort Leaton	R-S	P	N	N	N	I	N	N	I
19	Fort Martin Scott	S-W	I	N	N	I	P	I	N	P
20	Fort Mason	S-W	I	N	N	N	I	N	N	I
21	Fort McIntosh	U-S	P	P	P	P	P	P	P	P

(continued)

Table 4.3 (continued)

Fort ID	Fort Name	Land Use-Cover	PROVISIONING				REGULATING			
			Habitat	Food	Raw Materials and Energy	Fresh Water	Climate/Air Quality	Water Quality and Flood Control	Soil Protection/ Fertility	Pollination
22	Fort McKavett	R-W	I	P	N	I	I	P	I	I
23	Fort Parker	R-W	P	P	N	P	P	I	P	P
24	Fort Phantom Hill	R-S	I	N	N	N	I	N	N	I
25	Fort Reno	R-G	P	P	P	P	P	P	P	P
26	Fort Richardson	R-W	I	P	P	P	I	I	N	N
27	Fort Ringgold	U-S	I	I	N	I	I	I	N	I
28	Fort Sherman	R-F	P	P	N	P	P	I	P	P
29	Fort sill	S-G	P	P	P	P	P	P	P	P
30	Fort Stockton	U-S	I	N	N	N	I	N	N	P
31	Fort supply	S-G	I	N	N	N	I	N	N	I
32	Fort Towson	R-F	I	N	N	I	I	I	N	I
33	Fort Washita	R-W	I	I	N	P	I	I	N	I

Fort ID	CULTURAL Recreation for Mental and Physical Health	Aesthetics and Inspiration	Identity and Spirituality	Tourism	Annual average visitors	Current fort protected area (ha)	Nature trails (km)
1	P	I	N	P	30,000	6	0
2	P	P	P	I	4500	49,397	1.0
3	P	I	I	P	10,000	747	5.6
4	N	I	N	P	11,000	10	0
5	P	P	P	P	2500	12,141	366.9
6	P	P	P	P	3500	1093	99.8
7	P	P	I	P	54,800	16	5.4
8	P	P	P	P	2250	16	1.6
9	P	P	P	P	56,000	212	6.4
10	P	P	I	P	2000	121	2.2
11	P	I	I	P	80,000	6	0.8
12	P	I	N	P	55,000	35	0
13	I	I	N	I	100	36	0
14	P	P	I	P	11,000	205	4.0
15	P	P	P	P	1500	19	3.2
16	N	I	N	P	1750	1	0
17	N	I	N	P	3000	33	0
18	P	P	P	P	4500	9	1.6
19	N	I	N	P	16,000	10	0
20	N	I	N	I	2000	1	0
21	P	P	P	P	11,000	84	2.4

(continued)

Table 4.3 (continued)

Fort ID	CULTURAL				Annual average visitors	Current fort protected area (ha)	Nature trails (km)
	Recreation for Mental and Physical Health	Aesthetics and Inspiration	Identity and Spirituality	Tourism			
22	P	I	I	P	5000	32	2.4
23	P	P	P	P	53,000	620	21.1
24	N	I	N	I	5000	15	0
25	N	P	P	P	22,000	2711	0
26	P	P	I	P	36,100	184	20.0
27	P	P	I	P	200	98	2.0
28	P	I	I	P	62,500	259	7.2
29	P	P	P	I	16,000	37,901	1.3
30	I	P	N	P	12,500	11	0
31	N	I	N	I	600	2	0
32	I	P	N	P	9000	89	1.0
33	P	P	I	P	22,000	61	3.2

See Table 4.2 for description of ecosystem services. For each ecosystem service for each fort, I assigned a rating of purposeful (P), incidental (I), or nonexistent (N). Land uses are Rural (R), Suburban (S), and Urban (U). Land covers are Grassland (G), Shrubland (S), Woodland (W), and Forest (F)

McIntosh, McKavett, Parker, Ringgold, and Sill (Fig. 4.4). The organic garden on Fort McIntosh provides food for its animals, college personnel, farmers markets, and food banks, and is used for botany, culinary arts, and environmental classes at Laredo CC. Fort Gibson hosts several Bake Days every year where a restored brick oven is used to bake goods using corn and other crops from the garden. Some of their corn is also used for chicken feed at a nearby historical site. Fort Sill leases 2428 ha annually for hay, alfalfa, corn, winter wheat, sorghum, sesame, and livestock grazing. Serving as a USDA research facility, Fort Reno also does farming on a larger scale with more than 400 ha in periodic crop production (mostly winter wheat), most of which is used for livestock forage and feed (Fig. 4.4). Some of the cattle fed by Fort Reno crops include those at *Braum's* dairy farm just down the road in Tuttle, OK, which supplies dairy products to much of the Southern Plains. Some of the beef cattle fed by Fort Reno crops are used for human consumption. Other forts that raise livestock, where excess is sold or donated for slaughtering, are Chadbourne, Cibolo, and Griffin.

Gardens, farms, and ranching were all identified as purposeful food ecosystem resources (Table 4.3) because they require active management. Fishing on the other hand was listed as incidental, or purposeful if stocking occurred. Non-stocked fishing occurs on eight forts (Bliss, Clark, Concho, Duncan, Fisher, McIntosh, Ringgold, Washita), while stocked fishing occurs on ten forts (Boggy, Cibolo, Croghan, Graham, Griffin, Inge, Parker, Richardson, Sherman, Sill). Hunting (and limited trapping) occurs on five forts (Boggy, Bliss, Cibolo, Clark, Sill). I will note, how-

Fig. 4.4 Farming is still practiced on many of the forts, like in the small post gardens at Fort Gibson (top-left) and Fort McKavett (top-right). Fort Reno (bottom) is home to a USDA research facility that raises livestock and has more than 400 ha in periodic crop production, mostly winter wheat as shown here (Historic Fort Reno is in the background near the water tower)

ever, that fishing in the Rio Grande along Forts Duncan, McIntosh, and Ringgold is not productive because of poor water quality, limited access, and discouragement by the U.S. Border Patrol.

4.3.2.3 Raw Materials and Energy

Texas and Oklahoma have a wealth and wide variety of natural, mineral, and energy resources. The energy development of the region, whether oil, natural gas, wind, or solar, is particularly prominent. But by virtue of being protected places, extraction of raw materials and energy on the fort properties is limited. There are no vertical oil wells and only one vertical gas well on any of the properties (Richardson). Fort Richardson also has a directional gas well (i.e. wellhead not located on property), and Fort Griffin has a directional oil/gas well. Two forts had multiple small-scale wind turbines, Cibolo and McIntosh. Fort Cibolo also has a few solar panels. Fort Reno is a USDA Biomass Research Center and accordingly produces biofuels, but only periodically. The two U.S. Army bases have a federal mandate to increase renewable energy production, and have made some strides. Fort Sill has several geothermal wells to heat water and has installed solar panels on numerous buildings. There is also limited wood cutting allowed by the Fort Sill community. More ambitious, Fort Bliss has a "Net Zero" plan where it will produce as much energy as it consumes on an annual basis. So far, they have installed small-scale wind turbines, a 1.4-MW solar farm, and 13.4 MW in distributed rooftop solar panels on hundreds of buildings and houses. Both forts have a wealth of potential energy sources (via wind, solar, geothermal, natural gas) and are currently exploring options to further develop these resources.

4.3.2.4 Fresh Water

Given the semi-arid and drought-prone climate of the Southern Plains, all the forts were established near [presumably] reliable water sources. Water sources ended up not being reliable at Belknap and Chadbourne, resulting in their closure; and accordingly, there is no current use of surface or groundwater on their property. For the other forts where water resources are currently nonexistent (Table 4.3), this is a consequence of the property boundaries contracting since their establishment, away from their original water source. The rest of the forts contain rivers (impounded reservoirs in eight cases), springs, lakes/ponds, or wetlands; meaning they have at least incidental use of water resources. The larger forts, like Fort Sill mentioned in Sect. 4.3.2.1, contain most or all of these features.

Half of the forts (n = 17) are purposefully managing their water resources for a variety of benefits (Table 4.3). The fort water source benefitting the most people is the 27.5-million-gallon-per-day (mgd) solar-powered Kay Bailey Hutchison Desalination Plant on Fort Bliss, reportedly the largest inland desalination plant in the world. Not only does it serve the 164,000 people of the Fort Bliss military community, but it also contributes to El Paso Water Utilities' supply, which serves over 750,000 people. At a much smaller scale, Fort Parker Lake is the backup water supply for the City of Groesbeck, with a population of 4300. The other purposeful uses of water on the forts include local water supply, irrigation, boating, fishing, swimming, and tourism (Fig. 4.5). Obviously, these purposeful uses of water are benefitting other ecosystem services, some already described (habitat; irrigation for food crops) and others detailed below. One great example is how Fort Davis uses a groundwater well to keep its cottonwood grove alive during droughts, which benefits habitat, air quality, soil protection, aesthetics, and tourism, and has cultural significance (Myers 2000).

Fig. 4.5 Water resources on the forts are used for many benefits. At the Fort Clark Springs resort (top), Las Moras Springs, fed by the Edwards Aquifer, forms the headwaters of Las Moras Creek (bottom-left), which empties into the Rio Grande. In addition to local water supply, irrigation, fishing, recreation, and tourism, the 20 °C (68 °F) water from the springs is used to fill a 0.2 ha swimming pool on fort grounds (bottom-right) used for swimming and body temperature regulation

4.3.3 Regulating Services

4.3.3.1 Climate and Air Quality

When the forts were established, trees were highly valued as building materials and fuel. Nowadays, trees and other vegetation within the forts are valued more for their benefits to air quality. All the forts contain vegetation and thus incidentally benefit climate and air quality; however, 14 forts have allocated resources to plant new trees and preserve old ones (Table 4.3), like Fort Davis mentioned above. Fort Bliss alone has planted thousands of trees (DPW 2015). Many of the forts have planted native trees or restored native prairie (see Sect. 4.3.2.1), which also filters the air, stores/sequesters carbon, contributes to the water cycle, and regulates air temperature (Fig. 4.6). As the location of the Southern Plains Regional Climate Hub, Fort Reno is instrumental in demonstrating these benefits, as well as researching strategies to mitigate climate change. The air quality benefit of vegetation in forts is particularly important for the ten forts located in urban environments with artificial surfaces, high emissions, and higher levels of air pollution; however, fossil fuel fired power plants, smelters, industrial boilers, petroleum refineries, manufacturing, increased vehicular traffic, and wind-blown dust from around the two states and northern Mexico impact air quality of the entire region. The noticeable decrease in air quality since the mid-twentieth c. has led to hazes that impact the aesthetic views from the forts.

4.3.3.2 Water Quality and Flood Control

Water quality and flood control are additional environmental issues some of the forts have to address. Three quarters of the forts (n = 25) are located within active floodplains, and therefore incidentally mitigate downstream flooding and filter water pollutants, at least during floods, which are frequent in the Southern Plains. A few of the forts purposefully manage for floods and water quality (Table 4.3). Forts Bliss, Cibolo, Clark, Concho, Davis, McIntosh, McKavett, Reno, and Sill all maintain and restore riparian corridors throughout their property. Ten of the forts are

Fig. 4.6 Native grassland restoration is being carried out at several of the forts, including Fort Lancaster (left) and Fort Martin Scott (right)

located along or have creeks that flow into the Rio Grande, which in 1993 was designated by *American Rivers* as the most endangered river in the U.S. due to overpumping and excessive pollution. Consequently, these ten forts have the opportunity to improve its water quality, which Fort McIntosh has seized. With millions of gallons of raw sewage from Mexican tributaries and a plethora of toxic chemicals and trash from Laredo factories and warehouses, the Laredo section of the Rio Grande that runs along Fort McIntosh has been found to be the most polluted. In response, the Lamar Bruni Vergara Environmental Science Center (LBVESC) on Fort McIntosh has made many and considerable contributions to benefit the river by working with U.S. Border Patrol to mitigate its environmental impacts to the river and its riparian zone, including saving hundreds of trees from being felled, restoring 4 ha of native riparian habitat, restricting ATV usage on riverside trails, and preventing the construction of a border 'wall' in the floodplain which could increase downstream flooding. Fort Bliss has also benefited the Rio Grande by rehabilitating numerous incised arroyos which not only improves water quality, but also reduces soil erosion.

4.3.3.3 Soil Protection/Fertility

Of all the ecosystem services, soil protection/fertility was the least purposefully managed (Table 4.3). The argument could be made that all the forts are incidentally protecting soil by having vegetation, but I assessed whether the vegetation, structure, or activities were specifically designed to prevent soil erosion or promote soil fertility. Thus, forts with only mowed fields received a nonexistent for this category. I also did not count maintenance of the walking trails around fort buildings and ruins as incidental or purposeful. The forts that I identified as incidental were one that allows natural composting by not mowing or clearing vegetation (Boggy), ones that indirectly protect soil through riparian restoration (Bliss, Cibolo, Clark, Concho, Davis, McIntosh, McKavett, Reno, Sill), and ones that indirectly reduce soil erosion by hunting overpopulations of deer and feral hogs (Boggy, Bliss, Cibolo, Clark, Sill). Only seven forts purposefully manage for soil protection/fertility, one through riverbank stabilization (Concho) and three through composting (McIntosh, Parker, Sherman). Fort Parker State Park has also constructed several terraces to reduce soil erosion. Forts Bliss and Sill, according to their INRMPs, actively address all soil erosion on their property including nonpoint sources, plan land uses that minimize erosion, reroute roads out of arroyos or other eroding landscapes, reseed bare areas, limit hay cuttings to at least 10 cm above the soil, and construct stream crossings to prevent in-channel erosion. Fort Sill also fertilizes its agricultural fields with sludge from its wastewater treatment plant. Also noteworthy is that Fort Sill has 10,144 ha of 'prime farmland soils' which are protected under the Farmland Protection Policy Act of 1981. As a USDA research facility, Fort Reno conducts many activities to reduce soil erosion (no-till agriculture, cover crops) and promote soil fertility (manuring and other natural fertilizers).

4.3.3.4 Pollinaton

Pollination as it is used here (*sensu* TEEB 2010) is specific to cultivated crops. In order for a fort to receive a purposeful rating for this category, they have to (1) be using resources to plant or maintain wildflowers or other vegetation that attracts pollinators, (2) have at least one of the purposes of this activity being to attract pollinators, and (3) be located within 3 km (typical forage distance of honey bees) of 'cultivated crops' as identified using the 2011 National Land Cover Database (Homer et al. 2015) or have their own gardens (e.g. Clark, McIntosh). This third requirement resulted in some forts like Boggy receiving 'nonexistent' even though they are restoring a portion of their property to native vegetation potentially used by pollinators. Similarly, Forts Chadbourne, Davis, Duncan, Lancaster, and Richardson are all more than 3 km from any cultivated crops, and therefore received nonexistent (Table 4.3). Several of the forts are planting wildflowers, but are doing so merely for aesthetic purposes, and therefore receive an incidental rating. Of the forts that purposefully manage for pollination, most restore native prairie, manage butterfly gardens, and plant wildflowers with the purpose to attract pollinators. Fort McIntosh, for example, has two designated wildflower gardens and one butterfly garden which has already attracted over 40 species. Forts Bliss and Sill manage for specific plant and animal species that are important pollinators (DPW 2015). And Fort Reno conducts research and outreach on pollination.

4.3.4 Cultural Services

4.3.4.1 Recreation for Mental and Physical Health

Recreational opportunities on the forts that improve mental and physical health are multifarious. For 11 forts, outdoor recreation is their primary mission, including the Texas State Parks (Boggy, Parker, Richardson, Sherman), city/county parks (Belknap, Duncan, Fisher, Graham, Inge), and two resorts (Cibolo, Clark). The two schools, McIntosh and Ringgold, have outdoor recreational complexes and host a variety of activities that promote physical and mental health. The two U.S. Army bases have very active outdoor recreation programs and numerous parks, notably Biggs Park (Bliss) and Lake Elmer Thomas Recreation Area (Sill). Further, two of the heritage sites have parks on original fort property, which are included in this analysis: Hamilton Creek Park on Croghan and Rio Concho Park on Concho. Fort Graham received an incidental rating for this category because no resources are currently used for its maintenance, on account of Hill County recently terminating the lease with USACE. The other incidental rating is Fort Stockton, which is used for leisure walking and occasionally for city events like races, but does not use its resources for recreational purposes. The rest of the forts are primarily heritage sites that do not offer recreational opportunities (Table 4.3).

The most popular recreational activity on the forts with a purposeful rating is hiking (Fig. 4.7). Among all the forts, there are 559 km (347 mi) of nature hiking trails (Fig. 4.2); however, this value is skewed by Cibolo, which has 367 km alone. The other resort, Fort Clark Springs, has the next longest trail system with a total of 100 km, but only 27 km is available for everyday public use. The third longest trail system is 20 km, which Fort Richardson and Fort Parker State Parks each have. The median nature trail length among all forts is 1.6 km. The walking trails around fort buildings and ruins are not included in any of these figures, only trails that involve interaction with nature. Many of these trails are also used for biking, and some for horseback riding (Cibolo, Clark, Griffin, Inge, Richardson). Given the historical and cultural significance of the forts, many fort grounds and trails are used for prominent running and adventure races like the Port to Fort Adventure Race at Fort Gibson. Fort Washita hosts an annual track meet for about 3000 people.

The second most popular recreational activity is fishing, available at all four state parks (Fig. 4.8), both resorts, six city/county parks, both U.S. Army bases, both schools, and two of the heritage sites (Griffin, Washita), with most of these fish-stocking. Seasonal hunting is allowed on five forts (Boggy, Bliss, Cibolo, Clark, Sill). Boating, swimming, and camping are allowed at many of the sites as well. Most of the city/county parks have playgrounds and athletic fields, and four forts have golf courses (Bliss, Clark, Duncan, Sill). Among all the forts, there is a recreational activity for practically anyone's taste, even if it is just leisure walking around the heritage sites.

Fig. 4.7 Nature hiking is a popular recreational activity at many of the forts. At Fort Davis, hikers can explore 6.4 km of trails through mountainous terrain overlooking the fort

Fig. 4.8 Fishing is another popular recreational activity at many of the forts, especially the state parks like Fort Richardson (top). The quarry pond at Fort Richardson (bottom-left) is stocked with fish annually. Note how the fishing sign at the quarry pond (bottom-right) brings together various ecosystem service concepts

4.3.4.2 Aesthetics and Inspiration

All of the protected forts have historical and cultural significance, and all but a few are archaeological landmarks where cultural remains have been protected. Further, two-thirds of the forts have 'Living History Days' where they demonstrate cultural and technological aspects of the forts during the mid-1800s. Thus all these forts have incidental aesthetic and inspirational value, but some of the forts are purposefully dedicating resources [beyond just Living History Days] to promote appreciation of arts, culture, and science (Table 4.3). These activities include nature centers, art shows/classes, and modern cultural events. Fort Concho collaborates with the San Angelo Museum of Fine Arts for multiple events and programs. Similarly, Fort Leaton works with multiple organizations, including the Mexican Consulate, to host many and various arts and cultural events/programs/conferences. Native American cultural programs are held at Gibson, Towson, and Old Fort Parker Historic Site. Fort Richardson has a nature center (mostly preserved animals) on the fort grounds. Fort McIntosh has a more sophisticated environmental science center, natural history exhibit hall, and living laboratory that includes live native animals (Fig. 4.9). Forts Bliss, Clark, Duncan, McIntosh, Ringgold, and Sill all have multi-purpose centers used regularly for arts and cultural events.

Several forts take advantage of their unique, scenic landscape to host arts, cultural, and scientific activities. Fort Reno (with their 1200-ha native prairie) and Fort

Fig. 4.9 Fort Ringgold (top-left) is the site of numerous cultural activities, classes, and centers, including the community Multipurpose Center (top-right). Fort McIntosh provides a wide variety of community services (bottom-left) and is home to the Lamar Bruni Vergara Environmental Science Center & Natural History Exhibit Hall (bottom-right)

Davis (with their biodiverse mountainous desert terrain) both use their aesthetic landscapes for scientific research and classes in film, photography, botany, technology, and conservation. Fort McIntosh is home to the Rio Grande International Study Center, which not only conducts scientific research, but also leads a binational collaboration that preserves and protects the Rio Grande, its watershed and environment for the purpose of improving the environmental health of the community (Fig. 4.9). Regular scientific research also occurs on Forts Bliss, Lancaster, Leaton, Ringgold, and Sill. Fort Cibolo uses its picturesque landscape with abundant wildlife to provide guided trips for nature photography, horseback riding, and hiking. Fort Griffin offers several types of nature hikes and programs. Fort Griffin is also well-known for its Star Parties (i.e. stargazing events), which benefit from its unusually dark skies (Fig. 4.10). Other forts that have dark skies and regularly host stargazing events are Cibolo and Inge.

4.3.4.3 Identity and Spirituality

All of the forts create a sense of place/pride, but in keeping with the TEEB (2010) framework, I limit this classification to natural features such as springs, rivers, vegetation, and mountains. The Rio Grande, for example, is culturally significant, has sacred/religious meaning, and creates a sense of place/pride for multiple cultures. The four forts located along the Rio Grande receive at least an incidental rating for this category, but Leaton and McIntosh purposefully dedicate resources to promote

Fig. 4.10 Stargazing events are held regularly at three of the forts, with the most well-known being at Fort Griffin. (Image courtesy of the Texas Historical Commission)

this identity and spirituality. Fort McIntosh hosts the Dia del Rio festival every October which includes a river art exhibit, kayaking excursions, sustainability workshops, and a ceremony where ministers of multiple faiths bless the river and adjacent trail. A Native American ceremony is also held at this rare, natural crossing of the Rio Grande, where 5000 years of their history is acknowledged and blessed. The site of Fort Leaton is particularly significant and commemorated with multiple events because it is located along the *Camino Real* (of Mexico) and at *La Junta de los Rios*, the confluence of Rio Grande and Rio Conchos. Fort Ringgold, located on the Rio Grande, receives a purposeful rating for a different reason; it has three volcanic ash mounds (currently the sites for the Robert E. Lee House, Telegraph House, and Observation Tower) which were used by prehistoric Native Americans for stone tools and likely religious ceremonies. Created during the same geologic epoch about 30 million years ago is the 43-m high volcanic plug known as Mount Inge (Fig. 4.11), which was used by prehistoric and modern Native Americans. Mount Inge now serves another ecosystem service, blocking the light pollution from the city of Uvalde so that Fort Inge can still have stargazing events. The entire 12,140 ha property of Cibolo Creek Ranch is on a volcanic landscape, also from the same geologic epoch (Oligocene), and its three forts are located in a caldera. Fort Cibolo also benefits from its views over the Chisos Mountains, caves with fossils and Native American art, and springs.

Springs, important for identity and spirituality, are a common natural feature among the forts. Accordingly, cultural artifacts, some prehistoric, have been found

Fig. 4.11 Natural and cultural features at Fort Inge

around these springs. The most prominent of the springs is Fort Clark Springs, which has been transformed into a pond habitat, swimming pool, and park that hosts a wide variety of activities (Fig. 4.5). Other forts with notable springs include Bliss, Inge, McKavett, Parker, Richardson, Sherman, and Sill. Fort Sill also contains part of the Wichita Mountains, named for the Native American tribe that inhabited this region prior to European settlement, but also an important site to the Apache and Comanche tribes.

Three of the forts have culturally significant vegetation: the native prairie on Fort Reno that has never been plowed, the cottonwood grove on Fort Davis (detailed in Myers 2000), and multiple native species on Fort Bliss. Fort Bliss still allows multiple tribes to harvest natural resources used in religious ceremonies or that have cultural significance. One tribe is the Mescalero Apache who use agaves native to Fort Bliss for many different cultural/spiritual purposes.

4.3.4.4 Tourism

All of the forts are tourist attractions, attracting more than 600,000 people every year overall. The mean fort visitation is 18,370 and the median is 10,000. Only six of the forts received an incidental rating because they are not currently using resources to increase tourism (Table 4.3). The most visited fort is Fort Fisher due to it being situated along a riverwalk next to a major university (Baylor) in a large city (Waco) that is located along one of the busiest interstates in the nation (I-35), and also because it is the site of the Texas Ranger Hall of Fame and Museum (Fig. 4.12). In fact, the 80,000 annual visitors I documented are only for the Museum. The hundreds of people that use the Waco tourist information center, Riverwalk, and Fisher

Fig. 4.12 Fort Fisher Park (top) is home to the Texas Ranger Hall of Fame and Museum, which receives more than 80,000 tourists each year. Not pictured in this photo are the Riverwalk and Waco tourist information center, which are used by hundreds of people each day. Fort Sherman (middle), located in Lake Bob Sandlin State Park, is one of the most used campgrounds in Texas and the Southern Plains. Fort Davis National Historic Site (bottom) receives more than 100,000 tourists in some years

Park every day were not taken into account. Fort Sherman, as part of Lake Bob Sandlin State Park, is the second most visited with a range of 50–75 k over recent years. The popularity of this park is due to its proximity to Dallas and a major interstate (I-30), having a large lake with excellent fishing and other outdoor amenities, and functioning as a Texas State Park with one of the largest and most scenic campgrounds in the region (Fig. 4.12). Texas State Parks are well-advertised, affordable, and appeal to a large population of outdoor enthusiasts. The other two State Parks with campgrounds, Parker and Richardson, were also some of the most visited forts, 6th and 7th respectively. The other State Park, Boggy, does not have a campground and is relatively far from a major city. The third most visited fort, with 56,000 annually, is Fort Davis which is a National Historic Landmark and advertised broadly by the National Park Service. Visitation here has been as high as 135,800, in 1966 (Fig. 4.12).

The high attendance at some of the forts is due in part to their location in popular tourist destinations. This is definitely the case for Fort Martin Scott which is located in the new national tourist hotspot of Fredericksburg, known as one of the best wine regions in the nation. The fort property also contains the Texas Rangers Heritage Center. Fort Concho, the fifth most visited fort, is located in the metropolitan area of San Angelo, which is a popular regional tourist destination. Fort Concho is also highly engaged in tourism promotion and cultural events, plus a wide range of public service activities. Located along Route 66, Fort Reno experiences higher visitation than most forts, especially since it is primarily a research facility. Fort Gibson, one of the most visited of all forts despite being far from any major cities or highways, benefits from its location on the Trail of Tears National Historic Trail and being a Cherokee Nation cultural tourism destination. Likewise, McIntosh benefits from its location along El Camino Real de los Tejas National Historic Trail.

The protected natural areas of the forts and their diverse habitats (Sect. 4.3.2.1) also attract ecotourism. In terms of wildlife viewing, birding is by far the most popular activity on the forts. About half of them have birding events or provide field guides, and twenty are *eBird* hotspots (ebird.org). As described previously, the Rio Grande forts are prime birding destinations. Fort McIntosh, in particular, is one of the main stops for the annual Laredo Birding Festival, attended by birders from all over the U.S. Fort McIntosh also contains the Paso del Indio Nature Trail, which provides opportunities to experience, learn about, interpret, and conserve the natural environment. Other forts with popular and educational nature trails include Boggy, Clark, Davis, Griffin, Leaton, McKavett, Parker, Richardson, and Sherman. These nature trails enhance environmental awareness of fort visitors, in a manner that promotes all of the ecosystem services.

4.4 Concluding Remarks: Forts Then and Now

The forts established during the Southern Plains Indian Wars of 1821–1890 forever changed the physical and cultural landscape of Texas and Oklahoma. During this 70-year period, the forts allowed Texas to grow from just three settlements with a few thousand people in 1820 to 2.2 million people scattered across the state by 1890. Oklahoma (then part of the Arkansas Territory) was experiencing the beginnings of Indian relocation in 1820, but by 1889, Oklahoma was open to all settlers with the first of many Land Runs. The forts were set in places that provided a combination of ecosystem services: a clean and reliable water supply, timber or stone construction materials, arable soil, and productive grazing lands (Wooster 1987). Their location was also influenced by economic considerations. In fact, these economic (and political) considerations often outweighed military strategy (Wooster 1987). When the forts were established in Texas and Oklahoma in the nineteenth century, they provided numerous economic benefits. They were the impetus for infrastructure such as roads, rail, and water wells; they became centers of trade; they attracted many enterprises; and they were a major source of appropriations and other government funding (Smith 1999; Wooster 1987, pp. 13–14). In Texas alone, the federal government spent over $83 million between 1849 and 1900 on army infrastructure, operations, and soldier pay (Smith 1999). Although created with military, political, and economic intentions, these forts were more instrumental in creating societies, societies with an improved level of well-being.

Today, these forts continue to improve human well-being through their ecosystem services (Tables 4.2 and 4.3). They provide the provisioning services of habitat, food, raw materials, energy, and fresh water. A few of the forts even provide permanent human habitat (Bliss, Clark, Sill). They provide regulating services such as climate/air quality, water quality, flood control, soil protection/fertility, and pollination for crops. And they provide the often-overlooked cultural services of recreation for mental and physical health, aesthetics, inspiration, identity, spirituality, and tourism. Tourism is particularly important because the provision of ecosystem services is largely dependent on the amount of funding and resources of the fort.

Indeed, my 'purposeful' classification requires that resources are used to manage the ecosystem service. Many of the forts are limited in this regard. Managing for ecosystem services requires (wo)manpower. Several of the forts had only one full-time staff member or less (Duncan, Graham, Inge, Mason, Supply, Washita), and many others only had 2–3 staff members. Many of the forts have also experienced a decrease in volunteers over the past decade, which has forced them to reduce their operating hours, program offerings such as Living History Days, and educational outreach. Attendance has also decreased at some of the forts due to changes in infrastructure (e.g. decreased traffic by Fort Lancaster due to Interstate 10 replacing Highway 290 as the major route to western Texas) or simply changes in tourism patterns. This decrease in tourism reduces funding of the forts and their ability to provide services. Consequently, some forts have had to move their museums or headquarters to nearby cities to increase awareness (Phantom Hill, Supply).

Other forts have seen increases in attendance. With Fredericksburg becoming a major tourist destination, Fort Martin Scott is experiencing increases in attendance and funding, and accordingly is adding ecosystem services to their property such as prairie restoration, nature trails, and more educational programs. With the re-enchantment of Route 66, Fort Reno has also experienced increases in tourism, programs, and services. If ecosystem services are to increase, then more funding and resources will be needed.

One of my findings from this research is that many of the 'purposeful' ecosystem services are recent activities, as the forts have transitioned from heritage centers to multi-use facilities. Further, this concept of ecosystem services is relatively new to several of the fort directors/managers; they are in the process of incorporating ES into their property management, tourism, and education. At Fort Belknap, for example, a wildflower and butterfly garden was recently planted. Fort Towson is in the process of planting a native educational garden and expanding their trail system. Fort Griffin plans to expand their campground. Fort Martin Scott is in the process of restoring native prairie and plans to construct nature trails in the future. These are just a few of the examples of the initiatives of the protected forts, and thus my inventory of ecosystem services here is likely to expand over the coming years. And hopefully new forts like Fort Bird will be protected, commemorated, and restored. In the past few centuries, historical places such as forts have become increasingly protected, whether a national park, state park, city park, or through private entities. As William Cronon so elegantly conveyed in Ken Burns' (2011) documentary on *The National Parks: America's Best Idea*: "We come from nature. But we also come from our own past. And so the interpretation of nature and history, together, is not a distraction that the parks face. It is the very core of the enterprise. They're all about where we come from."

Acknowledgments Many fort/park site directors/managers provided invaluable information on the forts and their numerous ecosystem services and societal benefits. These included Mike Alexander (Fort Phantom Hill), Jim Argo (Fort Washita), David Bianchi (Fort Martin Scott), Bob Bluthardt (Fort Concho), Crista Bromley (Hamilton Creek Park, City of Burnet), Chris Caswell (Fort Sherman, Lake Bob Sandlin State Park), Manuela Cataño (Fort Leaton State Historic Site), John Davis (Fort Towson), Chris Deurmyer (Fort Sill), Bill Dillahunty (Fort Inge), Tom Fisher (Fort Parker State Park), Fort Sill National Historic Landmark and Museum, David Fowler (Fort Gibson, Oklahoma Historical Society), Robert Frie (Fort Richardson State Park), Prasanna Gowda (Fort Reno, USDA), Erica Guerra (Fort Ringgold, Rio Grande City CISD), Wesley Hamilton (Fort Boggy State Park), Jim Hammond (Fort Belknap), Roger Havlak (Rio Concho Park, City of San Angelo), John Heiner (Fort Davis National Historic Site), Fran Hoerster (Fort Mason), Taylor Hunter (Fort Supply), Monique Jensen (Fort Clark Springs), Byron Johnson (Fort Fisher, Texas Ranger Hall of Fame and Museum), Susan Johnston (Fort Graham chapter of the National Society of the Daughters of the American Revolution), Wanda Kienzle (Old Fort Bliss), Jane Lenoir (Fort Griffin), Sarah McReynolds (Old Fort Parker Historic Site), Ann Marshal (Fort Reno, BlueSTEM AgriLearning Center), Tom Miller (Fort McIntosh, Laredo Community College), Cody Mobley (Fort McKavett), Melba Montoya (Historic Fort Stockton), Brian Northup (Fort Reno, USDA), Russell Nowell (Fort Clark Historical Society), Wendy Ogden (Fort Reno, U.S. Cavalry Association), Sarah Overholser (Historical Fort Reno), Bob Rea (Fort Supply), Omar Reed (Fort Gibson), Garland & Lana Richards (Fort Chadbourne), Steven Ridlehuber (Fort Graham, USACE), Brian Severn (Cibolo Creek Ranch), Russell Skowronek (CHAPS Program, University of

Texas-Rio Grande Valley), Jefferson Spilman (Fort Lancaster), Glen Taylor (Fort Inglish), Jeff Taylor Sr. (Fort Duncan), Laura Whitehurst (Fort Phantom Hill), Darrell Williams (Fort Richardson), Milli Williams (Fort Croghan). Emily McBroom helped research locations and history of the forts. A special thanks to Victoria, Aaralyn, and Griffin Julian, who traveled to all the forts and helped me collect information, photographs, and memories.

References

Adams, C. E., Higginbotham, B. J., Rollins, D., et al. (2005). Regional perspectives and opportunities for feral hog management in Texas. *Wildlife Society Bulletin, 33*(4), 1312–1320. https://doi.org/10.2193/0091-7648(2005)33[1312:RPAOFF]2.0.CO;2.

Alexander, T. E., & Utley, D. K. (2012). *Faded glory: A century of forgotten Texas military sites, then and now*. College Station: Texas A&M University Press.

Alexander, T. E., & Utley, D. K. (2015). *Echoes of glory: Historic military sites across Texas*. College Station: Texas A&M University Press.

Archer, S. (1994). Woody plant encroachment into southwestern grasslands and savannas: Rates, patterns and proximate causes. In M. Vavra, W. A. Laycock, & R. D. Pieper (Eds.), *Ecological implications of livestock herbivory in the west* (pp. 13–68). Denver: Society for Range Management.

Awbrey, B. D., & Dooley, C. (1992). *Why stop?: A guide to Texas historical roadside markers*. Houston: Gulf Pub Co.

Burns, K. (2011). *The national parks: America's best idea – Part 6, "the morning of creation" (1946–1980)*. Washington, DC: Florentine Films.

Directorate of Public Works (DPW). (2015) *Environmental assessment implementation of the 2015 Integrated Natural Resources Management Plan Fort Bliss, Texas and New Mexico*. Environmental Division, Fort Bliss.

Field, R. (2006). *Forts of the American frontier 1820–91: The Southern Plains and southwest*. New York: Osprey.

Fischer, J., & Lindenmayer, D. B. (2007). Landscape modification and habitat fragmentation: A synthesis. *Global Ecology and Biogeography, 16*(3), 265–280. https://doi.org/10.1111/j.1466-8238.2007.00287.x.

Frantz, J. B. (1970). The significance of frontier forts to Texas. *The Southwestern Historical Quarterly, 74*(2), 204–222.

Griffith, G., Bryce, S., Omernik, J., & Rogers, A. (2007). *Ecoregions of Texas. Texas Commision on environmental quality* (p. 125). Austin.

Gwynne, S. C. (2010). *Empire of the summer moon*. New York: Scribner.

Haines-Young, R. (2016). *Report of results of a survey to assess the use of CICES, 2016. Support to EEA tasks under the EU MAES process*. Negotiated procedure No EEA/NSS/16/002.

Hannings, B. (2006). *Forts of the United States: An historical dictionary, 16th through 19th centuries*. Jefferson, NC: McFarland & Company, Inc.

Homer, C. G., Dewitz, J. A., Yang, L., et al. (2015). Completion of the 2011 National Land Cover Database for the conterminous United States – Representing a decade of land cover change information. *Photogrammetric Engineering and Remote Sensing, 81*(5), 345–354.

Johnsgard, P. A. (2012). *Wings over the Great Plains: Bird migrations in the Central Flyway*. Lincoln: Zea Books.

Michno, G. (2011). *The settlers' war: The struggle for the Texas frontier in the 1860s*. Caldwell: Caxton Press.

Millennium Ecosystem Assessment (MEA). (2005). *Ecosystems and human Well-being: Synthesis*. Washington, DC: Island Press.

Myers, D. K. (2000). *Fort Davis National Historic Site, Texas: Cultural landscape study* (p. 168). Santa Fe: National Park Service.
National Park Service (NPS). (1993). *Buffalo Soldiers: The African-American contribution to guarding the frontier*. The Museum Gazette, Jefferson National Expansion Memorial, https://www.nps.gov/jeff/learn/historyculture/upload/buffalo_soldiers.pdf. Last accessed 27 Aug 2017.
National Park Service (NPS). (2016). *NPSpecies for Fort Davis National Historic Site. National Park Service Integrated Resource Management Applications*, https://irma.nps.gov/NPSpecies/Search/SpeciesList/FODA. Last accessed 21 Dec 2016.
Pierce, G. S. (1969). *Texas under arms: The camps, posts, forts, & military towns of the Republic of Texas* (pp. 1836–1846). Austin: Encino Press.
Smith, T. T. (1999). *The U.S. Army and the Texas frontier economy: 1845–1900*. College Station: Texas A&M University Press.
Stout, G. (2014). *Integrated natural resources management plan and environmental assessment 2014–2018: Fires center of excellence and Fort Sill, Fort Sill, Oklahoma* (p. 252). Fort Sill: Natural Resources and Enforcement Branch, Environmental Quality Division, Directorate of Public Works, U.S. Army.
TEEB. (2010). *The economics of ecosystems and biodiversity: Mainstreaming the economics of nature: A synthesis of the approach, conclusions and recommendations of TEEB* (p. 36). Malta: United Nations Environment Programme. http://www.teebweb.org/our-publications/teeb-study-reports/synthesis-report/.
Woods, A. J., Omernik, J. M., Butler, D. R., et al. (2005). *Ecoregions of Oklahoma*. Reston: U.S. Geological Survey.
Wooster, R. (1987). *Soldiers, sutlers, and settlers: Garrison life on the Texas frontier*. College Station: Texas A&M University Press.

Chapter 5
The Conflict Landscape of Verdun, France: Conserving Cultural and Natural Heritage After WWI

Rémi de Matos Machado and Joseph P. Hupy

Abstract Known to have been the scene of one of the greatest battles of history, the battlefield of Verdun (France) is now a vast forest area of 10,000 ha. Created in the aftermath of the World War One, the forest of Verdun is the result of a long process led by the state since the 1920s. Formerly used to reconstitute damaged land, the forest has many functions today, including that of a place of memory, which draw nearly 250,000 visitors each year. The forest of Verdun also constitutes a high environmental value site, featuring a large mosaic of environments and a remarkable range of flora and fauna. Thanks to a LiDAR mission conducted in 2013 within the *"Forêt d'exception®"* project, nearly 115 km^2 of woodland were surveyed, allowing the National Forests Office (ONF) to reconstruct 3-D topography of the battlefield and map hundreds of kilometers of trenches and thousands of shelters still visible despite the century that had passed since the war. The revegetated forest land cover has helped to maintain the underlying scars on the landscape. After outlining the major phases of restoration of the Verdun battlefield, this chapter explores the links between the geomorphological traces left by WWI and present-day biodiversity in the forest of Verdun.

5.1 Introduction

In today's world, which is so influenced by modern high-tech cinema, it is easy to imagine that battle in warfare was always a highly destructive epic event capable of widespread destruction. Images rendered by the movie industry, depicting exploding cannon rounds in nineteenth-century warfare that dislodged fountains of earth

R. de Matos Machado (✉)
UFR Géographie, Histoire, Economie et Sociétés, Université Paris-Diderot, Paris, France
e-mail: remi.machado@univ-paris-diderot.fr

J. P. Hupy
School of Aviation and Transportation Technology, Purdue University, Indiana, West Lafayette, USA
e-mail: jhupy@purdue.edu

T. R. Lookingbill, P. D. Smallwood (eds.), *Collateral Values*, Landscape Series 25,
https://doi.org/10.1007/978-3-030-18991-4_5

and blasted soldiers and trees skyward, inaccurately describe the technological capabilities of first-generation warfare at the time. Most artillery rounds were simple, round cannon balls that did nothing more than bounce along the ground – intended to rip through massed soldiers in formation. Artillerymen who experimented with exploding cannon balls took a large risk, and the use of such artillery was limited to ship fighting, and some siege warfare. Prior to the twentieth century warfare environmental damage was, with some exceptions, quite limited in its scope and magnitude. Military campaigns were short, and preference was given to open agricultural fields where the armies could rely on decent roads and an ample food supply from both the surrounding countryside and a supply line that was close to their base of operations. Environmental damage often stemmed from armies pillaging the resources of the landscape to support themselves when their supply line was either broken or unreliable.

The Industrial Revolution and technological innovations of the twentieth century changed the face of warfare (Hupy 2005). Every aspect of modern war is of greater magnitude than that of warfare prior to the Industrial Age; armies and battlefields are larger, munitions are more powerful, and the disturbances are more widespread. Besides continual advances in explosive munitions technology, modern warfare contributes to environmental disturbance in many other forms, *e.g.*, heavy vehicle traffic, chemical defoliants, and atmospheric and water pollution (Westing 1972; Rose and Nathanail 2000; Brauer 2009; Closmann 2009; Hupy 2011). How the environmental damage rendered by warfare has progressively evolved relates to changes in technology, and ideas on how war is waged based on those technological innovations. Many landscapes around the world bear the vestiges and scars of warfare, yet several stand out in terms of the sheer amounts of devastation rendered upon them. One such landscape is the Western Front of the 1914–1918 Great War (WWI in the United States). Battlefields such as these attest to the fact that the damages rendered by warfare should be viewed as a catalyst to change, and that post-war economic activity also plays a great role in shaping the divergent pathways of landscape development in the post-war era.

This chapter examines the WWI battlefield of Verdun, France (Fig. 5.1) to identify how disturbances rendered on that battlefield have had a lasting impact on the modern landscape. After providing a historical overview of the battlefield, we analyze the post war landscape through an economic and ecological perspective. We conclude with a discussion of the ecological potential of landscapes disturbed by war.

5.2 Disturbance and Divergent Paths of Recovery: The Western Front of the Great War

Although several wars, such as the Franco-Prussian, the Russo-Japanese, and the Spanish-American, allowed armies to 'test' and develop munitions that utilized the weapons of modern war, it was not until WWI that these developments were fully implemented at an industrial scale (Bailey 2004; Keegan 1993). In WWI, the same

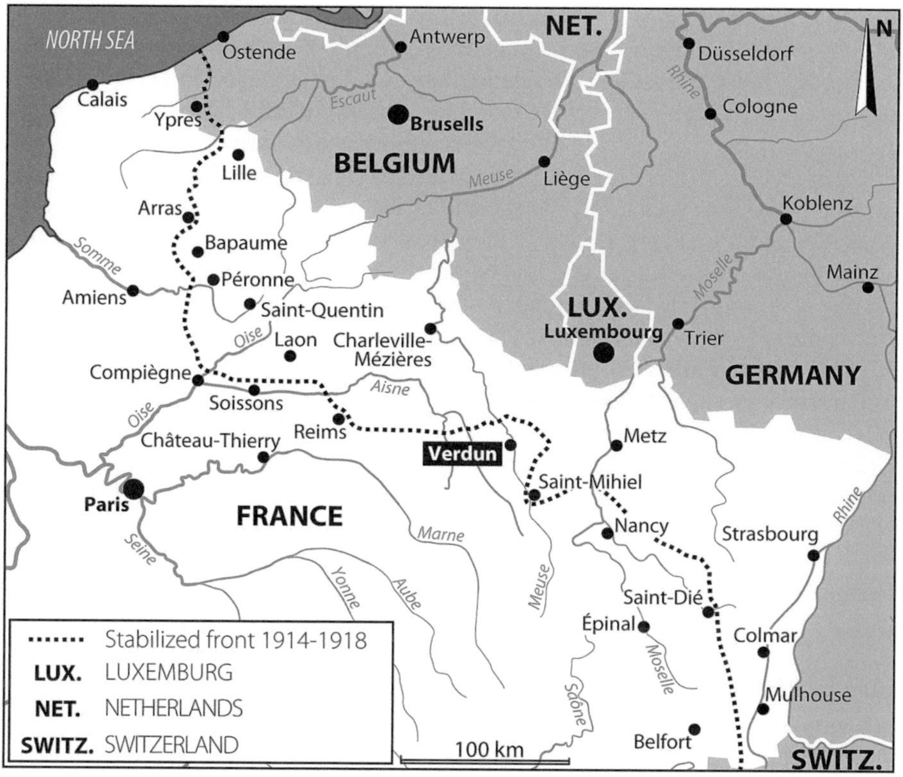

Fig. 5.1 Location of the Verdun battlefield on the Western Front (Designed and produced by R. De Matos Machado)

concepts associated with the Industrial Age were introduced into the philosophy of war. Instead of armies that numbered in the thousands of troops, a nation needed millions of troops in order to be a powerful, warring nation state. A nation required a well-built infrastructure and massive industrial complex just to support its massive armies. For example, by several months into the 'Great War' (as WWI would soon be known), it was realized that those nations capable of out-producing the other nation would have a distinct edge. Commanders also realized that the days of dashing cavalry charges and brightly colored uniforms, used so armies could communicate in the thick smoke of battle, were over; new tactics needed to be implemented.

The extremely long range of rifled infantry weapons and the rapidly firing machine gun forced commanders to take artillery off the front lines after several devastating losses in the beginning stages of WWI. Breech-loading rifles and rapid-firing weapons such as the Gatling gun had been in existence since the American Civil War, but army commanders were slow to adopt the changes brought about by these weapons, mainly because most countries did not upgrade their arsenals with these weapons until the advent of WWI. Artillery took up positions in the rear and perfected the art of indirect fire, based on the calls of forward observers. The role of artillery was to heavily shell an area in order to destroy enemy defenses and shatter its morale. Terms such as the 'straight barrage', 'rolling barrage', 'piled up barrage', and 'creeping barrage' were coined to refer to curtains of artillery fire placed directly in front of advancing troops to obliterate anything on the surface.

Before WWI, artillery units attached to armies were allotted, at most, 100 rounds per day for combat operations. By the end of the war, artillery units were assigned several hundred rounds per hour. In the age of smooth bore artillery, artillery units often brought one round per gun into battle; siege guns firing more than five rounds per day was considered exceptional. At the start of the war, artillery was seen as an arm to directly support the infantry and wars were won by élan, or courage of the infantry; by the war's end, the mantra of all commanders was, 'Artillery conquers and infantry occupies' (Gudmundsson 1993). Artillery therefore emerged from WWI as the deciding factor in battle.

The environmental consequences of this type of warfare obliterated forests and significantly cratered the landscape, thus creating wide swaths of destruction, limited only by the range that artillery could fire, which was well beyond the visible range of the gunners (Hogg 1985, 1987). Perhaps the best-known example of this swath of destruction is the Western Front, which was an average of 20 km wide and stretched from the English Channel to the border of Switzerland (Keegan 1998). Disturbance along this front varied, despite popular depictions and opinions. Four main variables related to differing magnitudes of disturbance along the front: (1) Location of armies in relation to the front line, with disturbance tapering off from the areas of stalemate; (2) Degree and duration of stagnation of the front line; (3) Position of the armies in relation to the topography of the landscape, with ridge fronts and tops receiving more damage than the backsides; (4) Characteristics of the underlying parent material or bedrock (Hupy 2005).

Today, the remnants of the Great War are very much at the behest of the cultural and economic landscape. Battlefields such as Verdun have limited public access and are treated more or less as mass graves at the battlefield level. Others, such as the Somme, have been placed back into agriculture, with portions preserved as memorials, monuments, and museums. Along the former front line and outside the memorials, vestiges of war remain in the form of German *Stollen* (underground bunkers), UXOs (unexploded shells), and woodlots filled with intact craters and trenches. In the areas that remain pocked with craters, the landscape is revegetated, yet soil development is different than outside of the craters, and areas of standing water exist where none stood before. Viewed in this context, the war did not so much permanently disturb the landscape, as to set it off on a divergent path of landscape evolution.

5.3 The Verdun Battlefield as an Object of Study

5.3.1 *Verdun: Study Area*

The Verdun battlefield is an excellent case study for examining the vestiges of war and divergent pathways in landscape evolution. Located at a little over 200 km east of Paris, in the vicinity of the Meuse (Lorraine region), the battlefield benefits from a particular geographical position on the heights of the Meuse valley and the town of Verdun (Fig. 5.2). The site occupies the Hauts-de-Meuse plateau, ranging between 300 and 350 meters a.s.l. The plateau surface is rugged and frequently cut by several valleys with varied morphologies, largely employed by the armies at the time of the conflict of 1914–1918 to shelter and to penetrate enemy defenses (Hupy 2005; Amat 2015).

Westward of the plateau is the incised Meuse valley, with an approximate depth of 150–200 m, occupied by the town of Verdun in one of its meanders. To the east, the study area is characterized by a wetland depression: the plain of Woëvre, with an altitude of 200–250 m a.s.l. The connection between the plateau and the plain comprises a slope: the cuesta of Meuse, commonly called 'Côtes-de-Meuse', which constituted at the end of the nineteenth century a natural barrier against attacks from the east (Amat 2015; Hupy 2005).

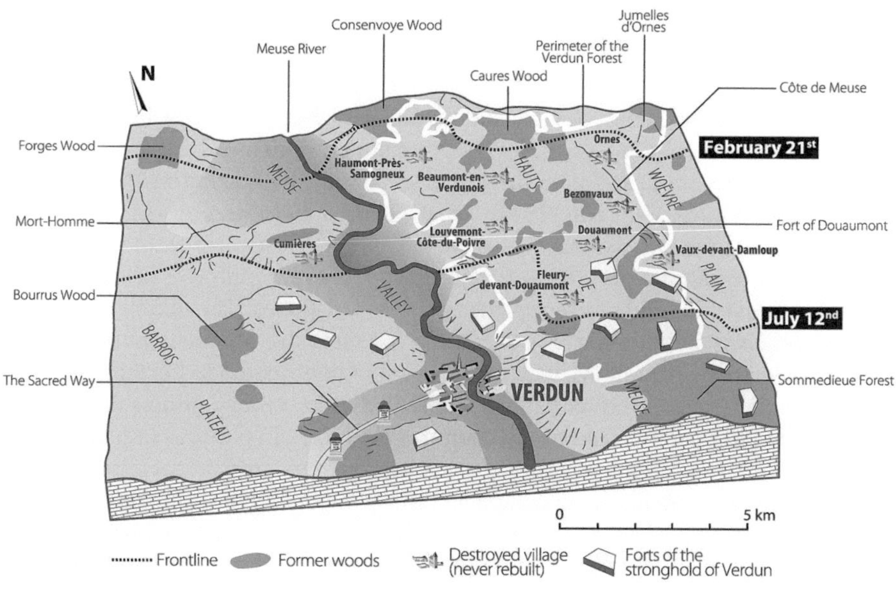

Fig. 5.2 3-D view of the Verdun battlefield in 1916 (Designed and produced by R. De Matos Machado)

The lithology of the battlefield is carbonate materials, alternatively resistant (massive limestone) and soft (clay and marl) from the Middle to Upper Jurassic period (Hupy 2005; Amat 2015). The plateau is sprinkled with many resurgences resulting from the presence of perched water tables. A first level appears between the series of Lower and Middle Oxfordian composed of resistant limestone and marl. This aquifer involves gravity depression springs on the Côtes-de-Meuse. A second level, which is less active, is above the clayey and marly series of Upper Oxfordian and is where most of the villages are located (Millarakis and Wagner 1999).

The Lorraine climate is characterized by a strong interannual variability of precipitation and temperature (Joly et al. 2010). The area tends towards a very altered oceanic climate, close to continentality, and sometimes takes the form of mountain climate (Joly et al. 2010). During the coldest month (January), climatic conditions are cold with median values ranging from 0 to 2 °C in the Meuse valley and from 0 to 5 °C on the plateau, contributing to the formation of a snow cover. Spring is marked by a rise in temperatures (average of 10 °C for April) and summers are moderately warm, with an average of 19.5 °C in July (the hottest month). Also, the area is characterized by a high amount of precipitation, with mean annual precipitations of 758 mm.

5.3.2 Verdun: History

From a human society point of view, the Verdun area was very much an agricultural landscape at the beginning of the twentieth century, dominated by open fields of wheat, oat, barley, rye, and potato, with a smattering of grazing cattle on gentle slopes (Malte-Brun 1845; Joly 1911; Amat 2015). Fields were organized around villages settled in the Meuse valley and the Hauts-de-Meuse plateau. In the margins of municipal territories, where landscape relief is more prohibitive of agricultural economic activities, fields were replaced by wooded slopes. The number of woodlots was relatively high in the area of Verdun but their size remained rather modest (Fig. 5.2). On the Côtes-de-Meuse, solar exposure contributed to the cultivation of fruits, especially grapevines and plum orchards. Lastly, valley bottoms promoted a configuration of marshy meadows, sometimes organized in pastures (Amat 2015).

In regard to traditional military geography, Verdun presented Germany with some of the worst conditions to launch an offensive operation: heavy precipitation; long, cold winters; steep east-facing escarpments; and soils that bogged down equipment and men (Winters et al. 1998). By 1915, WWI had succumbed to the realities of technology (Horne 1993). The Central and Allied powers were locked in a stalemate along the Western Front (Keegan 1998). Offensive movements were extremely limited due to extensive trench networks and the deadly efficiency of machine guns. Late in 1915, however, the German high command came up with a contingency plan that was designed to break the stalemate and ultimately knock France out of the war (Holstein 2002). The plan was unconventional in that it contained no strategic objectives and was designed merely to break the will of France to fight through a battle of attrition by 'Bleeding her to death'. Germany reasoned that by eliminating France from the war, Russia could be defeated in the east, then Germany could turn around and conquer her true enemy in the west, England. In essence, Germany wanted to 'eliminate the sword from England's right hand' by demoralizing France to the point of surrender and ultimate defeat.

Germany chose the Verdun region to launch the offensive because they knew this was a highly symbolic region of France and the French would throw everything they had into defending it solely for the sake of patriotism. The reason behind this was that during the Franco-Prussian War of 1870–1871, France had suffered a humiliating defeat at Sedan and was forced to cede control of its Alsace-Lorraine province to Germany (Ousby 2002). France, to protect her borders from further losses, had set up a massive ring of forts to protect the region. The area became a symbol of pride for France and a bulwark against German invasion. Thus, WWI began with Verdun bordering German territory. An additional incentive for the Germans to choose Verdun was supply line logistics. While France had few roads to supply the front, which could easily be destroyed (and were) by well-placed artillery, the German side of the front was blessed with a plethora of supply lines to feed the offensive (Michelin travel guide 1919; Winters et al. 1998; Bernède 2002).

The battle of Verdun began on February 21st, 1916 with the German army launching a 'steel storm' (Jünger 1961) 10 km north to Verdun (Fig. 5.2). The opening artillery barrage lasted for two straight days and was the largest artillery bombardment of the war up to that point (Martin 2001). Estimates from French soldiers placed incoming shells at rate of 25–30 rounds per minute in the Caures Wood, the center of the Verdun salient (Fig. 5.2; Service Historique de l'État Major de l'armée n.d.). At first, the Germans made large offensive gains, but eventually this battle, like the war itself, became locked in stalemate with both sides pouring hundreds of thousands of men into a futile struggle (Horne 1993; Bernède 2002). Both sides relied heavily upon artillery to break the morale of the enemy and 'soften up' the other side for an offensive effort to retake a fort, pillbox, trench, ridge or high point on the battlefield. These objectives changed hands daily with fresh craters filling in old craters and trenches excavated into trenches dug only hours before (Cowley 2004). For nearly a year, this exchange of artillery continued to pulverize the landscape into what many soldiers described as 'something from another world' (Bagnold 1990) or what pilots flying above would describe as like 'the surface of the moon'.

5.4 Effects of the Fighting on Local Environment

Although the principal impact of warfare on the Verdun battlefield was on the geomorphic landscape, with the creation of conflict-induced landforms such as trenches and bomb craters (Fig. 5.3; Ilyès 2006), those disturbances also involved an ecological significance, still visible today. These disturbances have their origin in the

Fig. 5.3 View looking south of the Verdun battlefield (Verdun forest) from the Abri 320 (Photo by De Matos Machado 2015)

massive use of explosive and chemical devices (Masson-Loodts 2014; Hubé 2016). The extensive shelling at Verdun caused numerous species of flora and fauna to disappear over the course of the battle and in the immediate years following these disturbances (Westing 1980; Parent 2004). From a faunistic perspective, Parent's study (2004) showed that all reptile populations that occupied the Verdun area before war had disappeared. The phenomenon also relates to the majority of amphibians. Nevertheless, the majority of amphibian communities recolonized the post-conflict landscapes with the exception of Pelodytes punctatus, now supposed to be extinct (Parent 2004).

For plant communities, the damage rendered by the explosion blasts created a destruction of the canopy and trunks after opening the forest cover. In extreme cases, vegetative cover on the ground was completely removed (Amat 2015). Along with vegetative removal and damage, humus and soils were also subjected to 'bombturbation' in a range of disturbance magnitudes, characterized by a compaction and mixing of soil horizons (Hupy 2006). In one of the more extreme examples, on the Thiaumont platform at the center of the battlefield, soil disturbance exceeds 10 m of depth. After the war, when the bedrock was exposed, chasmophytic species appeared, such as *Galeopsis angustifolia*, *Teucrium botrys*, *Teucrium montanum*, *Teucrium scorodonia*, *Thymus praecox*, and *Thymus pulegioides*. Such species sometimes persisted until the 1960s, before weathering and soil development covered exposed bedrock (Parent 2004).

Unlike mechanical effects, chemical disturbances are not easily visible but are particularly important in the conflict landscapes. Upon impact, millions of explosive and gas ordnance released chemical agents into the ground. Some of these particles have the capacity to impact the local environment for the long-term (Souvent and Pirc 2001; Bausinger and Preuß 2005; Bausinger et al. 2007; Van Meirvenne et al. 2008; Hupy 2011; Prestidge 2013; Hubé 2016). This is the case for perchlorate salt, widely used in munitions during WWI and relatively harmful. In 2012, its presence was noted within several aquifers close to the Western Front. Consideration must also be given to heavy metal pollution resulting from the corrosion of metal objects, especially shells. Indeed, some studies recorded an abnormally high rate of metal particles in the grounds of the Western Front. This mainly concerns copper, lead and zinc (Bausinger and Preuß 2005; Bausinger et al. 2007; Meerschman et al. 2011; Hubé 2016). Unfortunately, the situation is not improving since it is estimated that between 200 and 300 million unexploded ordnance (UXOs) still remain unrecovered (Fig. 5.4; Hubé 2016). Thus, it is probable that the chemical weathering of soils by munition deterioration involved a flora modification at a local scale, causing the disappearance of species not adapted to the new edaphic conditions or contributing to the introduction of specific species. Perhaps that would explain the low number of herbaceous species recorded by Parent (2004) in several areas of the modern Verdun battlefield (as low as 10 species on a 1 ha surface area).

Fig. 5.4 An unexploded shell discovered during a field mission in the Verdun forest (Photo by Jacquemot 2014)

5.5 Land Management in the Aftermath of the War

After WWI, the Verdun battlefield was abandoned; bodies were left to decay and the remnants of war were strewn everywhere. The devastation along the Western Front, particularly at Verdun, was by all means catastrophic. In addition to the million hectares of devastated fields, woods of the Verdun region were significantly diminished. Compared to the pre-battle landscape, 65% of Meuse forest land was damaged, translating to approximately 120,000 ha of forest destroyed during the battle (Amat 2015). The agricultural economy was also decimated. Large expanses of agricultural land were never replowed due to the tens of millions of craters and unexploded shells lying on or just beneath the surface. The war destroyed 325 villages, and of that number, nine have never been rebuilt (Amat 2015). Eventually the barren, cratered surface was covered by a thick mass of shrubby vegetation. French officials believed the area was forever devastated and abandoned any plans for restoration (Holstein 2002; Amat 2015).

Revegetation occurred quickly in the post-conflict years (Fig. 5.5), but the type of regrowth did not match the pre-war landscape. In the first few months of recovery following the war, a grass cover made up of messicole plants colonized the devastated lands. When veterans' groups in the mid-1920s complained to the French government that they could no longer visit their former positions due to the dense vegetative cover, the government first bought the land and designated it as a *Zone Rouge* (Red Zone), which means that it was too dangerous for normal public access (Webster 1994) and too damaged to be restored within current time and budget constraints (Amat 2015) – and then began clearing the thick vegetative cover, corpses, and unexploded shells from the surface.

Next, an effort was undertaken to replant the 9,292 ha of the Red Zone with a managed forest cover (Clout 1993; Amat 2015). Pre-war woods were cleaned and restored when canopy gaps were too large. Former fields were first replanted with fast-growing Scotch pine *(Pinus sylvestris)*, black pine *(Pinus nigra)* and Norway spruce *(Picea abies)* seedlings because they were able to tolerate nutrient-poor conditions (Fig. 5.5). The year 1935 marked the end of the first stage of the Red Zone reforestation, estimated at 36 million seedlings. Operations continued until the 1950s in the sectors where previous plantings failed. From the mid-1960s, the pine forests were eventually thinned and many areas were then replanted to European beech *(Fagus sylvatica)* for the reappearance of regional species. Today, some areas remain covered with conifers although the majority of the battlefield contains a beech-dominated, deciduous forest (Amat 2015).

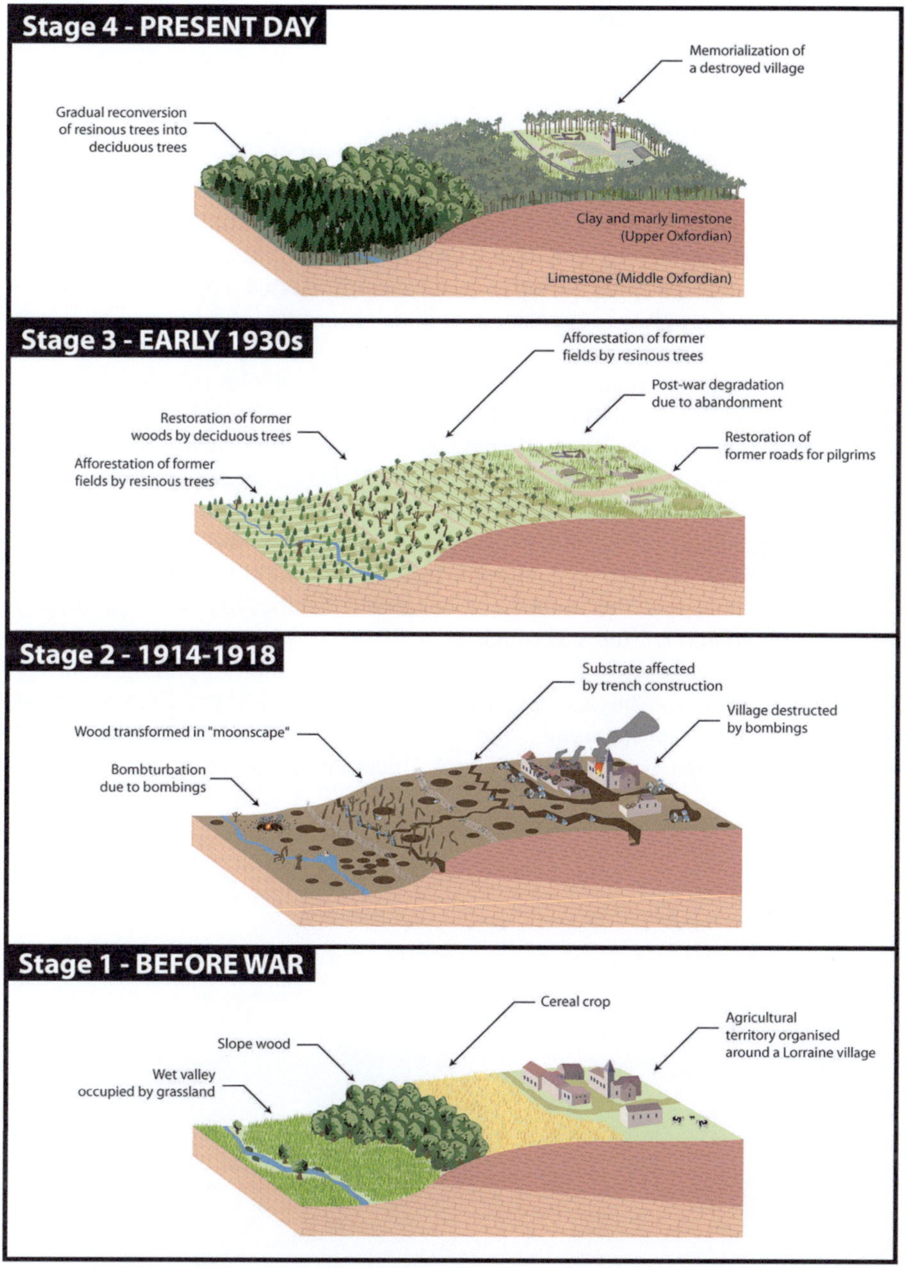

Fig. 5.5 Landscape history of the Verdun battlefield from the prewar period to the present day (Designed and produced by R. De Matos Machado)

5.6 Present-Day Landscapes Seen by Airborne LiDAR

By overlaying the battlefield with a large forest cover (Fig. 5.6), the management operated by the National Forests Office (ONF) has led to the conservation of many vestiges and topographic features inherited from WWI (Jacquemot and Legendre 2011; Schnitzler and Landolt 2013; De Matos Machado et al. 2016, 2019). Indeed, the classification of the lands as red zone has allowed the conflict-induced landforms to escape leveling by agricultural mechanized activities, as seen in the lands of the Verdun forest periphery. Moreover, the canopy exerted a protective action on the soil surface through the interception and the absorption of raindrops by leaves, branches, litter, and roots (Ziegler and Giambelluca 1998; Rey 2003). Thus, the forest of Verdun preserves memory of one of the wider battlefields of WWI.

In 2013, an airborne LiDAR (Light Detection and Ranging) mission was carried out on the 10,000 ha of the Verdun forest, allowing the reconstruction of the ground surface topography in 3-D images (Chase et al. 2011; Georges-Leroy et al. 2011; Wulder et al. 2012; Opitz and Cowley 2013). The high-resolution nature of these data (spatial resolution of 20 cm) has allowed scientists and forest managers to gain a better idea of the state of the geomorphic traces left by the conflict (Fig. 5.7). Also, LiDAR data show that the number of conserved conflict-induced landforms is demonstrably high (De Matos Machado et al. 2019).

Moreover, these images provide information about the morphological diversity of the landforms. They identify two categories of landform: 1) impact landforms, circulars or ellipsoidals, due to bombings (*i.e.*, shell holes); and 2) manmade landforms (*i.e.*, war remnants), linears or polygonals, which consti-

Fig. 5.6 View looking south of the Verdun forest from the town of Verdun (Photo by De Matos Machado 2016)

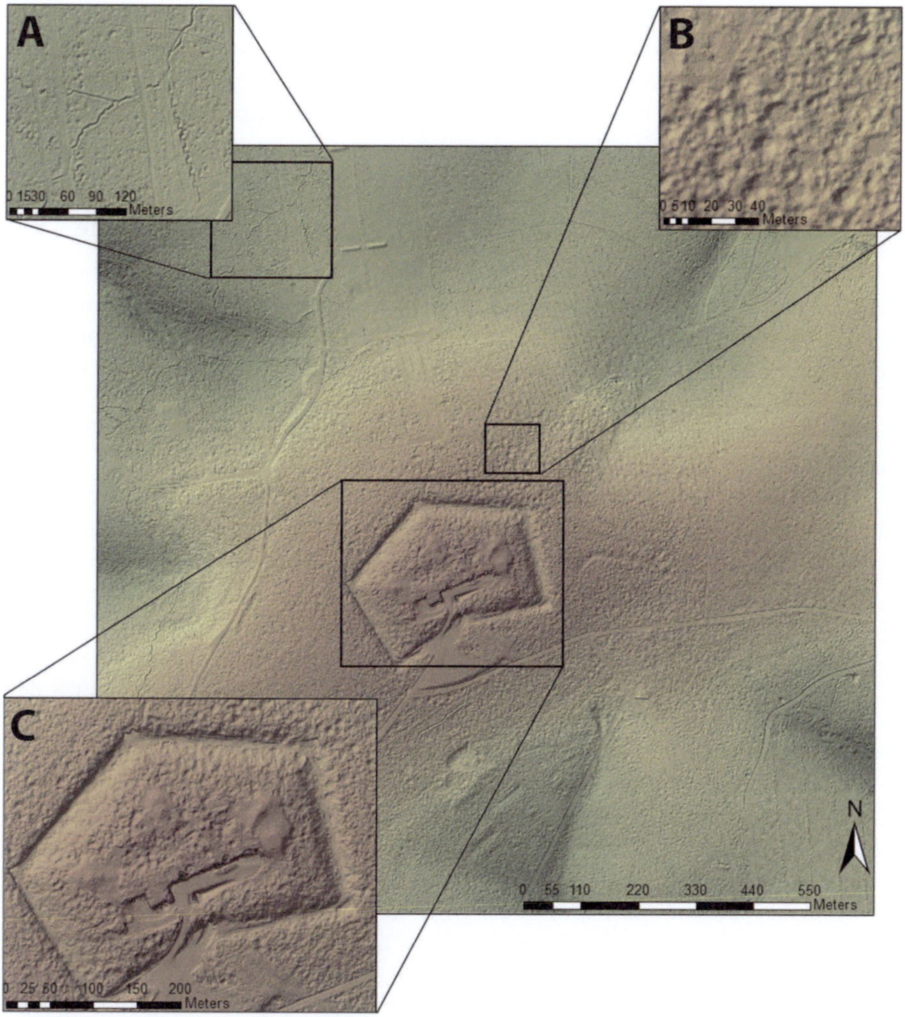

Fig. 5.7 LiDAR digital elevation model of the Verdun battlefield at 0.28-meter resolution: (**a**) Remnants of the French trenches, (**b**) Shell holes resulting from the bombings, and (**c**) The fort of Douaumont (designed and produced by J. Hupy from the LiDAR data of the National Forests Office of Verdun and the Regional Archaeology Department of Lorraine)

tute the remains of built facilities (*i.e.*, trenches, shelters, gun emplacements) (De Matos Machado et al. 2016, 2019).

Also, LiDAR data contribute to the analysis of the battlefield organization. Indeed, by mapping conflict induced-landforms, it is possible to understand the influence of geographical conditions in shaping defense positions. An analysis conducted in several parts of the Verdun forest revealed the essential role of pre-war woods in masking and reinforcing military facilities (Amat 2015; De Matos

Machado et al. 2016, 2019). Geological conditions also appeared to explain the location of war remnants and morphological diversity. Maps made from LiDAR data raise many questions and serve to underline the exceptional nature of the Verdun forest, through its heritage and historical and archaeological richness.

5.7 Modern Recognition of Verdun's Environmental Value

More known for its history of war than its environmental value, the Verdun forest of today is of high biogeographical and ecological interest. Indeed, the modern battlefield enjoys a double sacredness: the symbolism of the Verdun battle and the impressive biodiversity of the conflict landscapes. The forest planting performed in the 1920s, and continuing over the past century, have resulted in a diverse Verdun forest habitat.

Valiant conservation efforts of former woodlands have saved the 'survivor' trees. These relic trees are still numerous in the Verdun forest and provide suitable habitat for birds, bats, insects, and mushrooms (ONF 2005). They are easily recognizable by their winding morphology (numerous tree hollows, twisted branches), their great size, and the presence of war material (barbed wire, cables, shrapnel; ONF 2005; Amat 2015).

In planted lands, the reconversion of the old fields by pioneer, post-pioneer and dryad species allowed the introduction of exotic species such as *Laburnum anagyroides*, *Larix kaempferi*, *Pinus nigra*, *Pseudotsuga menziesii*, and *Quercus rubra*, which have the ability to develop on nutrient-poor soils (Parent 2004; Amat 2015). The planting of these 'substitution forests', especially in spruce stands, has also caused the emergence of orchids *(Cephalanthera damasonium, Cephalanthera longifolia, Epipactis helleborine, Neottia nidus-avis, Platanthera bifolia)* and saprophytic plants *(Monotropa hypopitys, Pyrola rotundifolia)*. It is also interesting to note the importance of lianas in these resinous forests such as *Clematis vitalba* and *Tamus communis* (Millarakis and Wagner 1999; Parent 2004; ONF 2005).

Between forest and clearing, contact zones appear, populated by edge species such as *Epipactis muelleri*, *Equisetum telmateia*, *Polygonatum odoratum*, *Vicia pisiformis*, and *Vincetoxicum hirundinaria*. This floristic richness can also be observed in opened environments surrounding the places of memory. Near the Verdun battlefield forts, monuments, destroyed villages, and cemeteries, we see numerous grass and wild lands particularly rich in orchids *(Epipactis helleborine, Neottia nidus-avis)*, in meadow species *(Bunias orientalis, Crepis sancta subsp. nemausensis, Globularia bisnagarica)*, and in naturalized decorative plants *(Buxus sempervirens, Dianthus chinensis, Lysimachia punctata)*. Maintenance work of these lands by the foresters strongly contributed to the introduction of this specific flora (Parent 2004).

Lastly, there is a spontaneous flora in the center of the Verdun forest that has developed without human intervention since the end of the war. This zone, called 'Plateau of Douaumont,' extends to 193 ha and hosts very different vegetation formations. Thus, we find fruit trees, mainly pear, apple, and plum trees, resulting from the old orchards of the Fleury destroyed village (Parent 2004; Amat 2015).

The presence of a rich natural heritage within the Verdun forest is also due to the military actions conducted during WWI. By modifying their local environment, the armies unconsciously produced new landscape patterns, sometimes with benefits to fauna and flora. One notable example is the case of the flooded shell holes, which appear seasonally when the ground water levels are highest and when they cause overflow. These landscapes can be observed locally, mainly on the top of the Hauts-de-Meuse plateau, where there are outcrops of impermeable clay and marly limestone of Upper Oxfordian. Their presence is also noted in the wet valley bottoms of the plateau (Millarakis and Wagner 1999; Parent 2004; Hupy 2005; Hupy and Schaetzl 2008).

Unlike many other conflict-induced landforms, shell holes are closed depressions. Because of this morphology, runoff waters are systematically trapped. Thus, during winter and spring, small puddles seed the forest. New hygrophilous species have colonized those landforms that did not exist before WWI. Several remarkable species have been recorded in the flooded shell holes of the Verdun forest, mainly amphibians. Among these are newts *(Lissotriton helveticus, Lissotriton vulgaris)* including one benefiting from effective protection measures *(Triturus cristatus)*, frogs, and toads, such as *Bombina variegata* (Fig. 5.8), an iconic species protected at the national level and supposed to be extinct in Belgium, which lies 30 km north. The flooded shell holes also provide botanical interest since they host a hygrophi-

Fig. 5.8 A yellow-bellied toad *(Bombina variegata)* photographed during a census campaign conducted by the ONF (De Matos Machado 2016)

Fig. 5.9 A colony of greater horseshoe bats *(Rhinolophus ferrumequinum)* photographed inside the fort of Souville (De Matos Machado 2016)

lous flora. More than 10 species of plants have been detected in association with these puddles, such as *Epipactis palustris*, which constitutes a species of regional interest and has been placed on the list of the French threatened orchids (Parent 2004; ONF 2005; Amat 2015).

Underground facilities also are biodiversity 'hotspots'. Indeed, colonies of bats find refuge within these abandoned buildings (Fig. 5.9). Located away from human society and near a huge reserve of food, bats proliferate in the Verdun forest. In their period of hibernation, their population exceeds 10,000 individuals. Their diversity must be underlined since 19 of the 23 species recorded in the Lorraine region are populating the Verdun forest. In addition to the two protected species *Rhinolophus ferrumequinum* and *Myotis bechsteinii*, war remnants host *Myotis emarginatus*, *Myotis myotis* and *Rhinolophus hipposideros* (Masson-Loodts 2014; Tillon et al. 2017).

Finally, concrete and stone constructions host rupicolous flora. Grass installed on the walls of these facilities contributed to the introduction of hybrid orchids such as *Orchiaceras verdunensis*, *Orchiaceras bispuria* and *Orchiaceras melsheimeri*. Five species of horsetails and 15 species of ferns have also been identified on the Verdun forest forts, including *Asplenium scolopendrium*, *Asplenium viride* and *Asplenium fontanum* (Parent 2004).

5.8 Conservation and Valorization Issues

Although traces of the Great War are still numerous in the Verdun forest, they are threatened by erosion and degradation. This disappearance could be harmful to animals and plants installed in these reliefs. Currently, it is important to establish an inventory of conflict-induced landforms for improving their conservation, and at the same time, preserving the natural heritage of the battlefield.

A major concern of the Verdun forest managers is its attendance. Even immediately following the end of hostilities in 1918, hundreds of thousands of people flocked to the Western Front to witness first-hand the carnage, or perhaps to look for lost loved ones. A first illustrated tour guide of the Verdun battlefield was published in 1919 by André Michelin to accompany the pilgrims on their visit (Michelin travel guide 1919). In the 1950s, the tourist numbers exceeded 300,000 visitors, and some believe it may have exceeded 500,000 as part of the centenary celebrations of WWI (Brandt 1994; Amat et al. 2015). In addition, the area is visited by the people of the region who see in the Verdun forest a large space of leisure and recreation consisting of hiking and equestrian paths and ATV trail networks (Amat et al. 2015). The archaeological richness of the battlefield also attracts looters, particularly active during these celebration times, who collect or sell archaeological artefacts extracted from the soils (Jacquemot and Legendre 2011; Schnitzler and Landolt 2013). Nor should we forget the foresters of the ONF who regularly undertake cuts and silvicultural operations within the framework of the Verdun forest management.

To counteract this phenomenon, the establishment of a new forest management plan (2006–2020) and the labeling of the Verdun forest (2014) have allowed the ONF to undertake for the last few years an integrated forest management, closely coordinated with the Regional Archaeology Department of Lorraine (SRA), historians, and geographers (ONF 2005; Schnitzler and Landolt 2013; Amat 2015). First, an inventory of the remnants was conducted on the whole area based on the existing maps. The same inventory is carried out for the remarkable plant and animal species. For the yellow-bellied toad and bats, population census is regularly conducted in order to understand their movements and to locate favorable sites for their installation. These inventory documents are regularly updated using LiDAR imagery and pedestrian prospection. This procedure identifies 'sensitive' areas to be considered for concrete management practices (Košir et al. 2015). In the field, the ONF agents adapt to archaeological issues by creating a more respectful road network. In high historical value areas, no land leveling or gear traffic is allowed around the trenches and military facilities. Other initiatives are under study such as the use of animal traction instead of mechanical power for transporting timber. Shell holes are excluded from these provisional measures because they are not considered archaeological objects. However, special attention is paid to relic trees.

The natural heritage of the Verdun forest also benefits from protection measures aimed specifically at lawns close to the forts, bats, the yellow-bellied toad, and the crested newt (ONF 2005; Amat 2015). Indeed, the site has been part of the European Natura 2000 network since 2005 and hosts a Special Area of Conservation (SAC;

'Corridor de la Meuse'). The Verdun Forest is also classified as a Natural Zone of Interest for Ecology, Flora, and Fauna (ZNIEFF) type 1 ('Forêt de Verdun) and type 2 ('Côtes-de-Meuse'). Thus, the forest management plan takes into consideration this ecological zoning by offering management practices adapted to these issues: (1) for plants, use of biodegradable oils, maintenance of lawns by extraction of pioneer trees and invasive shrubs, late mowing; (2) for wildlife, land leveling unauthorized at certain times of the year, maintenance of the puddles (flooded shell holes and trenches), biennial or late mowing around bat roosts to increase the number of insects, early cuts outlawed around breeding sites (ONF 2005).

In order to welcome and support visitors in their visit, the forest management plan also includes restoration of communication routes and monuments (Douaumont Ossuary, forts, reconstituted trenches, stele, sculptures) and construction of new infrastructures. Thus, the Memorial of Verdun proposes a new visit tour with panoramic views and interactive panels allowing manipulation in space and time of aerial photographs, maps and LiDAR data (Fig. 5.10). Several publicly accessible hiking trails have been built in the southern part of the forest and designed from the LiDAR imagery. Trilingual educational panels concerning forest management, biodiversity, and Verdun battle history complement these itineraries. For the most adventurous, the ONF offers guided trips to discover war remnants from a new angle. 'Paths of biodiversity' are also available to contemplate the remarkable species of the Verdun forest. Finally, note that the mayors of destroyed villages and many local associations help to preserve the memory of the Great War by their safeguarding and recovery initiatives (ONF 2005; Amat 2015; Amat et al. 2015).

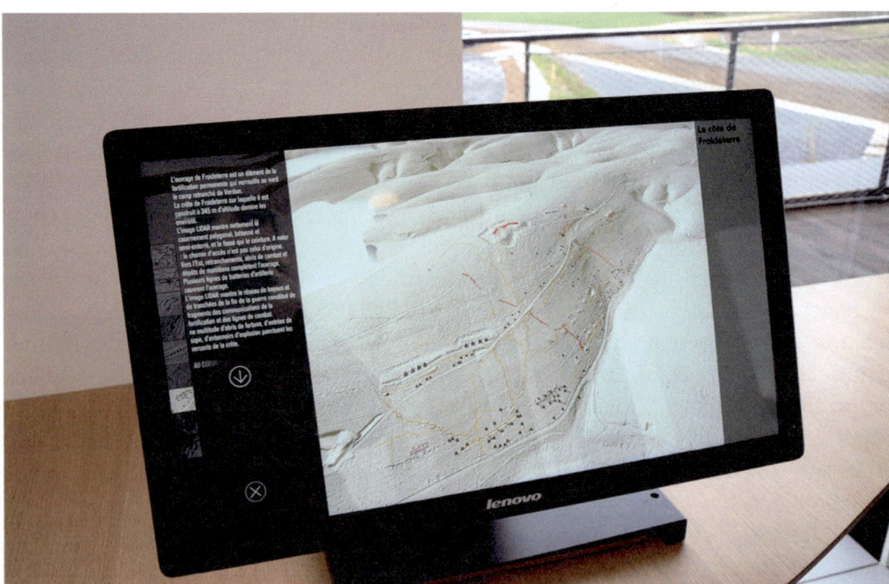

Fig. 5.10 LiDAR data displayed by an interactive screen of the Verdun Memorial (De Matos Machado 2016)

5.9 Conclusions

Thus, the battlefield of Verdun constitutes a practically unique example of massive destruction of the landscape by the war. Prior to the battle of Verdun, the landscape of the Loraine province of France was predominantly agricultural, covered with fields of grain, grazing cattle, and small woodlots separating the hundreds of villages. In early 1916, hundreds of millions of artillery left the area devoid of vegetation and pock-marked with craters of varying sizes.

Verdun is also the revival of life. Unlike in other parts of the Western Front, the State decided to replace the devastated lands by a forest because fields were too devastated to be recultivated. Thanks to the remarkable work conducted by the foresters since the 1920s, vegetation has claimed its rights. This forest has also favored the introduction of exotic flora and fauna, benefiting both from the diversity of habitats created by foresters to prosper. This biodiversity is also closely linked to the landforms inherited from the fighting. For even if a century has passed since the end of the conflict, the number of conflict-induced landforms is still very important as shown by LiDAR images acquired in 2013 on the Verdun forest. Thus, many animal and plant species benefit from the wet conditions offered by the flooded shell holes or the former underground networks, which they use for shelter.

After being the home place of the soldiers for several months, the Verdun battlefield became the habitat of many animal and plant species that should be preserved in the same way as the vestiges of war. For some years, foresters, archaeologists, historians, and geographers have been working together to maintain this historic and natural heritage. Conservative management of the battlefield has been facilitated since the acquisition of LiDAR data in 2013, promoting inventory and mapping of war remnants and potential ecological habitats.

Finally, unlike most battlefields of the Great War, the site of Verdun has managed to preserve the memory of the first world conflict through the safeguard of the vestiges of war on 10,000 ha. Moreover, few places of memory on the Western Front have conciliated historical heritage and biodiversity. The battlefield of Verdun can boast of having done it.

Acknowledgements The authors would like to thank Stéphanie Jacquemot of the DRAC Grand Est – Regional Archaeology Department, Gersende Gérard, Gérald Colin, René-Marc Pineau, and Juliette Foltier of the National Forests Office of Verdun, for LiDAR data of the Verdun forest. We also thank Gilles Arnaud-Fassetta and François Bétard – University Paris-Diderot, for their useful and constructive remarks which helped to improve the manuscript.

References

Amat, J. P. (2015). *Les forêts de la Grande Guerre: histoire, mémoire, patrimoine*. Paris: Presses de l'Université Paris-Sorbonne.

Amat, J. P., Filippucci, P., & Savouret, E. (2015). 'The cemetery of France': Reconstruction and memorialisation on the battlefield of Verdun (France). In M. L. Sørensen & D. Viejo-Rose (Eds.), *War and cultural heritage* (pp. 46–68). New York: Biographies of place. Cambridge University Press.

Bagnold, R. A. (1990). *Sand, wind, and war: memoirs of a desert explorer*. Tucson: University of Arizona Press.

Bailey, J. B. A. (2004). *Field artillery and firepower*. Annapolis: Naval Institute Press.

Bausinger, T., & Preuß, J. (2005). Environmental remnants of the First World War: Soil contamination of a burning ground for arsenical ammunition. *Bulletin of Environmental Contamination and Toxicology, 74*, 1045–1052. https://doi.org/10.1007/s00128-005-0686-z.

Bausinger, T., Bonnaire, E., & Preuß, J. (2007). Exposure assessment of a burning ground for chemical ammunition on the Great War battlefields of Verdun. *Science of the Total Environment, 382*, 259–271. https://doi.org/10.1016/j.scitotenv.2007.04.029.

Bernède, A. (2002). *Verdun 1916: le point de vue français*. Le Mans: Éditions Cénomane.

Brandt, S. (1994). Le voyage aux champs de bataille. *Vingtième Siècle, revue d'histoire, 41*, 18–22. https://doi.org/10.3406/xxs.1994.3262.

Brauer, J. (2009). *War and nature: The environmental consequences of war in a globalized world*. Lanham: AltaMira Press.

Chase, A. F., Chase, D. Z., Weishampel, J. F., et al. (2011). Airborne LiDAR, archaeology, and the ancient Maya landscape at Caracol, Belize. *Journal of Archaeological Science, 38*, 387–398. https://doi.org/10.1016/j.jas.2010.09.018.

Closmann, C. E. (Ed.). (2009). *War and the environment*. College Station: Military destruction in the modern Age. Texas A&M University Press.

Clout, H. D. (1993). The revival of rural Lorraine after the Great War. *Geografiska Annaler, 75B*(2), 73–91. https://doi.org/10.2307/490701.

Cowley, R. (2004). The unreal city: the trenches of World War I. In J. M. Morris (Ed.), *Readings in American military history* (pp. 187–198). Upper Saddle River: Pearson/Prentice Hall.

De Matos Machado, R., Amat, J. P., Arnaud-Fassetta, G., & Bétard, F. (2016). Potentialités de l'outil LiDAR pour cartographier les vestiges de la Grande Guerre en milieu intra-forestier (bois des Caures, forêt domaniale de Verdun, Meuse). *EchoGéo 38* [online]. https://doi.org/10.4000/echogeo.14791.

De Matos Machado, R., Toumazet, J. P., Bergès, J. C., Amat, J. P., Arnaud-Fassetta, G., Bétard, F., Bilodeau, C., Hupy, J. P., & Jacquemot, S. (2019). War landform mapping and classification on the Verdun battlefield (France) using airborne LiDAR and multivariate analysis. *Earth Surface Processes and Landforms, 44*(7), 1430–1448. https://doi.org/10.1002/esp.4586.

Georges-Leroy, M., Bock, J., Dambrine, E., & Dupouey, J. L. (2011). Apport du lidar à la connaissance de l'histoire de l'occupation du sol en forêt de Haye. *ArchéoSciences, 35*, 117–129. https://doi.org/10.4000/archeosciences.3015.

Gudmundsson, B. I. (1993). *On artillery*. Westport: Praeger.

Hogg, I. V. (1985). *The illustrated history of ammunition*. Secaucus: Chartwell Books.

Hogg, I. V. (1987). *The illustrated encyclopedia of artillery*. Secaucus: Chartwell Books.

Holstein, C. (2002). *Fort Douaumont: Verdun*. Barnsley: Pen & Sword.

Horne, A. (1993). *The price of glory*. London: Penguin.

Hubé, D. (2016). *Sur les traces d'un secret enfoui: enquête sur l'héritage toxique de la Grande Guerre*. Paris: Michalon.

Hupy, J. P. (2005). *Assessing landscape disturbance and recovery across a WWI battlefield: Verdun, France*. Dissertation: Michigan State University.

Hupy, J. P. (2006). The long-term effects of explosive munitions on the WWI battlefield surface of Verdun, France. *Scottish Geographical Journal, 122*(3), 167–184. https://doi.org/10.1080/00369220618737264.

Hupy, J. P. (2011). Khe Sanh, Vietnam: Examining the long-term impacts of warfare on the physical landscape. In E. Palka & F. Galgano (Eds.), *Modern military geography* (pp. 312–326). New York: Routledge.
Hupy, J. P., & Schaetzl, R. J. (2008). Soil development on the WWI battlefield of Verdun, France. *Geoderma, 145*(1), 37–49. https://doi.org/10.1016/j.geoderma.2008.01.024.
Ilyès, Z. (2006). Military activities: Warfare and defense. In J. Szabo, L. David, & D. Loczy (Eds.), *Anthropogenic geomorphology: A guide to man-made landforms* (pp. 217–231). Dordrecht: Springer. https://doi.org/10.1007/978-90-481-3058-0.
Jacquemot, S., & Legendre, J. P. (Eds.). (2011). *Vestiges de guerres en Lorraine: le patrimoine des conflits mondiaux*. Metz: Éditions Serpenoise.
Joly, H. (1911). *Géographie physique de la Lorraine et de ses enveloppes*. Nancy: Albert Barbier.
Joly, D., Brossard, T., Cardot, H., et al. (2010). Les types de climats en France, une construction spatiale. *Cybergéo: European Journal of Geography, Cartographie, Imagerie, SIG*. https://doi.org/10.4000/cybergeo.23155.
Jünger, E. (1961). *In Stahlgewittern. 2. Bände*. Stuttgart: Klett-Cotta.
Keegan, J. (1993). *A history of warfare*. New York: Vintage.
Keegan, J. (1998). *The First World War*. New York: Vintage.
Košir, U., Mlekuž, D., & Črešnar, M. (2015). The use of modern technologies for documenting and interpreting conflict landscapes. Case study of the Bovec Area (Posočje Region, Slovenia). In A. Jankovič Potočnik, M. Zupančič, & A. Marn (Eds.), *Handbook of typical historic technologies of fortress constructions* (pp. 62–68). Ljubljana: Saving of European cultural heritage fortified monuments in Central Europe. Zavod za varstvo kulturne dediščine Slovenije.
Malte-Brun, V. A. (1845). *Précis de la géographie universelle*. Paris: Imprimerie de Bourgogne et Martinez.
Martin, W. (2001). *Verdun 1916: 'They shall not pass'*. Oxford: Osprey Publishing.
Masson-Loodts, I. (2014). *Paysages en batailles. Les séquelles environnementales de la Grande Guerre*. Brussels: Éditions Nevicata.
Meerschman, E., Cockx, L., Monirul Islam, M., et al. (2011). Geostatistical assessment of the impact of World War I on the spatial occurrence of soil heavy metals. *Ambio, 40*, 417–424. https://doi.org/10.1007/s13280-010-0104-6.
Michelin travel guide. (1919). *The Battle of Verdun 1914–1918*. Clermont-Ferrand: Michelin & Cie.
Millarakis, P., & Wagner, I. (1999). *Étude de la végétation de la forêt domaniale de Verdun*. Verdun: Office National des Forêts (ONF).
Office National des Forêts. (2005). *Forêt domaniale de Verdun. Révision d'aménagement forestier 2006–2020*. Verdun: Office National des Forêts (ONF).
Opitz, R. S., & Cowley, D. C. (Eds.). (2013). *Interpreting archaeological topography: Airborne laser scanning. 3D data and ground observation*. Oxford: Oxbow books.
Ousby, I. (2002). *The road to Verdun: World War I's most momentous battle and the folly of nationalism*. New York: Doubleday.
Parent, G.H. (2004). Trois études sur la zone rouge de Verdun, une zone totalement sinistrée. I. L'herpétofaune. II. La diversité floristique. III. Les sites d'intérêt botanique et zoologique à protéger prioritairement. *Ferrantia*, Travaux scientifiques du Musée national d'histoire naturelle de Luxembourg. https://ps.mnhn.lu/ferrantia/publications/Ferrantia38.pdf.
Prestidge, O. (2013). Forêt de guerre: Natural remembrances of the Great War. *Exchanges: The Warwick Research Journal, 1*(1), 16–34.
Rey, F. (2003). Influence of vegetation distribution on sediment yield in forested marly gullies. *Catena, 50*, 549–562. https://doi.org/10.1016/S0341-8162(02)00121-2.
Rose, E. P. F., & Nathanail, C. P. (Eds.). (2000). *Geology and warfare: Examples of the influence of terrain and geologists on military operations*. Bath: Geological Society.
Schnitzler, B., & Landolt, M. (Eds.). (2013). *À l'est, du nouveau ! Archéologie de la Grande Guerre en Alsace et en Lorraine*. Strasbourg: Éditions des Musées de la Ville de Strasbourg.

Service Historique de l'État Major de l'armée. (n.d.). *Journal des marches et opérations du 56 ème bataillon de chasseurs à pied pendant la campagne du 2 août 1914 au 6 juillet 1916*. Paris: État Major de l'Armée (EMA), Imprimerie-Librairie Universelle Laurent Fournier.

Souvent, P., & Pirc, S. (2001). Pollution caused by metallic fragments introduced into soils because of World War I activities. *Environmental Geology, 40*(3), 317–323. https://doi.org/10.1007/s002540000156.

Tillon, L., Raynaud, J.-C., Le Roux, M., & Darnis, T. (2017). *Programme « Forêt d'exception » : inventaire des chiroptères, forêts domaniales de Verdun et de Morthomme (55)*. Paris: ONF Réseau Mammifères.

Van Meirvenne, M., Meklit, T., Verstraete, S., et al. (2008). Could shelling in the First World War have increased copper concentrations in the soil around Ypres? *European Journal of Soil Science, 59*, 372–379. https://doi.org/10.1111/j.1365-2389.2007.01014.x.

Webster, D. (1994). The soldiers moved on. The war moved on. The bombs stayed. *Smithsonian, 24*(11), 26–37.

Westing, A. H. (1972). The cratering of Indochina. *Scientific American, 226*(5), 21–29. https://doi.org/10.1038/scientificamerican0572-20.

Westing, A. H. (1980). *Warfare in a fragile world: Military impact on the human environment*. London: Taylor & Francis.

Winters, H. A., Galloway, G. E., Jr., Reynolds, W. J., & Rhyne, D. W. (1998). *Battling the elements: Weather and terrain in the conduct of war*. Baltimore: Johns Hopkins University Press.

Wulder, M. A., White, J. C., Nelson, R. F., et al. (2012). Lidar sampling for large-area forest characterization: A review. *Remote Sensing of Environment, 121*, 196–209. https://doi.org/10.1016/j.rse.2012.02.001

Ziegler, A. D., & Giambelluca, T. W. (1998). Influence of revegetation efforts on hydrologic response and erosion, Kaho'Olawe Island, Hawai'i. *Land Degradation & Development, 9*, 189–206. https://doi.org/10.1002/(SICI)1099-145X(199805/06)9:3<189::AID-LDR272>3.0.CO;2-R.

Chapter 6
The Ecosystem Service Impacts from Invasive Plants in Antietam National Battlefield

Todd R. Lookingbill, Emily S. Minor, and Lisa A. Wainger

Abstract Following their memorialization as protected landscapes, battlefield parks can provide a blend of cultural and other ecosystem services. Among the many threats to providing these services are non-native invasive plants. In this chapter, we assess the threats imposed by biological invasions of non-native plants in battlefield parks and discuss management strategies. We use evidence from the scientific and economic literature and the expert judgment of biologists, economists, and park managers to identify the harms caused by invasives and to characterize their effects on park ecosystem services. Based on this evidence, we propose four generic stressor-response relationships to describe the relationships between invasion extent and ecological endpoints such as park vegetation structure and diversity. Using Antietam National Battlefield as a case study, we tailor the general stressor-response curves to four specific species representing different functional groups of invasive plants: trees, shrubs, vines, and herbaceous forbs. We next link the ecological response of changes in vegetation structure and diversity to relevant ecosystem service impacts using interviews with national park service personnel and the economic literature. We identify four broad categories of parks users who might be affected by these losses of services: causal visitors, avid recreationalists, park neighbors, and non-use beneficiaries. Our findings reveal a general lack of experimental evidence quantifying the ecosystem service impacts of invasive plants. This lack of evidence, combined with the likely non-linear effects of non-native plant invasions on ecological endpoints, could catch managers unaware of dangerous thresholds in long-term resource management of battlefield landscapes.

T. R. Lookingbill (✉)
Department of Geography and the Environment, University of Richmond, Richmond, VA, USA
e-mail: tlooking@richmond.edu

E. S. Minor
Department of Biological Sciences, Institute for Environmental Science and Policy, University of Illinois at Chicago, Chicago, IL, USA

L. A. Wainger
Chesapeake Biological Laboratory, University of Maryland Center for Environmental Science, Cambridge, MD, USA

T. R. Lookingbill, P. D. Smallwood (eds.), *Collateral Values*, Landscape Series 25,
https://doi.org/10.1007/978-3-030-18991-4_6

6.1 Introduction

Battlefield parks provide a blend of cultural and ecological benefits that can extend beyond their boundaries. These benefits can be considered *ecosystem services* (Costanza et al. 1997), which for battlefield parks include recreational opportunities, preservation of historic viewsheds, aesthetic enjoyment for visitors, conservation of biodiversity, and increases in neighboring property values (Wainger et al. 2012). Like most parks, battlefield parks face wide-ranging threats to these services, many of which are associated with their geography. For strategic reasons, battles have historically often been fought on the outskirts of cities. In the years following war, many of the landscapes surrounding these battlefield sites have experienced high rates of suburbanized development as nearby metropolitan centers expand (Lookingbill et al. 2014a). The expanding population can bring opportunities for increases in the ecosystem services provided by the park, but also increases in the anthropogenic impacts to the site.

This trajectory of landscape change is illustrated by Antietam National Battlefield. Located in the Appalachian Ridge and Valley province of western Maryland, USA (Fig. 6.1), Antietam was the site of the highest number of military casualties in a single day in U.S. history. On September 17, 1862, in one of the defining conflicts of the U.S. Civil War, a total of 23,000 soldiers were lost in the hostilities. A major turning point in the war, the outcome of the battle paved the way for the issuance of

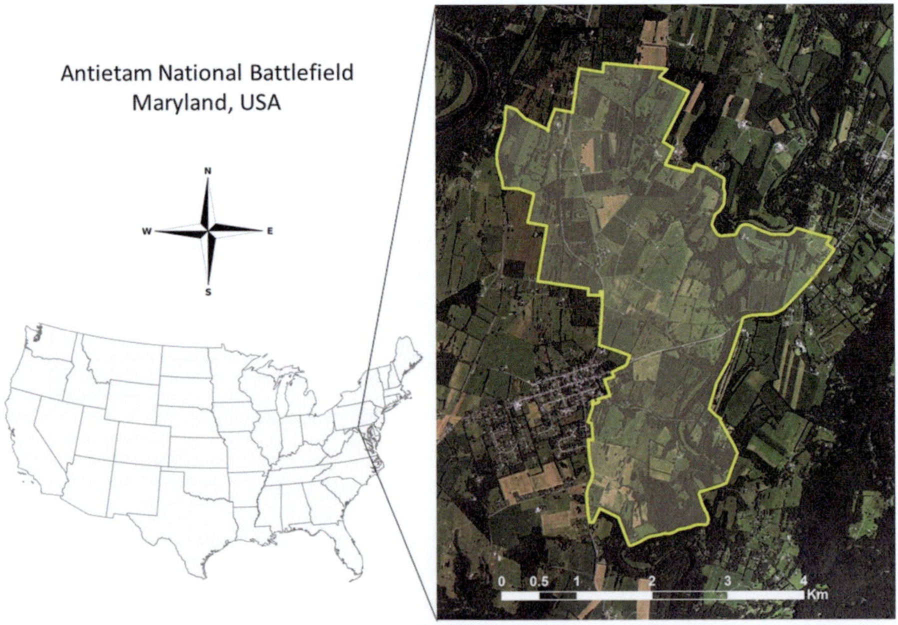

Fig. 6.1 Antietam National Battlefield located in the Mid-Atlantic region of the USA and surrounded by mixed forest, agriculture and urban development

the Emancipation Proclamation, making the eradication of slavery an explicit objective of the war (McPherson 2002).

The memorialization of the site began almost immediately following the Confederate retreat from the battlefield. A private cemetery was created on site within five years and was transferred to the U.S. War Department in 1879. As Congress turned its attention to preserving Civil War landscapes during the "Golden Age" of battlefield preservation, Antietam (along with Gettysburg, Chickamauga and Chattanooga, Shiloh, and Vicksburg) was one of the first Civil War parks to be established in 1890 (Smith 2008). In 1933, control of the 65 acres of park land was passed from the War Department to the National Park Service. The size of the park increased by an order of magnitude to 600 acres in the 1960s, as part of the Civil War Centennial commemoration. A series of additional acquisitions in the 1980s increased the acreage another fivefold to its current 3200 acres (Madron and Tilton, Chap. 2 of this book). The park today receives approximately 350,000 visitors per year, which comes in at the bottom of the list of the original five Civil War parks. However, the increase in visitation of over 75% in the past five decades tops the list for these parks (https://irma.nps.gov/Stats/Reports/Park).

Forest cover comprises approximately 14% of the park, mostly in small woodlots but also as contiguous corridors along Antietam Creek on the east side of the park, and the Potomac River, just to the west of the park (Table 6.1; Fig. 6.1). The remaining vegetation in the park is predominantly open fields and agricultural leases. Vegetation succession in the fields and other parts of the park requires constant and sometimes-extensive management to maintain the historical landscape. The land surrounding the park is a mixture of agricultural and urbanized areas (Fig. 6.1). The park is typical of Civil War battlefield parks of the Mid-Atlantic USA that are located in suburbanizing, mixed-use landscapes. As shown for Antietam, agricultural abandonment of small family farms has resulted in less agriculture and greater forest cover adjacent to the parks than in the parks themselves, where fields are preserved for their historical value (Table 6.1). However, these adjacent forests are being rapidly lost to suburban and exurban development (Suarez-Rubio et al. 2012).

Antietam National Battlefield is also demonstrative of regional environmental stressors in its invasive plants problems. Non-native invasive plants often accompany human encroachment around parks (Allen et al. 2009). These plants invade from surrounding homes and are carried into the parks by visitors, animals, and other vectors (Minor et al. 2009). As these parks mature, invasive plants can threaten the ecosystem functions and services that they provide. Once established, invasive plants can have negative ecological or cultural impacts on the landscape, including

Table 6.1 Percentage of dominant land cover classes in Antietam National Battlefield and in 5-km buffer surrounding the park (source: 2006 National Land Cover Dataset)

	Forest	Agriculture and Fields
Inside park	14.0%	75.5%
Within 5-km buffer of park	32.5%	55.7%

destroying historic structures (Celesti-Grapow and Blasi 2004), altering insect (Bezemer et al. 2014) and bird communities (Skórka et al. 2010), and degrading viewsheds and general sense of place (Barendse et al. 2016). These changes, in turn, affect park visitors, neighbors, and other stakeholders.

Managing invasive plants on historic battlefields is operationally challenging, as many of these undesirable species have become well-established, are likely to reinvade following treatment, and require ongoing control due to plant persistence and regional propagule pressure (Lookingbill et al. 2014b). The reduction of invasive plants must also be balanced against competing cultural and natural resource priorities. Understanding how to allocate management efforts is a multi-faceted problem that requires identifying the level of control effort that generates net benefits to the park.

The relationship between invasive plants and diminished ecosystem services is indirect and linked by the effect that invasive plants have on ecological endpoints. It is helpful to consider these relationships in two parts (Fig. 6.2). In the first part, changes in invasive species (for example, measured as spatial extent or density) generate changes to ecological endpoints. We call these relationships *stressor-response functions*. After determining stressor-response relationships, the changes in ecological endpoints must next be related to user preferences to quantify the economic value of affected ecosystem services. *Economic damage functions* are used to relate changes in one or more ecological endpoints to the benefits users derive from ecosystem services. Damage functions are created by quantifying how much people would be willing to trade off other goods and services to get more (quality or quantity) of a particular ecosystem service, usually by measuring willingness-to-pay.

In this analysis, we address the first half of this equation to build stressor-response functions for four plant species that are commonly invasive to battlefield parks within the Mid-Atlantic region of the United States. Of the multiple ecological endpoints identified in our earlier work (Wainger et al. 2012), we chose one type of ecological endpoint affected by invasive plants: impacts to vegetation structure and diversity. The remainder of this chapter describes a literature review and expert elicitation workshop conducted to build stressor-response functions for these impacts. We conclude with a summary of the ecosystem service users or beneficiaries likely to be affected by reducing vegetation quality in our focal park. However,

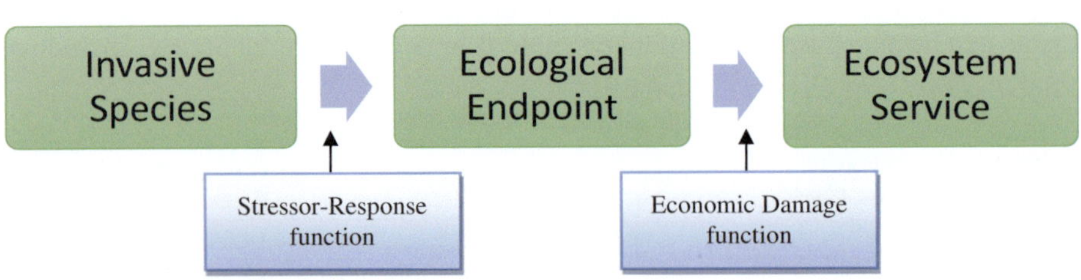

Fig. 6.2 Relationship between the spread of invasive plant species, ecological change, and associated impacts to ecosystem services

we do not generate quantitative economic damage functions here, which would be needed to value the economic impacts of changes in invasive species.

6.2 Stressor-Response Functions

In this chapter, we use changes in vegetation structure and diversity as the primary ecological endpoint of interest. Invasives also impact other ecological endpoints relevant to human well-being, including changes in insects, birds, water, air, or landscape features. We selected vegetation characteristics as our endpoint because of their pervasive effect on many other ecological conditions and observable ecosystem services. In this battlefield park, vegetation structure has a specific cultural interpretation because the park has an explicit mandate to maintain the landscape (including vegetation) to be consistent with the historical period of the Civil War. Therefore, changes to vegetation structure and diversity would lead to direct social harms for users who wish to experience historical accuracy, and these changes would be inconsistent with park management goals. Vegetation structure also influences other cultural and ecological properties of the park such as the overall aesthetics, wildlife viewing opportunities, and habitat quality.

To quantify the impact of invasive plants on vegetation structure and diversity, we surveyed the literature that demonstrated measurable effects of invasive presence or density on ecological endpoints. To fill the considerable data gaps that we found in the literature, we also consulted an expert panel of biologists, economists, and National Park Service management personnel with knowledge of Antietam National Battlefield. Our first goal was to develop a set of general curves depicting potential stressor-response relationships, where invasive plant abundance was the stress and an ecological endpoint relevant to ecosystem services was the response. The shape of these theoretical curves would then be refined on a case-by-case basis to describe specific stressor-response relationships within Antietam National Battlefield.

Briefing materials were made available via a website to participants in the expert panel workshop, including a project summary and supporting papers from the scientific literature (additional details in Wainger et al. 2012). Participants were asked to complete a questionnaire at the beginning of the workshop about the potential form of the stressor-response function for one or more species-endpoint combinations (Box 6.1). Then the group jointly discussed the evidence for different forms of the stressor-response functions. Thus, the workshop used both individual and group approaches to extract expert knowledge on impacts of invasive plants. The workshop was not aimed at achieving consensus but at eliciting expert judgment in unbiased ways to inform the eventual development of the generalized stressor-response functions.

The expert panel concurred with our finding that the empirical literature was inadequate to fully characterize stressor-response functions and suggested that the best approach to fill the data gap was to apply a family of theoretical curves. The

Box 6.1 Example questions from the expert workshop. These questions are intended to link changes in cover of a particular invasive species *(Ailanthus altissima)* to a change in an ecological endpoint (vegetation structure and diversity)

Question 1: What do you believe to be the major effects of Ailanthus on vegetation structure and diversity?

Question 2: What do you believe to be the causal mechanisms of those changes?

Question 3: Please use the table and/or chart below to fill in the relationship that you consider to be the most probable between % cover of Ailanthus and % of native vegetation structure and diversity.

% cover Ailanthus	% vegetation structure/diversity
0	
20	
40	
60	
80	
100	

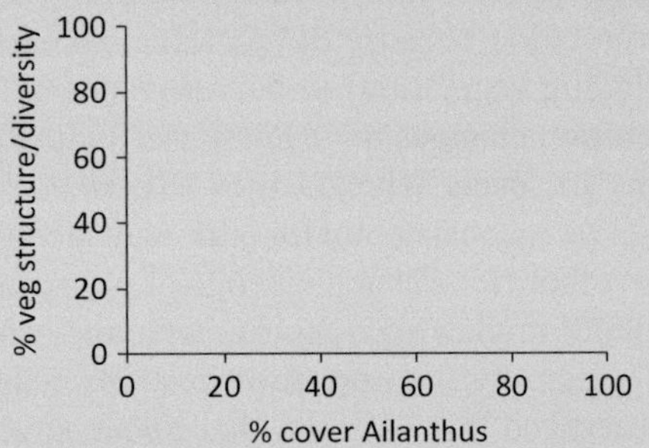

Please list the assumptions/evidence supporting the relationships depicted:

curves could either be parameterized with data when available or applied as a general model of invasive plant impacts using qualitative information from the literature, when quantitative data were lacking. For example, the ecological endpoint affected by the increase in invasive species could be scaled along the y-axis to represent the best and worst-case conditions possible using a reference ecosystem to act as a benchmark for the non-degraded system. Similarly, the invasive abundance (or percent stressor) could be scaled from zero to a theoretical, maximum possible abundance along the x-axis.

We settled on four generalized, theoretical curves to represent the suite of species impacts (Fig. 6.3). Curve I represents a highly sensitive endpoint where low abundance of invasive plants has a large impact on the ecological endpoint. In contrast, curve IV represents a low-sensitivity endpoint where impact remains low until the invasive plant reaches high density. Curve II represents an endpoint with intermediate sensitivity, where invasive plants have minimal impact until an intermediate abundance level is reached, after which the impact increases rapidly. Curve III represents a linear relationship between density of the invasive plant and impact on the ecological endpoint.

We next considered the species-specific forms of these stressor-response relationships for four different invasive plants found within Antietam National Battlefield. The four species represent a mix of functional forms (tree, vine, shrub, and herb), and all have been labeled as species of concern to the park (Table 6.2).

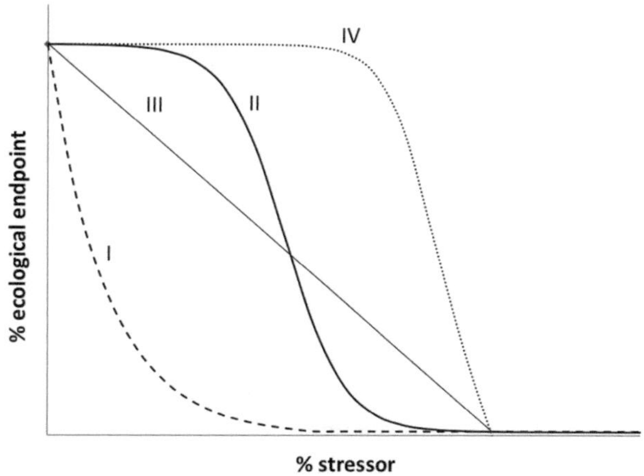

Fig. 6.3 Theoretical stressor-response curves (Adapted from Yokomizo et al. 2009)

Table 6.2 Case study species

Functional form	Common Name	Scientific Name
Tree	Tree-of-heaven	*Ailanthus altissima*
Shrub	Multiflora rose	*Rosa multiflora*
Vine	Oriental bittersweet	*Celastrus orbiculatus*
Herb	Garlic mustard	*Alliaria petiolata*

For each species, we searched the literature for information about its effect on ecosystems, focusing on potential impacts to vegetation structure and diversity as the ecological endpoint. In the following sections, we provide an overview of the relevant literature for each of the four focal invasive plant species. We use this information to select and apply appropriate theoretical stressor-response curves to depict the relationship between the percent cover of the invasive species and changes in the ecological endpoint of concern.

6.2.1 Ailanthus altissima

Ailanthus altissima, sometimes called "tree-of-heaven," is a common invasive tree in Antietam National Battlefield. Native to China, it has subsequently spread to all other continents except Antarctica (Kowarik and Säumel 2007). In its non-native range, tree-of-heaven can have major impacts on native vegetation structure and diversity.

Tree-of-heaven has several competitive advantages that alter its local environment and contribute to its success in invading new areas. Seeds are dispersed by wind but also are buoyant and remain viable after long-distance transport via water (Landenberger et al. 2007; Kowarik and Säumel 2008). Once established, individuals mature quickly, develop root networks that form dense clonal stands, and are able to resprout when cut (Miller 1990; Kowarik and Säumel 2007). The roots, leaves, and stems of tree-of-heaven exude chemicals that can negatively affect neighboring plants (Lawrence et al. 1991; Gómez-Aparicio and Canham 2008a; Heisey 1996). Greenhouse studies have shown that these allelopathic compounds reduce germination of seeds and damage or kill seedlings of multiple plant species (Heisey 1996). The impacts are greatest on heterospecific individuals previously unexposed to the species (Lawrence et al. 1991), suggesting that impacts of tree-of-heaven may be largest in newly invaded areas. Furthermore, its seedlings have fast-developing root systems and can be strong competitors for below-ground resources (Call and Nilsen 2005), altering the availability of soil resources in its proximity (Gómez-Aparicio and Canham 2008a; Vilá et al. 2006; Constán-Nava et al. 2015).

Together, tree-of-heaven's fast growth, allelopathy, and altered soil resources affect recruitment and growth of native plant species and often lead to changes in community composition and phylodiversity (Gómez-Aparicio and Canham 2008b; Vilá et al. 2006; Constán-Nava et al. 2015). Several field studies have shown a significant decrease in native plant species richness in areas invaded by tree-of-heaven. A field study near Paris, France, compared diversity under tree-of-heaven and native tree species in four different habitat types (Motard et al. 2011). In each habitat, understory vegetation under tree-of-heaven was significantly lower in species richness and species rarity than vegetation under native trees. The 15–30% loss of diversity under tree-of-heaven was linked to an increase in root suckers (Motard et al. 2011). Similarly, Vilá et al. (2006) found a 24% decrease in native plant species richness in invaded plots on Mediterranean islands compared to nearby non-invaded plots. These studies suggest that native vegetation structure and diversity may be relatively sensitive to tree-of-heaven invasion, perhaps resulting in a type I stressor-response curve (Fig. 6.4).

6.2.2 Rosa multiflora

Rosa multiflora, or multiflora rose, was intentionally introduced to the United States in the 1800s as a horticultural plant (Rehder 1936). The non-native shrub was further promoted in the mid-1900s as a way to reduce soil erosion and create living hedges for agriculture (Steavenson 1946; Reichard and White 2001). In subsequent years, multiflora rose has become an invasive species. In 2008, it was the most common introduced species in the Northeast and Midwest USA, found in over 27% of 1302 forest inventory plots (Schulz and Gray 2013). Today it is regulated or classified as a noxious weed in 12 states (USDA Federal and State Noxious Weeds; https://plants.usda.gov/java/noxiousDriver).

Multiflora rose spreads via a combination of sexual and clonal reproduction and can form large patches of monocultures up to 32 m in circumference (Jesse et al. 2010). It exhibits shade-avoiding traits (Dlugos et al. 2015) and is often most dense in successional habitats, open areas, and roadsides (Christen and Matlack 2009; Yates et al. 2004). However, multiflora rose also appears capable of establishing in a closed-canopy forest (Matlack and Schaub 2011) and can invade systems dominated by longer-lived species with slower turnover (Yurkonis et al. 2005). Its extended growing season allows understory shrubs to photosynthesize while canopy trees are bare (Dlugos et al. 2015).

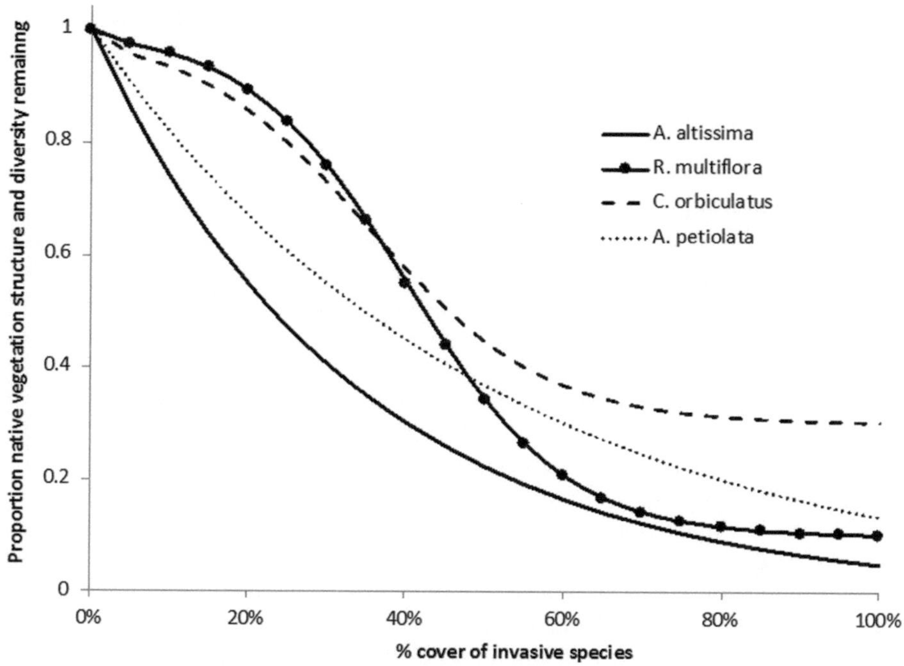

Fig. 6.4 Projected stressor-response curves for four case study species based on literature review and expert panel workshop

This large shrub can provide greater cover for foraging seed predators, therefore leading to greater discovery and removal of heterospecific seeds (Meiners and LoGiudice 2003; Meiners 2007), potentially leading to changes in community structure or species composition. A long-term study of abandoned agricultural land found that invasive shrub species, including multiflora rose, were associated with declines in both plant species richness (Meiners et al. 2001) and colonization of new species (Yurkonis et al. 2005). However, this negative impact was seen only at intermediate and high levels of invasion. Therefore, we expect vegetation structure and diversity to display a type II stressor-response curve to multiflora rose invasion (Fig. 6.4).

6.2.3 Celastrus orbiculatus

Celastrus orbiculatus, Oriental bittersweet, is a deciduous woody vine that was brought to the United States for cultivation as an ornamental species. Oriental bittersweet closely resembles a native con-generic species, *C. scandens*, but appears to outperform the native species across a broad range of light and temperature conditions (Leicht-Young et al. 2007). The congenerics also hybridize, which potentially contributes to the decline of the native species (Pooler et al. 2002; Zaya et al. 2015). Oriental bittersweet is commonly found in agricultural areas, natural forests, planted forests, grasslands, riparian zones, disturbed sites, and urban areas (Sundarapandian et al. 2015). Its range extends along the eastern seaboard from Maine to North Carolina, and west to Tennessee, Missouri, and Minnesota.

Traits leading to rapid colonization and spread of Oriental bittersweet include its high survivorship and ability to "sit and wait" for conditions that allow rapid growth (Greenberg et al. 2001). Oriental bittersweet apparently does not undergo density-dependent thinning under high densities (Leicht-Young et al. 2011). In full and partial sun, Oriental bittersweet grows quickly and may be able to overtop 1–2 m tall vegetation by the end of one growing season (Ellsworth et al. 2004). However, the plant is also tolerant of low light conditions (Leicht-Young et al. 2007), and seedlings can establish and survive under closed canopies at rates comparable to shade-tolerant trees (Ellsworth et al. 2004; Greenberg et al. 2001). Tolerance to low light probably explains why Oriental bittersweet colonized four different seral stages with equal frequency in the northern Piedmont of the United States (Robertson et al. 1994). Oriental bittersweet damages hardwood stands by stem girdling, increasing risk of ice damage, and eventually causing death by shading the crown foliage (McNab and Meeker 1987). Finally, Oriental bittersweet has alelopathic chemicals in its leaves that inhibit germination, for example of radish seeds (Pisula and Meiners 2010), and may also alter soil composition and soil processes in invaded areas, which could impact the re-introduction of native plants to these areas (Leicht-Young et al. 2009).

In their review of the top ten invasive climbing vines in the world, Sundarapandian et al. (2015) describe the primary ecosystem impact of Oriental bittersweet as overtopping native flora. This overtopping effect was seen in a four-decade study of old-field development in the northeastern United States when Oriental bittersweet invaded a portion of the field; two distinct plant communities developed based on the presence or absence of the invasive vine (Fike and Niering 2009). Bittersweet had much less of an effect in an established interior floodplain forest, where plots containing bittersweet had only slightly lower diversity, richness, and total abundance compared to plots without the invasive plant (Browder 2011). As much of Antietam National Battlefield consists of established floodplain forests, we expect that vegetation structure and diversity will show relatively low sensitivity to bittersweet invasion and follow a type II stressor-response (Fig. 6.4).

6.2.4 Alliaria petiolata

Alliaria petiolate, or garlic mustard, is a biennial plant in the Mustard family. The species is native to Europe and Asia and was introduced to North America in the mid-1800s as a culinary herb. It is currently listed as noxious in states throughout the eastern and mid-western United States. Garlic mustard is one of the few invaders that is able to grow in undisturbed woodland communities, where it is found in the understory of a variety of deciduous forests and woodlands (Munger 2001).

Garlic mustard has several traits that contribute to its success as an invader. Unlike many invasive species, garlic mustard can form dense monocultures in heavily shaded and semi-shaded habitats (Cavers et al. 1979). It produces allelopathic chemicals that suppress mycorrhizal fungi (Stinson et al. 2007), which subsequently decreases growth and survival of native mycorrhizal plants (Callaway et al. 2008). Garlic mustard tissues contain cyanide at levels considered toxic to many vertebrates (Cipollini and Gruner 2007), which probably explains why it is grazed little by mammalian or avian herbivores. In fact, in areas of high deer abundance, deer facilitate invasion by avoiding garlic mustard in favor of eating native plants instead (Knight et al. 2009). Additionally, research suggests that garlic mustard invasions may change soil nutrient availability in a way that promotes continued proliferation (Rodgers et al. 2008).

Impacts on native vegetation include mortality of existing trees and changes in understory composition. Several observational studies have shown a negative correlation between abundance of garlic mustard in the forest understory and diversity of native plant species (tree seedlings in particular), indicating that invasion by garlic mustard may lead to changes in native ecosystem structure and loss of canopy-forming trees (Stinson et al. 2007; Knight et al. 2009). In combination with deer browsing, garlic mustard can have particularly negative effects on growth of red oak (*Quercus rubra*) seedlings (Waller and Maas 2013). However, other studies show conflicting results (Rose et al. 2013; Davis et al. 2012), and some researchers suggest that garlic mustard invasion may be driven by native plant declines rather than the reverse (Phillips-Mao et al. 2014), or that the effects of garlic mustard at a site may change over time (Davis et al. 2012) or under different circumstances (Cipollini and Cipollini 2016). Therefore, while we expect garlic mustard invasion to have a relatively rapid effect on vegetation diversity because of attributes such as its allelopathy, we do not expect the final impact on the ecological endpoint to be as severe as for some other invasive species (Fig. 6.4).

It is worth noting that none of the four species are projected to have a linear stressor-response relationship. Tree-of-heaven and garlic mustard impacts are represented as type I curves, depicting their relatively large effects on vegetation at low abundance due to their fast growth, allelopathy, and ability to alter soil chemistry (especially for tree-of-heaven). There is less evidence of damages at low densities for the other two species, which follow type II theoretical curves. However, the available evidence indicates that multiflora rose can cause quite large impacts as it approaches its maximum possible abundance. From a management perspective, it would therefore be important to implement treatment for this species while it is still in its early phases of establishment before it can cause substantial harm. None of the curves were able to be parameterized from quantitative data. Instead all curves relied on qualitative evidence from the scientific literature and our expert panelists.

6.3 Ecosystem Services Impacted by Invasive Plants

The stressor-response functions are a step towards economic analyses of impacts on ecosystem services since they capture the degree to which invasives change ecosystem structure and function. These changes can be valued by translating them into ecosystem service benefits that affect people (the second step in Fig. 6.2). To value ecosystem services requires assessing people's degree of concern for any change in terms of what they would be willing to pay to avoid change (or accept change) (Freeman et al. 2014).

In the absence of the considerable work required to measure values of ecosystem service changes, ecological endpoints can be used as leading indicators of values. In this case, changes in vegetation structure and diversity can be related to their impact on the ecosystem services derived from battlefield landscapes. For example, the endpoint "vegetation structure" affects landscape character, length of views, light penetration below canopy, and habitat, which in turn affect aesthetic enjoyment of visitors interested in history, hiking, and birdwatching. Changes to ecological endpoints triggered by non-native plant invasions also can lead to changes in non-use values (Smith 1987), which affect people who never visit the park but who value preserving historic character, wildlife habitat, and biodiversity (Wainger et al. 2018).

From interviews and a review of recreational, social science, and economic literature, we identified the ecosystem services that we expected to be impacted by invasive plant-induced changes in vegetation structure and diversity. These ecosystem services are linked with four different groups of park users, each of which might differ in its sensitivity to ecological or physical changes (Table 6.3). For

Table 6.3 Ecosystem services influenced by changes in vegetation structure and diversity induced by invasive plants

Ecosystem services by user group
Casual visitors
Aesthetics of visitor experience
Convenient road/water access
Walking, hiking, biking opportunities
Safety of outdoor recreation
Avid recreationalists
Birdwatching
Native plant/wildflower viewing
Insect watching (e.g., butterflies)
Amphibian/reptile watching
Nature photography
Historic/cultural tourism
Neighbors
Safety and convenience of travel
Aesthetics from roads and viewpoints
Property values
Buffer incompatible uses
Maintenance costs (energy use, yard maintenance)
Distant and Non-use beneficiaries
Climate regulation
Native ecosystem preservation
Charismatic species preservation
Maintenance of significant natural areas

example, casual visitors might be less sensitive than avid recreationalists to visual aesthetics changes.

The set of ecosystem services that we identified emphasize the recreational benefits of battlefield parks. The services also recognize the benefits of battlefield parks on nearby property values and consider the collateral values associated with people knowing that natural areas are protected (i.e., non-uses). We recognize that parks provide a multitude of other services that are not quantified here. Further, we have not captured the values to scientific and educational user groups, even though we know that Antietam National Battlefield is well-used by both types of users. Instead, we focus on services for which there exists sufficient published literature to form the basis of a discussion.

6.3.1 Casual Visitors

Casual visitors, including joggers, hikers, dog-walkers, horse-back riders, and recreational drivers, tend to visit the park looking for exercise or relaxation. Users in this category are assumed to be less sensitive to changes in vegetation than avid recreationists but might still be affected by major changes in flora and fauna that could occur at high densities of invasive plants (Ioja et al. 2011; Zhang et al. 2015). In addition, at lower densities of invasive plants, their enjoyment might be affected by changes in aesthetic qualities, safety, or convenience. Invasive plants might affect their recreational experience by altering the character of the vegetation (e.g., making the understory more dense), reducing the probability of encounters with charismatic species such as birds and butterflies, increasing the probability of tree falls, and increasing health risks to people or pets from direct plant contact (e.g., cuts, skin irritations, and burns).

Driving for pleasure and recreation is one of the most common ways that casual visitors experience battlefield parks (NSRE n.d.). The value of this activity depends on the drivers' ability to enjoy scenic vistas and is enhanced by having well-maintained parking areas and pull-offs for enjoying the scenery and taking photographs (Hallo and Manning 2009). Some NPS units have explicit goals of offering recreational driving experiences. This service is sensitive to changes in vegetation structure that affect views and the risks to park users from tree falls.

6.3.2 Avid Recreationists

Avid recreationists include those who visit parks to experience specific species or ecosystems (i.e., nature viewers), or to experience the historical or cultural heritage of the park (i.e., historical and cultural tourists). This group is likely to be more sensitive to changes caused by invasive species than casual visitors.

Nature viewing includes birdwatching, native plant/wildflower viewing, insect watching (e.g., butterflies), amphibian/reptile watching, and nature photography.

Nature viewing will be affected if native plants induce changes in vegetation that ripple to higher trophic levels. For example, the introduction of garlic mustard is partially credited with decline of the rare West Virginia white butterfly (*Pieris virginiensis*) due to chemicals that appear to be toxic to the larval form (Davis et al. 2015). Richness of other arthropods may also be negatively affected by garlic mustard invasion (McCary 2016). Through impacts on insect communities and particularly on lepidopteran larvae, which are a disproportionately valuable source of food for multiple species of terrestrial birds (Tallamy 2004; Tallamy and Shropshire 2009), invasive plants can have indirect impacts on bird communities as well.

Invasive plants can also directly impact bird communities by changing vegetation structure and composition. In urbanizing landscapes, bird nests in exotic shrubs (including multiflora rose) experience higher daily mortality rates than those in native shrubs. This is likely due to reduced nest height and larger shrub volume surrounding the nests in exotic shrubs (Borgmann and Rodewald 2004). Furthermore, fruits of invasive shrubs like multiflora rose have been shown to be less appealing and less nutritious for migratory birds than fruits from native shrubs and thus may alter migratory stop-over dynamics (Bolser et al. 2013; Smith et al. 2013).

A sizable subset of recreationists visits parks, especially battlefield parks, with the primary goal of experiencing the historical or cultural heritage of the sites. The entire park can be a cultural or historical site, or components within the park may be the focus of the visit. For example, many Civil War battlefield parks aim to maintain vegetation appropriate to the period of conflict in the 1860s; this service would be threatened if invasive plants altered the historic vegetation. Similarly, invasive plants could threaten cultural tourism if vegetation damaged structures of cultural importance or significant aspects of historical vistas.

6.3.3 Neighbors

Battlefield parks provide amenities to neighbors because they provide open space, aesthetic benefits, and convenient recreational opportunities. Studies suggest that property adjacent to parks or other types of open space often has enhanced value relative to similar property not similarly situated (Crompton 2001; Geoghegan 2002; McConnell and Walls 2005). Hence, battlefield parks would be expected to have a positive effect on adjacent and nearby homes (Lutzenhiser and Netusil 2001). Due to their urban and suburban setting, battlefield parks often include major commuting roads for adjacent landowners that intersect or form park administrative boundaries.

However, with invasive plants, the value of this amenity can diminish if park aesthetics decline in the area adjacent to the property (Fox 1990) or if proximity becomes a "disamenity" due to threats of falling trees or a persistent source of noxious weeds. Large-scale changes in vegetation structure can further lead to changes in nearby heating and cooling cost (Nowak et al. 2006), although energy-saving benefits may already be captured in the property value premium.

6.3.4 Distant and Non-use Beneficiaries

Non-use or passive-use services are those associated with preserving a resource that will not be used in any tangible way. For example, people may value the existence of diverse vegetation communities even if they never plan to visit these communities. This service is associated with sense of place and intergenerational stewardship, which is the notion that people have an ethical responsibility to future generations to care for nature (Welburn 2014). Non-use services are typically divided into existence, option, and bequest values (Smith 1987), which correspond to benefits associated with knowing a resource exists, preserving the option to use it in the future, or providing the opportunity for future generations to use or enjoy that resource.

Non-use environmental values are generally underappreciated for battlefield landscapes. Non-use values for environmental preservation have the potential to outweigh those for cultural/historical preservation (Turner and Willmarth 2014), but may never be measured for parks that do not contain rare or endangered species. Nevertheless, more data on all types of non-use values, including cultural and historical services, provided by battlefield parks would be useful, as studies have shown that non-use values can greatly exceed use values (Grosclaude and Soguel 1993; Ruijgrok 2006).

6.4 Conclusions: Managing Biological Invasions to Promote Ecosystem Services in Battlefield Landscapes

Battlefield parks present an opportunity to provide numerous ecosystem services in close proximity to urban areas. However, as these parks mature, the management of their natural resources requires increasing attention. In particular, invasive plants represent a pervasive and costly challenge that will continue to grow over time. In this chapter, we presented a systematic approach to understanding how invasive plants can impact the ecosystem services provided by battlefield parks. This is a crucial step in identifying the level of control effort that will provide the greatest net benefits to the parks' diverse stakeholders.

Our approach starts by identifying stressor-response functions that connect invasion by non-native plants to ecological endpoints: here, changes in vegetation structure and diversity. We approached this task by conducting a literature review and a workshop to elicit expert opinion. This experience was illuminating in several ways. First, we quickly encountered limitations in terms of published literature about how invasive plants affect relevant ecological endpoints, such as native vegetation diversity and structure. Of the existing studies, which focus on a limited set of species (and are unevenly distributed across functional forms), the vast majority were observational rather than experimental, leaving uncertainty as to the potential causality of invasive-native plant relationships.

A second lesson learned was that while experts generally agreed that invasive plants have nonlinear effects on native vegetation, we found very little in the literature to either support or refute this idea. If our proposed stressor-response curves are reasonably accurate, lack of information combined with the nonlinear effects depicted by the curves could catch managers unaware of dangerous thresholds. Furthermore, these results would have implications for cost-effectiveness targeting of management actions. As one example, the existence of nonlinear relationships dictate that all restoration will not do the same amount of good. The magnitude of change in ecological endpoint is a function of the amount of change in invasive abundance as quantified by the slope of the stressor-response curve. The fact that the slope may differ at different invasive abundances dictates that restoration opportunities should be weighed carefully to target situations in which the greatest change in ecological endpoint is achieved per unit decrease in invasive abundance.

We offer a further caution that the impacts of non-native, invasive plants are not spatially uniform. First, the spread of invasive plants in these heterogeneous battlefield landscapes is highly variable depending on, among other factors, the location of introduction, the spatial patterning of the physical landscape, and the dispersal mode of the invading species (Holdenreider et al. 2004; Ferrari and Lookingbill 2009; Minor et al. 2009). In addition, the ecosystem services that are impacted are also spatially variable. For example, we would expect impacts to wildlife viewing to be concentrated in park management parcels of high habitat value for rare or charismatic species where wildlife viewers typically concentrate their activities (Fig. 6.5). The impact to historical visitors would spread along the main tour roads and trails of parks, as well as around park visitor centers. In light of this spatial variability, spatial targeting of management may be an approach to maximizing the benefits of management.

We see several potential next steps in this research. The stressor-response models developed in this chapter are only a first-cut and could be further refined with additional experimental data, for example. Lack of adequate information to better quantify these relationships is a common lament among invasive species managers. However, conceptual models like those presented here are valuable for prioritizing restoration activities and improving the return on investment from these actions, assuming they can approximate the shape of the function reasonably well.

An obvious omission from this chapter is the creation of damage functions — how do altered environmental endpoints quantitatively affect ecosystem service benefits? The evaluation of these functions requires consideration of how the biophysical changes in the environment influence the social or economic benefits that people derive from parks. For example, an invasive species may reduce the rate of groundwater recharge in a system – an ecological response. However, this change will create an economic harm only if the change in groundwater recharge rate is sufficient to reduce water levels where it is being pumped and may be a substantial harm only in areas where water is scarce and therefore valuable (e.g., Zavaleta 2000; Le Maitre et al. 2002). Assessments of the financial impacts and costs required to prevent and eradicate invasive plants are rare, but an important next step (Abella 2014). Based on general assumptions about the value premium associated with proximity to natural parks (e.g., Curtis 1993; Lutzenhiser and Netusil 2001), we might expect, for example, an overall invasive plant impact of nearly \$500,000 to the 440 homes within the zone of influence of Antietam National Battlefield (Wainger et al. 2012).

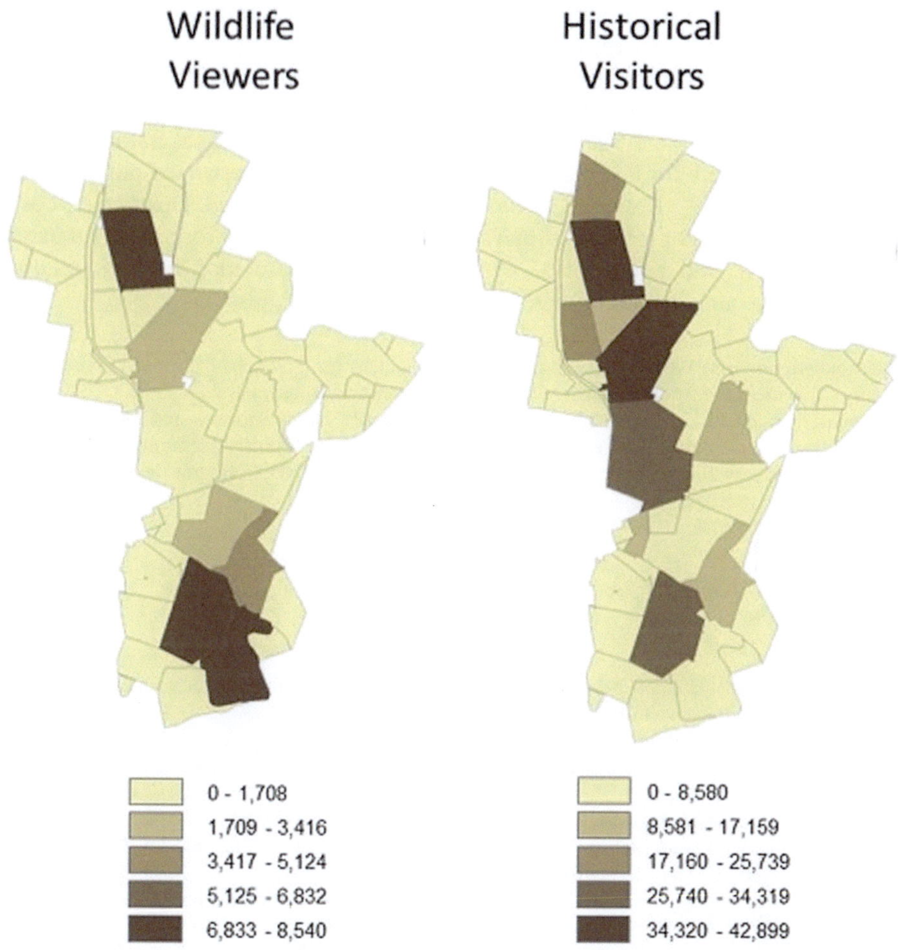

Fig. 6.5 Spatial patterning of two different ecosystem services at Antietam National Battlefield (annual number of users estimated per management unit)

Additional research is needed to examine effects of other invasive species and other ecological endpoints beyond vegetation structure and diversity. Invasive species are also certainly not the only threats to battlefield parks. New studies could consider the potential interactive effects of invasive plants with climate change, increasing isolation within an urbanizing matrix, and regional air-quality degradation. More detailed maps of the damages wrought by non-native, invasive plants also would be of great value to managers who must decide where to focus their eradication efforts.

We present this case study for Antietam National Battlefield, but the challenges are widespread for battlefield parks. For example, the cover of *Ligustrum sinense*, Chinese privet, increased five-fold from 1993 to 2008 in Chickamauga and Chattanooga National Military Park, contributing to a 70% decrease in native herbaceous plants, including rare and priority species (Sutter et al. 2011; Abella 2015).

National Park Service resources for managing non-native, invasive plants at most battlefield parks are scarce, often relying solely on one of the 16 Exotic Plant Management Teams spread throughout the country (Fraley et al. 2007). These management actions are expensive. A recent effort to reduce the density of *Lonicera morrowii*, Japanese honeysuckle, by half at Fort Necessity National Battlefield cost up to $9300/hectare for just the first treatment when plants were hand pulled (Love and Anderson 2009). However, these threats can not be ignored. Developing cost-effective strategies for treating invasive plants will be a priority for battlefield parks as they mature and embrace the ecosystem services provided by their natural resources as a complement to their historical and cultural missions.

Acknowledgements We are grateful to our expert panelists for their contributions to the development of the stressor-response curves: Frank Archuleta, Norm Bourg, Pat Campbell, Geoff Clark, Cynthia Huebner, Peter Marra, Bill McShea, John Parrish, Doug Samson, and Doug Tallamy. Alynne Bayard, Joe Ferrari, and Kristin Politano assisted with preliminary literature research. This study was partially funded under Cooperative Agreement H3992060006 with the National Park Service.

References

Abella, S. R. (2014). Effectiveness of exotic plant treatments on National Park Service lands in the United States. *Invasive Plant Science and Management, 7*, 147–163.

Abella, S. R. (2015). *Conserving national parks*. Charleston: CreateSpace.

Allen, J., Brown, C., & Stohlgren, T. (2009). Non-native plant invasions of United States national parks. *Biological Invasions, 11*, 2195–2207.

Barendse, J., Roux, D., Erfmann, W., Baard, J., Kraaij, T., & Nieuwoudt, C. (2016). Viewshed and sense of place as conservation features: A case study and research agenda for South Africa's national parks. *Koedoe, 58*, 1–16.

Bezemer, T. M., Jeffrey, A. H., & Cronin, J. T. (2014). Response of native insect communities to invasive plants. *Annual Review of Entomology, 59*, 119–141.

Bolser, J. A., Alan, R. R., Smith, A. D., Li, L. Y., Seeram, N. P., & McWilliams, S. R. (2013). Birds select fruits with more anthocyanins and phenolic compounds during autumn migration. *Wilson Journal of Ornithology, 125*, 97–108.

Borgmann, K. L., & Rodewald, A. D. (2004). Nest predation in an urbanizing landscape: The role of exotic shrubs. *Ecological Applications, 14*, 1757–1765.

Browder, J. R. (2011). *The effect of Celastrus orbiculatus, Oriental bittersweet, on the herbaceous layer along a western North Carolina creek*. Master's Thesis, Western Carolina University.

Call, L. J., & Nilsen, E. T. (2005). Analysis of interactions between the invasive tree-of-heaven (*Ailanthus altissima*) and the native black locust (*Robinia pseudoacacia*). *Plant Ecology, 176*, 275–285.

Callaway, R. M., Cipollini, D., Barto, K., Thelen, G. C., Hallett, S. G., Prati, D., Stinson, K., & Klironomos, J. (2008). Novel weapons: Invasive plant suppresses fungal mutualists in America but not in its native Europe. *Ecology, 89*, 1043–1055.

Cavers, P., Heagy, M., & Kokron, R. (1979). The biology of Canadian weeds *Alliaria petiolata* (M. Bieb.) Cavara and Grande. *Canadian Journal of Plant Science, 59*, 217–229.

Celesti-Grapow, L., & Blasi, C. (2004). The role of alien and native weeds in the deterioration of archaeological remains in Italy. *Weed Technology, 18*, 1508–1513.

Christen, D. C., & Matlack, G. R. (2009). The habitat and conduit functions of roads in the spread of three invasive plant species. *Biological Invasions, 11*, 453–465.

Cipollini, D., & Cipollini, K. (2016). A review of garlic mustard (*Alliaria petiolata*, Brassicaceae) as an allelopathic plant. *The Journal of the Torrey Botanical Society, 143*, 339–348.
Cipollini, D., & Gruner, B. (2007). Cyanide in the chemical arsenal of garlic mustard, *Alliaria petiolate*. *Journal of Chemical Ecology, 33*, 85–94.
Constán-Nava, S., Soliveres, S., & Torices, R. (2015). Direct and indirect effects of invasion by the alien tree *Ailanthus altissima o*n riparian plant communities and ecosystem multifunctionality. *Biological Invasions, 17*, 1095–1108.
Costanza, R., d'Arge, R., de Groot, R., Farber, S., Grasso, M., Hanno, B., Limburg, K., Naeem, S., O'Neill, R. V., Paruelo, J., Raskin, R. G., Sutton, P., & van den Belt, M. (1997). The value of the world's ecosystem services and natural capital. *Nature, 387*, 253–260.
Crompton, J. L. (2001). The impact of parks on property values: A review of the empirical evidence. *Journal of Leisure Research, 33*, 1–31.
Curtis, R. E. (1993). *Valuing open space in Maryland: An hedonic analysis*. College Park: University of Maryland.
Davis, M. A., Colehou, A., Daney, J., Foster, E., Macmillen, C., Merrill, F., O'Neil, J., Pearson, M., Whitney, M., Anderson, M. D., & Dosch, J. J. (2012). The population dynamics and ecological effects of garlic mustard, *Alliaria petiolata*, in a Minnesota oak woodland. *The American Midland Naturalist, 168*, 364–374.
Davis, S. L., Frisch, T., Bjarnholt, N., & Cipollini, D. (2015). How does garlic mustard lure and kill the West Virginia white butterfly? *Journal of Chemical Ecology, 41*, 948–955.
Dlugos, D. M., Collins, H., Bartelme, E. M., & Drenovsky, R. E. (2015). The non-native plant *Rosa multiflora* expresses shade avoidance traits under low light availability. *American Journal of Botany, 102*, 1323–1331.
Ellsworth, J. W., Harrington, R. A., & Fownes, J. H. (2004). Survival, growth and gas exchange of *Celastrus orbiculatus* seedlings in sun and shade. *The American Midland Naturalist, 151*, 233–240.
Ferrari, J., & Lookingbill, T. (2009). Initial conditions and their effect on invasion velocity across heterogeneous landscapes. *Biological Invasions, 11*, 1247–1258.
Fike, J., & Niering, W. A. (2009). Four decades of old field vegetation development and the role of *Celastrus orbiculatus* in the northeastern United States. *Journal of Vegetation Science, 10*, 482–492.
Fox, T. (1990). *Urban open space: An investment that pays*. The Neighborhood Open Space Coalition.
Fraley, N., Furqueron, C., Pernas, T., & Worsham, E. (2007). The National Park Service's exotic plant management teams in the southeast and Caribbean. *Natural Areas Journal, 27*, 232–235.
Freeman, A. M., Herriges, J. A., & Kling, C. L. (2014). *The measurement of environmental and resource values: Theory and methods* (3rd ed.). New York: RFF Press.
Geoghegan, J. (2002). The value of open spaces in residential land use. *Land Use Policy, 19*, 91–98.
Gómez-Aparicio, L., & Canham, C. D. (2008a). Neighbourhood analyses of the allelopathic effects of the invasive tree *Ailanthus altissima* in temperate forests. *Journal of Ecology, 96*, 447–458.
Gómez-Aparicio, L., & Canham, C. D. (2008b). Neighborhood models of the effects of invasive tree species on ecosystem processes. *Ecological Monographs, 78*, 69–86.
Greenberg, C. H., Smith, L. M., & Levey, D. J. (2001). Fruit fate, seed germination and growth of an invasive vine – An experimental test of 'sit and wait' strategy. *Biological Invasions, 3*, 363–372.
Grosclaude, P., & Soguel, N. (1993). *Contingent valuation of damages to historic buildings: A case study of road traffic externalities*. Centre for Social and Economic Research on the Global Environment.
Hallo, J. C., & Manning, R. E. (2009). Transportation and recreation: A case study of visitors driving for pleasure at Acadia National Park. *Journal of Transport Geography, 17*, 491–499.
Heisey, R. M. (1996). Identification of an allelopathic compound from *Ailanthus altissima* and characterization of its herbicidal activity. *American Journal of Botany, 83*, 192–200.
Holdenreider, O., Pautasso, M., Weisberg, P. J., & Lonsdale, D. (2004). Tree diseases and landscape processes: The challenge of landscape pathology. *Trends in Ecology & Evolution, 19*, 446–452.

Ioja, C. I., Rozylowicz, L., Patroescu, M., Nita, M. R., & Vanau, G. O. (2011). Dog walkers' vs. other park visitors' perceptions: The importance of planning sustainable urban parks in Bucharest, Romania. *Landscape and Urban Planning, 103*, 74–82.

Jesse, L. C., Nason, J. D., Obrycki, J. J., & Moloney, K. A. (2010). Quantifying the levels of sexual reproduction and clonal spread in the invasive plant, *Rosa multiflora. Biological Invasions, 12*, 1847–1854.

Knight, T. M., Dunn, J. L., Smith, L. A., Davis, J., & Kalisz, S. (2009). Deer facilitate invasive plant success in a Pennsylvania forest understory. *Natural Areas Journal, 29*, 110–116.

Kowarik, I., & Säumel, I. (2007). Biological flora of Central Europe: *Ailanthus altissima* (Mill.) Swingle. *Perspectives in Plant Ecology, Evolution and Systematics, 8*, 207–237.

Kowarik, I., & Säumel, I. (2008). Water dispersal as an additional pathway to invasions by the primarily wind-dispersed tree *Ailanthus altissima. Plant Ecology, 198*, 241–252.

Landenberger, R., Kota, N., & McGraw, J. (2007). Seed dispersal of the non-native invasive tree *Ailanthus altissima* into contrasting environments. *Plant Ecology, 192*, 55–70.

Lawrence, J. G., Colwell, A., & Sexton, O. J. (1991). The ecological impact of allelopathy in *Ailanthus altissima* Simaroubaceae. *American Journal of Botany, 78*, 948–958.

Leicht-Young, S. A., Silander, J. A., Jr., & Latimer, A. M. (2007). Comparative performance of invasive and native *Celastrus* species across environmental gradients. *Oecologia, 154*, 273–282.

Leicht-Young, S. A., O'Donnell, H., Latimer, A. M., & Silander, J. A., Jr. (2009). Effects of an invasive plant species, *Celastrus orbiculatus*, on soil composition and processes. *The American Midland Naturalist, 161*, 219–231.

Leicht-Young, S. A., Latimer, A. M., & Silander, J. A., Jr. (2011). Lianas escape self-thinning: Experimental evidence of positive density dependence in temperate lianas *Celastrus orbiculatus* and *C. scandens*. *Perspectives in Plant Ecology and Evolution Systematics, 13*, 163–172.

Le Maitre, D. C., van Wilgen, B. W., Gelderblom, C. M., Bailey, C., Chapman, R. A., & Nel, J. A. (2002). Invasive alien trees and water resources in South Africa: Case studies of the costs and benefits of management. *Forest Ecology and Management, 160*, 143–159.

Lookingbill, T., Schmit, J. P., Tessel, S., Suarez-Rubio, M., & Hilderbrand, R. (2014a). Assessing national park resource condition along an urban-rural gradient in and around Washington, D.C., USA. *Ecological Indicators, 42*, 147–159.

Lookingbill, T., Minor, E., Bukach, N., Ferrari, J., & Wainger, L. (2014b). Incorporating risk of reinvasion to prioritize sites for invasive species management. *Natural Areas Journal, 34*, 268–281.

Love, J. P., & Anderson, J. T. (2009). Seasonal effects of four control methods on the invasive Morrow's honeysuckle (*Lonicera morrowii*) and initial responses of understory plants in the southwestern Pennsylvania old field. *Restoration Ecology, 17*, 549–559.

Lutzenhiser, M., & Netusil, N. R. (2001). The effect of open spaces on a home's sale price. *Contemporary Economic Policy, 19*, 291–298.

Matlack, G. R., & Schaub, J. R. (2011). Long-term persistence and spatial assortment of nonnative plant species in second-growth forests. *Ecography, 34*, 649–658.

McCary, M. (2016). *Evaluating the impacts of invasive plants on the forest-floor food web*. Doctoral dissertation, University of Illinois at Chicago.

McConnell, V., & Walls, M. (2005). *The value of open space: Evidence from studies of nonmarket benefits*. Lincoln Institute of Land Policy Working Paper.

McNab, W. H., & Meeker, M. (1987). Oriental bittersweet: A growing threat to hardwood silviculture in the Appalachians. *Northern Journal of Applied Forestry, 4*, 174–177.

McPherson, J. M. (2002). *Crossroads of freedom: Antietam, the battle that changed the course of the Civil War*. New York: Oxford University Press.

Meiners, S. J. (2007). Apparent competition: An impact of exotic shrub invasion on tree regeneration. *Biological Invasions, 9*, 849–855.

Meiners, S. J., & LoGiudice, K. (2003). Temporal consistency in the spatial pattern of seed predation across a forest – Old field edge. *Plant Ecology, 168*, 45–55.

Meiners, S. J., Pickett, S. T. A., & Cadenasso, M. L. (2001). Effects of plant invasions on the species richness of abandoned agricultural land. *Ecography, 24*, 633–644.

Miller, J. H. (1990). *Ailanthus altissmia* (Mill.) Swingle. In R. M. Burns, & B. H. Honkala (tech cords) *Silvics of North America, Volume 2 Hardwoods*. Washington, DC: US Department of Agriculture, Forest Service, pp 101–104.

Minor, E., Tessel, S., Engelhardt, K., & Lookingbill, T. (2009). The role of landscape connectivity in assembling exotic plant communities. *Ecology, 90*, 1802–1809.

Motard, E., Muratet, A., Clair-Maczulajtyx, D., & Machon, N. (2011). Does the invasive species *Ailanthus altissima* threaten floristic diversity of temperate peri-urban forests? *Comptes Rendus Biologies, 334*, 872–879.

Munger, G. T. (2001) *Alliaria petiolata*. In *Fire effects information system*, [Online]. U.S. Department of Agriculture, Forest Service, Rocky Mountain Research Station, Fire Sciences Laboratory. Available: http://www.fs.fed.us/database/feis/

National Survey on Recreation and the Environment (NSRE). (n.d.). *Fifth report (2000–2002)*. Athens/Knoxville: The Interagency National Survey Consortium, coordinated by the USDA Forest Service, Recreation, Wilderness, and Demographics Trends Research Group/Human Dimensions Research Laboratory, University of Tennessee.

Nowak, D. J., Crane, D. E., & Stevens, J. C. (2006). Air pollution removal by urban trees and shrubs in the United States. *Urban Forestry & Urban Greening, 4*, 115–123.

Phillips-Mao, L., Larson, D. L., & Jordan, N. R. (2014). Effects of native herbs and light on garlic mustard (*Alliaria petiolata*) invasion. *Invasive Plant Science and Management, 7*, 257–268.

Pisula, N. L., & Meiners, S. J. (2010). Relative allelopathic potential of invasive plant species in a young disturbed woodland. *The Journal of the Torrey Botanical Society, 137*, 81–87.

Pooler, M. R., Dix, R. L., & Feely, J. (2002). Interspecific hybridizations between the native bittersweet, *Celastrus scandens*, and the introduced invasive species, *C. orbiculatus*. *Southeastern Naturalist, 1*, 69–76.

Rehder, A. (1936). On the history of the introduction of woody plants into North America. *National Horticulture Magazine, 15*, 245–257.

Reichard, S. H., & White, P. (2001). Horticulture as a pathway of invasive plant introductions in the United States. *Bioscience, 51*, 103–113.

Robertson, D. J., Robertson, M. C., & Tague, T. (1994). Colonization dynamics of four exotic plants in a northern Piedmont natural area. *Bulletin of the Torrey Botanical Club, 121*, 107–118.

Rodgers, V. L., Wolfe, B. E., Werden, L. K., & Finzi, A. C. (2008). The invasive species *Alliaria petiolata* (garlic mustard) increases soil nutrient availability in northern hardwood-conifer forests. *Oecologia, 157*, 459–471.

Rose, S. D., Endress, A. G., Frank, P. J., Kwit, M. C., & Helge, J. C. (2013). Increasing invasion of *Alliaria petiolata* (M. Bieb.) Cavara and Grande and change in the understory community across eight years in a fragmented Illinois woodland. *Journal of Biodiversity Management and Forestry, 2*, 1.

Ruijgrok, E. C. M. (2006). The three economic values of cultural heritage: A case study in the Netherlands. *Journal of Cultural Heritage, 7*, 206–213.

Schulz, B. K., & Gray, A. N. (2013). The new flora of northeastern USA: Quantifying introduced plant species occupancy in forest ecosystems. *Environmental Monitoring and Assessment, 185*, 3931–3957.

Skórka, P., Lenda, M., & Tryjanowski, P. (2010). Invasive alien goldenrods negatively affect grassland bird communities in Eastern Europe. *Biological Conservation, 143*, 856–861.

Smith, S. S., DeSando, S. A., & Pagano, T. (2013). The value of native and invasive fruit-bearing shrubs for migrating songbirds. *Northeastern Naturalist, 20*, 171–184.

Smith, T. B. (2008). *The golden age of battlefield preservation: The decade of the 1890s and the establishment of America's first five military parks*. Knoxville: University of Tennessee Press.

Smith, V. K. (1987). Nonuse values in benefit cost analysis. *Southern Economic Journal, 54*, 19–26.

Steavenson, H. A. (1946). Multiflora rose for farm hedges. *Journal of Wildlife Management, 10*, 227–234.

Stinson, K., Kaufman, S., Durbin, L., & Lowenstein, F. (2007). Impacts of garlic mustard invasion on a forest understory community. *Northeastern Naturalist, 14*, 73–88.

Suarez-Rubio, M., Lookingbill, T., & Elmore, A. (2012). Exurban development derived from Landsat from 1986 to 2009 surrounding the District of Columbia, USA. *Remote Sensing of Environment, 124*, 360–370.

Sundarapandian, S., Muthumperumal, C., & Subashree, K. (2015). Biological invasion of vines, their impacts and management. In N. Parthasarathy (Ed.), *Biodiversity of lianas. Sustainable development and biodiversity* (Vol. 5). Switzerland: Springer.

Sutter, R. D., Grovus, T. E., Lyons Smith, R., Nordman, C., Pyne, M., & Hogan, T. (2011). Monitoring change in a central U.S. calcareous glade: Resampling transects established in 1993. *Natural Areas Journal, 31*, 163–172.

Tallamy, D. W. (2004). Do alien plants reduce insect biomass? *Conservation Biology, 18*, 1689–1692.

Tallamy, D. W., & Shropshire, K. J. (2009). Ranking lepidopteran use of native versus introduced plants. *Conservation Biology, 23*, 941–947.

Turner, R. W., & Willmarth, B. (2014). *Valuation of cultural and natural resources in North Cascades National Park: Results from a tournament-style contingent choice survey*. Economics Faculty Working Papers. 38. http://commons.colgate.edu/econ_facschol/38

Vilá, M., Tessier, M., Suehs, C. M., Brundu, G., Carta, L., Galanidis, A., Lambdon, P., Manca, M., Medail, F., Moragues, E., Traveset, A., Troumbis, A. Y., & Hulme, P. E. (2006). Local and regional assessments of the impacts of plant invaders on vegetation structure and soil properties of Mediterranean islands. *Journal of Biogeography, 33*, 853–861.

Wainger, L. A., Lookingbill, T. R., & Minor, E. S. (2012). *To weed or not to weed? An economic decision support tool for National Capital Region Parks*. University of Maryland Center for Environmental Science Technical Report to National Park Service. TS-734-19.

Wainger, L. A., Helcoski, R., Farge, K. W., Espinola, B. A., & Green, G. T. (2018). Evidence of a shared value for nature. *Ecological Economics, 154*, 107–116.

Waller, D. M., & Maas, L. I. (2013). Do white-tailed deer and the exotic plant garlic mustard interact to affect the growth and persistence of native forest plants? *Forest Ecology and Management, 304*, 296–302.

Welburn, D. (2014). Rawlsian environmental stewardship and intergenerational justice. *Environmental Ethics, 36*, 387–404.

Yates, E. D., Levai, D. F., & Williams, C. L. (2004). Recruitment of three non-native invasive plants into a fragmented forest in southern Illinois. *Forest Ecology and Management, 190*, 119–130.

Yokomizo, H., Possingham, H. P., Thomas, M. B., & Buckley, Y. M. (2009). Managing the impact of invasive species: The value of knowing the density-impact curve. *Ecological Applications, 19*, 376–386.

Yurkonis, K. A., Meiners, S. J., & Wachholder, B. (2005). Invasion impacts diversity through altered community dynamics. *Journal of Ecology, 93*, 1053–1061.

Zavaleta, E. (2000). The economic value of controlling an invasive shrub. *Ambio, 29*, 462–467.

Zaya, D., Leicht-Young, S., Pavlovic, N., Feldheim, K., & Ashley, M. (2015). Genetic characterization of hybridization between native and invasive bittersweet vines (*Celastrus* spp.). *Biological Invasions, 17*, 2975–2988.

Zhang, W., Yang, J., Ma, L., & Huang, C. H. (2015). Factors affecting the use of urban green spaces for physical activities: Views of young urban residents in Beijing. *Urban Forestry & Urban Greening, 14*, 851–857.

Part III
Borderlands

Chapter 7
Valuing the Wounds of War: Korea's DMZ as Nature Preserve

Lisa M. Brady

Abstract Korea's Demilitarized Zone (DMZ) contains a wealth of what this volume calls "collateral values" ranging from undisturbed habitat and de facto protection for the peninsula's rare and endangered species to conservation of a variety of cultural sites, both ancient and more modern, that have deep meaning for Korea's people. This chapter examines the environmental history of the DMZ, illustrating how changes in both nature and culture since 1953 have helped to transform a relic of war into a symbol of peace. Although never intended to be permanent, the DMZ has radically altered the landscape of the Korean peninsula in ways that many inside both North Korea and South Korea, as well as across the globe, wish to make lasting. Because of its scientific, environmental, and cultural importance, the DMZ has become the object of preservation efforts even as reunification remains an important goal. This chapter argues that, in both its material and symbolic forms, the DMZ may reify the discontinuities between north and south but, in very important ways, it also serves as a green ribbon of hope, a place that commemorates the human sacrifices of military conflict and protects a variety of ecosystems and species that are found nowhere else on the peninsula.

7.1 Introduction

Clearly distinguishable in satellite images, a ribbon of green bisects the Korean peninsula at its narrowest point (Fig. 7.1). Depicted cartographically, the ribbon becomes a line, a transformation that simplifies and obscures its historical and ecological significance. On the ground, this meandering corridor has been many things at once: a de facto international boundary, an active war zone, a lively tourist destination, and an unofficial nature preserve (Brady 2012). This is the Korean Demilitarized Zone (DMZ), a visible, tangible boundary inscribed on Korea's landscape and an emotional, political divide embedded in Korean culture.

L. M. Brady (✉)
Department of History, Boise State University, Boise, ID, USA
e-mail: lisabrady@boisestate.edu

T. R. Lookingbill, P. D. Smallwood (eds.), *Collateral Values*, Landscape Series 25,
https://doi.org/10.1007/978-3-030-18991-4_7

Fig. 7.1 Demilitarized Zone (DMZ) between the Democratic People's Republic of Korea (north) and the Republic of Korea (south) as seen from space. Image courtesy Jesse Allen, NASA's Earth Observatory, Goddard Space Flight Center. Arrow added to indicate western half of the DMZ. https://earthobservatory.nasa.gov/IOTD/view.php?id=3660

The DMZ is, on average, 4 km wide, although additional buffer areas on either side (generally cleared of forests and in some cases under agricultural cultivation) broaden its girth by another 4–16 km. Roughly following the 38th parallel, it meanders 250 km east from the Han River estuary through the Taebaek Mountains to the East Sea/Sea of Japan (see Fig. 7.2). Its approximately 1000 square kilometers comprise about one half of one percent of the peninsula's total area (220,847 km^2). Although this narrow band of territory is one of the most heavily land-mined and militarized areas on the planet, it is also one of the safest—for nature. The DMZ has provided largely undisturbed habitat for a wide range of species including white-naped and red-crowned cranes, Amur gorals, and wild boar. Recent estimates suggest that at least 2000 species exist within the DMZ's fences (Kim 2013). Such

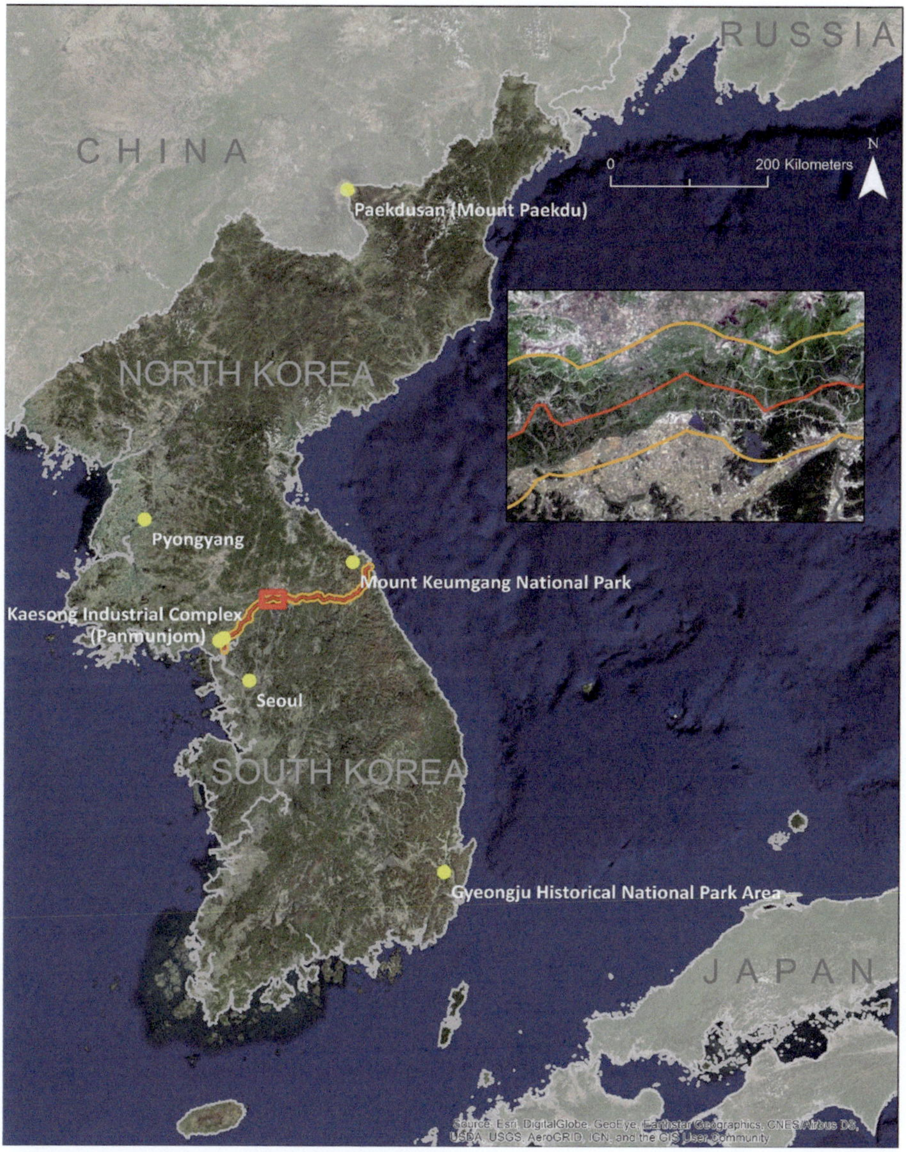

Fig. 7.2 DMZ and other points of interest on the Korean Peninsula. Inset: the buffer zone around the DMZ with detail illustrating the high percent forest cover

proliferation of wildlife, both plants and animals, has attracted a great deal of attention since at least the 1990s. As David Havlick and others have explained, the DMZ has become a tourist destination and an educational site for those interested in Korea's natural history and ecology (Chap. 9 of this book). Moreover, it has become a rallying cry for those wishing to protect what is arguably some of the last wild land left on the peninsula.

A case in point: In 2005, the DMZ Forum—an international, U.S.-based, 501(c)(3) organization—hosted a conference in Seoul, South Korea, featuring panel discussions and keynote addresses by scientists, politicians, and conservationists, including billionaire philanthropist and CNN founder Ted Turner, all of whom supported some kind of protection for the DMZ and its denizens. This was the fifth such conference sponsored and organized by the group. Previous meetings—held in either Seoul or New York City—included talks by world-renowned ecologist Edward O. Wilson, forest ecologist and peace activist Arthur H. Westing, and former U.S. Ambassador to South Korea Stephen Bosworth. Across the years, the gatherings consistently centered on the goal of setting aside the DMZ as a nature preserve, an idea that garnered the attention and support of world leaders like former U.S. president Jimmy Carter and the late president of South Africa, Nelson Mandela. The DMZ Forum, founded in 1997 by two Korean Americans, Dr. Seung-ho Lee and Dr. Ke-chung Kim,[1] began as an effort "to support conservation of the unique biological and cultural resources of Korea's Demilitarized Zone, transforming it from a symbol of war and separation to a place of peace among humans and between humans and nature" (DMZ Forum).

The DMZ Forum's founding in 1997 and its early conferences coincided with a thawing of relations between the Republic of Korea (ROK) and the Democratic People's Republic of Korea (DPRK). During the administrations of Kim Dae-jung (1998–2003) and Roh Moo-hyun (2003–2008), South Korea instituted a policy of engagement and cooperation with North Korea, led by Kim Jong-il from 1997 through 2011, to encourage more open diplomatic dialogue and broader economic partnerships between the two nations (Gelézeau et al 2013). Known as the "Sunshine Policy," the increased level of communication and cooperation between the Koreas facilitated the creation of joint development projects in the DPRK, among which were the Mount Keumgang Tourist Region on the eastern side of the peninsula in Kangwon Province, nine miles north of the heavily mined border, and the Kaesong Industrial Complex (KIC) located thirty miles northeast of Seoul and six miles above the DMZ (see Fig. 7.2). The industrial complex in particular aimed at strengthening ties between the two nations through mutual economic growth when it opened in 2004 (the DPRK received needed infusions of cash through wages, the South Korean companies got access to cheap labor, and the ROK benefitted from an increased tax base on goods produced there). Beyond the financial incentive, the Sunshine Policy and its affiliated projects intended to break down political barriers so that families separated by the 1950–1953 Korean War could reunite, if not permanently, then at least more easily and more frequently for brief periods. Together, supporters of the policy anticipated that it would help facilitate diplomatic conversations and negotiations, thus bringing the two Koreas closer to more open borders and, potentially, unification, thus making the DMZ obsolete.

[1] Most Korean names in this chapter follow traditional Korean practice of placing the surname first and given names after. Where they do not, name order reflects how the individuals list their names in publications or correspondence.

Although the cooperative projects garnered general support, some concerns about their broader implications arose, especially among those who had watched nature flourish within the confines of the DMZ. Conservation organizations such as the DMZ Forum advocated better relations between the two nations but feared what would happen to the flora, fauna, and ecosystems that enjoyed some protection inside the DMZ's borders should the commercial and industrial developments expand or, in the case of reunification, should the DMZ be erased from the map. They had valid reasons for such apprehension. Both the Keumgang Tourist Region and the KIC required major infrastructure projects, ranging from sewer systems to roadways, all of which could potentially compromise the area's environmental integrity. The transportation routes in particular had direct impact on the border region: getting people to the mountain, workers to the plant, and manufactured goods to market necessitated breaching the DMZ, thus fragmenting ecosystems and interrupting migration paths. On a small scale, this presented manageable disruptions to natural processes; but, if more widely implemented, it could spell disaster for some of the last remaining native and wild habitats on the peninsula.

Beyond infrastructural issues, extensive human activity in the two regions portended environmental degradation, including increased pollution and expanded exploitation of natural resources. Some numbers provide context for these fears: Over one million South Koreans (some estimates place it closer to two million) visited the resort at Mount Keumgang in the decade between its opening in 1998 and 2008, when its doors were shuttered after Park Wang-ja, a tourist from Seoul, was shot and killed by DPRK military personnel after she wandered into a restricted area (Watts 2008; Yonhap News Agency 2014). The KIC, at its height, employed some 54,000 North Korean workers and around 500 South Korean supervisors and generated U.S.$2.7 billion in inter-Korean trade in 2015 alone (these numbers vary according to source; see, for example, Choe 2016 and Kim 2016). The KIC, which produced watches, small electronics, and textiles, among other consumer goods, operated continuously for a dozen years, with the exception of a temporary work stoppage in 2013 when the DPRK pulled its workers from the factory in opposition to joint ROK-U.S. military training exercises. In early 2016, the ROK shut down the facility to protest the North's ballistic missile and nuclear weapons tests (McCurry 2016).[2]

The closures of Mount Keumgang and the KIC did not assuage environmentalists' fears, however; if anything, levels of concern rose as the chances for renewed military conflict flared. Diplomatic relations deteriorated under the Lee Myung-bak (2008–2013) and Park Geun-hye (2013–2017) administrations and became even more strained after Kim Jong-un ascended to the DPRK's Supreme Leadership upon the death of his father in December 2011. After several years of belligerent rhetoric and fast-tracking the DPRK's nuclear weapons program, Kim Jong-un

[2]As of this writing in early 2018, the Kaesong Industrial Complex remains closed. Its future is not clear.

made a surprising reversal in January 2018, opening up dialogue with the ROK and intimating that he was willing to both de-nuclearize the Korean Peninsula and negotiate a final peace. The two nations held an Inter-Korean Summit on April 27, 2018, to discuss the future of the peninsula and signed the Panmunjeom Declaration for Peace, Prosperity, and Unification of the Korean Peninsula.[3]

To a certain extent, the state of inter-Korean politics—blessed by sunshine or cursed by shadow—changes little but the rhetoric for those who wish to preserve the peninsula's natural heritage. In good times, nature in the DMZ is portrayed as fragile, susceptible to destruction from unchecked greed and exploitation; in bad times, its environmental health becomes common ground from which pan-Korean cooperation and understanding can grow. In either case, advocates assert, the DMZ must be protected.

That the DMZ exists only as a result of vicious conflict enhances its symbolic significance. Were it a remote forest or mountain valley, degraded by subsistence farmers desperate for food and fuel, or perhaps a wetland destroyed by corporate or state exploitation, the DMZ likely would have few defenders beyond committed environmentalists and local residents. But because it is the physical manifestation of the Korean people's greatest shared tragedy, and because nature there seems to ignore the ideological competition and animosity that created it, protecting the DMZ has gained significant support inside both Koreas and beyond. This is perhaps because it epitomizes the notion of what this volume calls "collateral value": unintended positive environmental outcomes of military conflict.

War may be a destructive force, arbitrary in its desolation, but it can also be indiscriminately generative, clearing the way for beneficial developments to occur. Which path a particular war will take—devastation or renovation—and what its ultimate human and environmental consequences will be, are impossible to predict. In 1953, Korea was shattered, with at least two million of its nearly 30 million people dead (military and civilian), millions more displaced and destitute, its rivers degraded and polluted, its farmlands turned into killing fields, and upwards of 65% of its forests destroyed (on population in Korea on eve of war, see Trewartha and Zelinsky 1955; on deforestation estimates, see Jeong et al. 2013). By the 1980s, the Korean people had begun to recover, with both the ROK and DPRK economies experiencing growth, but that came at a significant environmental cost. Today, climate change, industrial development, deforestation, and on-again-off-again tensions between the two Koreas continue to undermine the peninsula's environmental health.

[3] The reasons behind this stunning change in policy have yet to be revealed, but there is speculation that a collapse at the DPRK's nuclear testing facility at Punggye-ri under Mt. Mantap may have contributed (see, for example, Taylor 2018). Other factors may have included pressure from Xi Jin-ping, president of the People's Republic of China, economic promises from ROK president Moon Jae-in, and changes in U.S. foreign policy under the Donald J. Trump administration (see, for example, Choe 2018 and DeThomas 2018). For the official ROK perspective on the April 27, 2018, Inter-Korean Summit, see http://www.korea.net/Government/Current-Affairs/National-Affairs?affairId=656, especially the text of the Panmunjeom Declaration (in English) http://www.korea.net/Government/Current-Affairs/National-Affairs/view?subId=641&affairId=656&pageIndex=1&articleId=3354. Accessed May 11, 2018.

The DMZ took a different path after the conflict than the rest of Korea, however. Fenced off from direct development and settlement, it was the only area that presented some semblance of ecological health apart from the national parks that began to crop up in the late 1960s and 1970s. The DPRK currently has six national parks, all of them in mountain settings. South Korea established its first nature-based national park at Jirisan Mountain in 1967; since then, the nation has created a total of 22 parks, with all but one, Gyeongju (est. 1968), focused on natural heritage (Fig. 7.3). Seventeen of the ROK's parks promote mountain landscapes and the remaining four protect marine and coastal areas. According to the Korea National Park Service (KNPS) website, 6.7% of the nation's area is protected within the park system. These reserves, however, are under constant threat from tourism and other human-driven environmental pressures. Bukhansan National Park in the Seoul Capital Region, for example, is surrounded by Seoul's urban sprawl and, therefore, subject to industrial and urban pollution and intense use from over five million visitors each year. Nature in the DMZ, however, is protected from what the KNPS describes as "having too many visitors," a situation that has undermined ecological health at its many parks (KNPS n.d.). In the DMZ, nature enjoys a respite from constant human pressures.

7.2 Making the DMZ

The DMZ has two creators: nature and war. Both are equally pertinent to the debates over its preservation, with its cultural significance as important to arguments for its preservation as its environmental value (see Havlick, Chap. 9). Out of war's death and destruction came resiliency and rebirth; the dark past has given way to the potential of a bright future; and nature and culture unite to reveal a promising route toward hope and reconciliation. As a place apart—recognizable for its distinctiveness—the DMZ has a relatively brief history (Brady 2008). Officially designated on July 27, 1953, when the ROK and DPRK signed the cease-fire and armistice agreement that ended formal military hostilities (but not the war) on the peninsula, the DMZ not only represents Korea's tragic history, but also epitomizes an optimistic view to the future. It became a place of ecological regeneration upon which Koreans on both sides pinned their dreams of reunification and reconciliation. The DMZ evolved into a living laboratory where aspects of Korea's natural history can be studied, understood, and perhaps preserved. Although the DMZ is deeply political in its origins and function, Koreans across the peninsula and across the world see the DMZ and its ecological promise as a means by which politics can be transcended and where we can find a host of collateral values. In the case of the DMZ, those benefits are both ecological and cultural, categories that are too often depicted as mutually exclusive. There can be no separation between what has human value and what has natural value in the DMZ because, without continued strife, without human conflict, nature would have been as under threat there as it has been everywhere else on the peninsula.

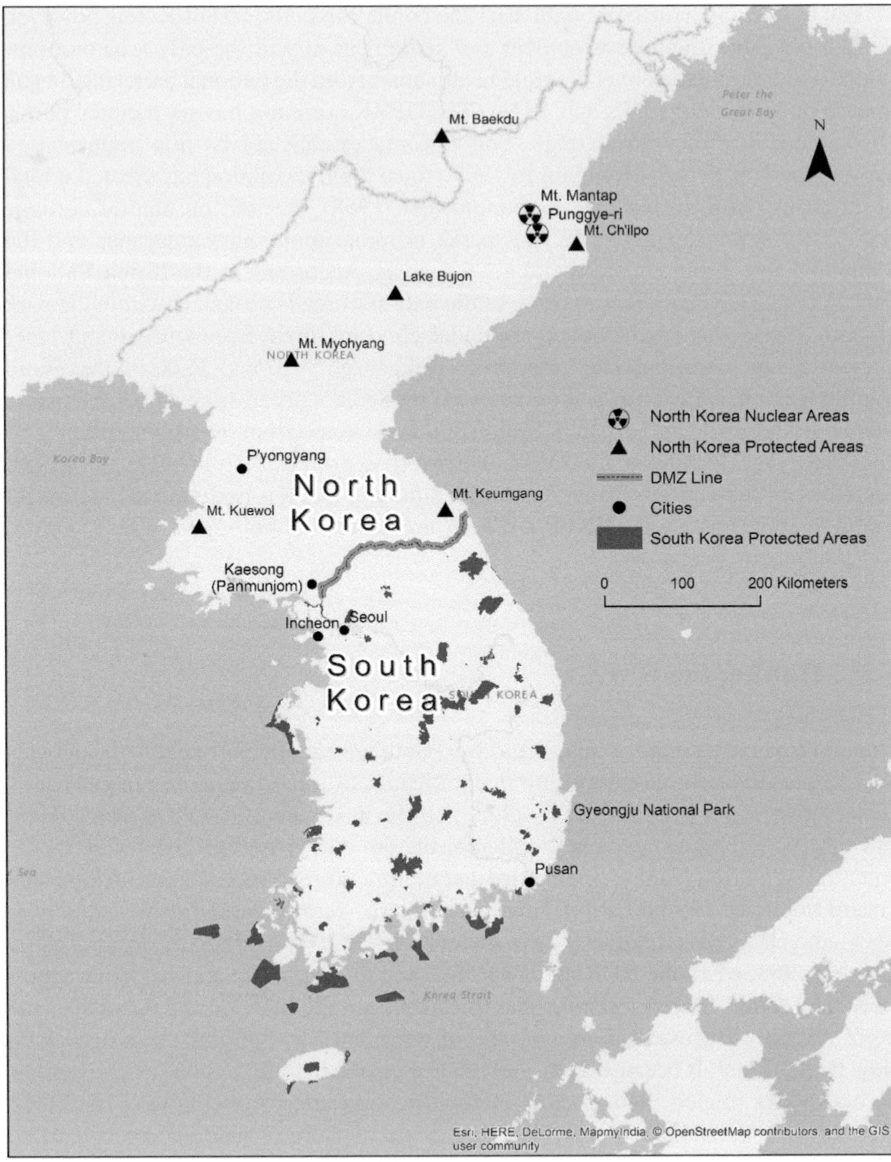

Fig. 7.3 Protected Areas of the Korean Peninsula. North Korean protected areas are indicated with triangles, because their borders are not available

Although today the DMZ is unique in its environmental makeup, such has not always been the case. Prior to becoming a politically significant region in the mid-twentieth century, the area along the 38th parallel was historically and environmentally unremarkable except for being the narrowest part of the peninsula. Indeed, it shared geological, ecological, and climatological characteristics with many other regions across Korea. It is in this similarity to other areas on the peninsula, however, where it derives its importance from a conservation perspective.

Providing an overview of Korea's physical environment helps to make this point more clearly (Fig. 7.2). Mountains cover the eastern two-thirds of the peninsula, a result of the Precambrian convergence of the Eurasian and Pacific tectonic plates. Throughout the Taebaeksan Range, which runs nearly the entire length of the peninsula north to south like a spine, gneiss and granite compose the basement rock with significant overlays of sedimentary rock deposited over millions of years as sea levels rose and fell. The remainder of the peninsula—that is, the western third—consists of generally flat plains traversed by rivers that typically flow westward to empty into the Yellow Sea, although some smaller streams, mostly in the northern half of the peninsula, debouch into the East Sea (Sea of Japan). Spanning the center of the peninsula east to west (and, incidentally, along the northern border of the DMZ) is the Imjingang Belt, a collisional suture zone of igneous rock created during the Mesozoic period in what some have called the most important event in the peninsula's geological history (Chough 2013, 16). Korea's geological and hydrological characteristics, in combination with its monsoonal climate (most of its precipitation occurs in July and August), leave its soils highly susceptible to erosion, especially where forest and other types of vegetative cover are thin (Woo 1982).

Korea is biologically diverse, with over 41,400 species recorded (and tens of thousands more as yet unrecorded), including at least 2200 native species, some unique to the peninsula and others that can be found in the wider East Asian region. Historically, Korea's fauna included approximately one hundred mammal species, such as the Asiatic black bear (*Ursus thibetanus*), the Amur leopard (*Panthera pardus orientalis)*, and the Amur or Siberian tiger (*Panthera tigris altaica*). Most of its larger mammals, especially predator species, are now rare, endangered, or extinct on the peninsula due to habitat loss or overhunting. Smaller mammals include the Amur or long-tailed goral (*Naemorhedus caudatus*), wild boars (*Sus scrofa*), water deer (*Hydropotes inermis*), and the leopard cat (*Prionailurus bengalensis*). The peninsula is also home to a variety of riverine and marine mammals such as otters, sea lions, and dolphins, and to small terrestrial creatures such as hares, voles, and bats. Birds are also prominent members of Korea's animal life, with at least 500 species ranging from herons and kestrels to plovers, owls, and—most famously—white-naped and red-crowned cranes (*Grus vipio* and *Grus japonensis*, respectively). Korea also boasts over 1200 species of fish and at least 15,600 species of insects. Its flora is equally diverse, with more than 5000 species of vascular plants alone. Korea's forests typically contain both evergreen and deciduous plants such as the Korean fir (*Abies koreana*), the Korean red pine (*Pinus densiflora*), and the palmate or Japanese maple (*Acer palmatum*) with understories that incorporate shrubs and herbs such as Korean blueberry (*Vaccinium koreanum*) and Korean ginseng (*Panax ginseng*) (ROK 2014).

Humans, of course, also claim a prominent role in Korea's geological and ecological history. Hominin remains dating to at least 46,000 years ago have been found in Ryonggok Cave, near Pyongyang, DPRK, with agricultural sites dating from between 10,000 and 8000 years ago (Bae and Guyomarc'h 2015; Totman

2004). Human changes to Korea's landscape have taken multiple forms, from terracing of hillsides for agriculture and erosion control purposes, to damming rivers for irrigation and hydroelectric use. Urbanization and industrialization since the 1960s have also taken their toll, ramping up pollution, resource extraction, and habitat destruction, and restructuring the physical landscape through rerouting streams, digging out mines in mountains, and covering vast areas of land with roads and high rises. By far the most extensive changes have been to Korea's forests, which largely coincide with its mountains. As the main source for building materials and fuel, forests long have been under significant pressure as Korea's human population grew and its social structures developed. Forests remain among the peninsula's most important natural resource and continue to face overexploitation, with serious implications for biodiversity and erosion concerns (Fedman 2015; Totman 2004; UNEP 2003).

Such a broad description of the peninsula's environment admittedly elides regional variations that affected the evolution of Korea's culture, society, and ecology over the past several millennia (see Totman 2004 for a more extensive treatment); nevertheless, it provides a useful baseline from which to begin discussion of the DMZ and why it is now the focus of conservation efforts. Importantly, the 4 km wide, 250 km-long swath represents a virtually complete cross-section of the peninsula's geology and ecology. Like the rest of Korea, the DMZ is predominantly mountainous, with its eastern reaches characterized by steep, rugged peaks of highly weathered bedrock and eroded slopes. Scores of rivers and streams wend their way through the zone, depositing silt and watering wetlands as they flow west toward the sea. The western terminus of the DMZ runs through the Han River Estuary, where the still-disputed Northern Limit Line extends into the Yellow Sea, serving as the maritime boundary between the ROK and DPRK. Its forests, marshes, grasslands, and waterways provide food and habitat to a wide variety of flora and fauna including, until 1953, humans (for an accounting of the species in the DMZ, see Kim 2013).

Complete, perhaps, but not pristine, despite assertions to the contrary. The DMZ's current ecology does not conform to ecological conditions on the peninsula prior to human habitation, nor does it reflect the state of nature just prior to the war that created this strange nature preserve. That is, it is not some kind of primordial place where we can see Korea's untouched natural history. Before the DMZ became a line on a map, before it became home to rare species of plants and animals, it supported farmers and villagers who grew crops from its soil, took fish and fresh water from its streams, and collected firewood and building materials from its forests. Its resources provided sustenance for generations of people who altered, managed, and, at times, overexploited the region's natural systems. Three years of war destroyed these long-term interactions between people and nature—and the ecologies such exchanges created—first through near total devastation of the area's physical and cultural landscapes and then by excluding all but military personnel from its newly drawn boundaries. Acknowledging as much does not undermine the value of the DMZ from an ecological standpoint, however; instead, recognizing the unique quality and history of its nature offers researchers and policy-makers crucial insights into the variety of ecological and cultural values the DMZ can provide.

7.3 Life in the DMZ

One of the major lessons the DMZ can teach us is that nature is resilient. Ecosystems and species can be fragile, even threatened or destroyed by the slightest changes in temperature, precipitation, biodiversity levels, or habitat loss, but nature—the physical and biological world we have come to define as not-human—adjusts, adapts, and evolves. What persists after such changes may or may not resemble what we find familiar, and may or may not be what we value, but it remains nature nevertheless. In the case of the Korean Demilitarized Zone, post-war nature has both real and symbolic worth to Koreans on both sides of the line, as well as to a global community that embraces environmentalist philosophies and appreciates what nature can teach us through science and other forms of knowledge and understanding.

We do not know, precisely, what lives in the DMZ, which is cordoned off by high fences topped with razor wire and guarded by highly trained military personnel. What we do have is a good idea of what *might* live there, based on anecdotal evidence, increasingly comprehensive data collected from the buffer zones on either side of the border, and a few, highly localized biological surveys done during the development phase of the Keumgang Tourist Region and the KIC. As it is in the DMZ, human activity is severely restricted in the buffer areas, with only a small number of farms and villages, and even fewer industrial sites, allowed to operate within their boundaries. Since the 1990s, interest in the DMZ for its ecological promise has resulted in increased access for researchers in the border region as both the DPRK and the ROK seek information about the state of their environmental health and solutions for environmental problems, but to date we do not have a complete accounting of the numbers, types, and diversity of species residing in the DMZ or its buffer zones.

As early as 1995, researchers began formal surveys of the biota in the DMZ corridor, documenting thousands of plant, animal, and insect species (Kim 2013). The majority of these studies have taken place in the southern civilian control zone (CCZ), a one- to fourteen-kilometer wide buffer zone along the entire southern edge of the border, where the central authorities grant research access more often than in the DMZ itself. Studies and projects take place in the analogous buffer zone to the north as well, with the International Crane Foundation's red-crowned and white-naped crane migratory reserve project in Cheorwon as the best example (ICF.org). There are also a small number of sites where ecological studies have been conducted within the DMZ, specifically near the Panmunjeom Joint Security Area, along the KIC train line, and in Cheorwon County, Kangwon Province (Kim 2013).

Among the first times scientists were allowed into the DMZ itself was during construction of the highway and railroad servicing the KIC. These infrastructure projects provided a much-anticipated opportunity to do field research and find out what actually existed between the razor-wire fences. Among those scientists was Kwi-gon Kim, professor of planning and author of *The Demilitarized Zone of Korea* (2013). According to Kim, researchers have identified around 2000 species of plants and animals, which he estimates to be approximately two-thirds of the total number that likely exist in the entire DMZ region. He notes, however, that only about 10% of the species in the DMZ have been formally documented (Kim 2013, xv–xvi). Among the species documented in the DMZ/CCZ research areas, over two dozen are legally protected by various South Korean governmental agencies and some are listed on the International Union for Conservation of Nature and Natural Resources (IUCN) Red List of Threatened Species (see Table 7.1).

The DMZ is more valuable than the sum of its parts, however. Although each species of plant and animal has inherent ecological value, combined they make significant contributions to the peninsula's overall environmental health. Thus, conservation of the DMZ's species, endangered, threatened, or otherwise, is only one goal for scientists and policy makers. Maintenance of ecosystems is equally important, perhaps even a greater priority, as individual species cannot thrive if their habitats are compromised or destroyed. Thus arises the impetus for protecting as much of the DMZ intact as possible.

7.4 DMZ to TBPP

So, what can a protected DMZ do? Who or what could it benefit? What concerns are associated with the creation of a bilateral, multilateral, or UN-supported conservation area between the two Koreas? What could happen if the Korean War officially ends and a unified nation is once again restored to the peninsula? These are among the many questions supporters (and opponents) of turning the DMZ into a nature reserve ask as they look toward the future.

Since at least the early 1990s, various entities within and outside of the Korean peninsula have clamored for the DMZ to not only more closely resemble its name—demilitarized—rather than its character—heavily armed and guarded—but to also become both a symbol of and first step toward inter-Korean reconciliation and reunification. Prominent among such calls are suggestions for turning to nature as common ground. The underlying premise is that mutual respect for the natural environment, as well as shared understanding that healthy ecosystems benefit human communities, could override the ideological impasse that has dominated the peninsula since the Korean War. This turn toward nature as a neutral arbiter between the DPRK and the ROK has resulted in a variety of proposals, many of which revolve around turning the DMZ into some kind of transboundary peace park (Kim 2007).

The notions that nature protection both requires and fosters cooperation and that nature parks located on national (or other territorial) boundaries can become peace parks predate efforts in Korea by at least a century (Chester 2006, 20). The second quarter of the twentieth century witnessed a number of boundary-area regions set aside for conservation or preservation purposes, with peace and cooperation as

Table 7.1 Examples of species in the DMZ/CCZ research areas

Species	ROK Designation and Protecting Agency[a]	IUCN Red List Status[b]	Notes
Northern Pipevine (*Aristolochia contorta*)	Rare; Korea Forest Service	Not evaluated	Important to traditional Korean medicine
Red-Crowned Crane (*Grus japonensis*)	Endangered Species I; Ministry of Environment	Endangered ver 3.1; population trend decreasing	Culturally important
White-Naped Crane (*Grus vipio*)	Endangered Species II; Ministry of Environment, National Heritage	Vulnerable ver 3.1; population trend decreasing	Culturally important
Bean Goose (*Anser fabalis*)	Endangered Species II; Ministry of Environment	Least Concern ver 3.1; population trend decreasing	
Eurasian Black or Cinereous Vulture (*Aegypius monachus*)	Endangered Species II; Ministry of Environment, National Heritage	Near Threatened ver 3.1; population trend decreasing	
Eurasian Buzzard (*Buteo buteo*)	Endangered Species II; Ministry of Environment	Least Concern ver 3.1; population trend stable	
Eurasian Otter (*Lutra lutra*)	Endangered Species I; Ministry of Environment, National Heritage	Near Threatened ver 3.1; population trend decreasing	
Leopard Cat (*Felis bengalensis euptilura*)	Endangered Species II; Ministry of Environment, National Heritage	Not evaluated	
Yellow-throated Marten (*Martes flavigula koreana*)	Endangered Species II; Ministry of Environment	Least Concern ver 3.1; population trend decreasing	Now considered to be the apex predator on the peninsula with the regional extinction of the Siberian Tiger (Panthera tigris spp. altaica)[c]
Long-tailed goral (*Naemorhedus caudatus*)	Endangered Species II; Ministry of Environment	Vulnerable ver 3.1; population trend decreasing	
Longhorn Beetle (*Callipogon relictus Semenov-Tian-Shansky*)	Endangered Species II; National Heritage	Not evaluated	

(continued)

Table 7.1 (continued)

Species	ROK Designation and Protecting Agency[a]	IUCN Red List Status[b]	Notes
Dragonfly (*Nannophya pygmaea Rambur*)	Endangered Species II; Ministry of Environment	Least Concerned ver 3.1; population trend stable	
Freshwater Gudgeon or Kogori (*Gobiobotia macrocephala*)	Endangered Species II; Ministry of Environment	Not evaluated	Found only in Korea[d]
Amur rat snake (*Elaphe schrenckii*)	Endangered Species I; Ministry of Environment, National Heritage	Not evaluated	

[a]Kim (2013, 186–189)
[b]Republic of Korea Ministry of Environment, Government Press Release, "The yellow-throated marten to be the apex predator," 14 Jan 2013. http://www.korea.net/Government/Briefing-Room/Press-Releases/view?articleId=1923. Accessed 16 March 2017
[c]*Gobiobotia macrocephala*. http://www.fishbase.se/summary/25580. Accessed March 16, 2017
[d]The IUCN Red List of Threatened Species. Version 2016–3. www.iucnredlist.org. Downloaded 16 March 2017

additional benefits. Among the first was the Waterton-Glacier International Peace Park between Alberta, Canada, and Montana, U.S.A., established in 1932. That same year, Poland and Czechoslovakia created a transboundary park in the Pieniny Mountains (Chester 2006, 21; Hoenig 2014). In the first case, distinct national parks dedicated to the conservation of nature antedated the international agreement identifying them as a joint peace park. In the latter, conservation and peace were proposed as mutually constitutive; there, natural scientists advocated for the creation of a transnational park that would preserve important landscapes while at the same time offer a peaceful solution to border disputes between the two nations after the dissolution of the Habsburg Empire (Hoenig 2014). Interest in transboundary parks—whether designated as peace parks or not—expanded greatly over the course of the twentieth century, so much so that by the second decade of the twenty-first century, over 200 such parks or conservation areas had been formed (Vasilijević et al. 2015, 5).

The most recent IUCN definition of a protected area is: "A clearly defined geographical space, recognized, dedicated and managed, through legal or other effective means, to achieve the long-term conservation of nature with associated ecosystem services and cultural values" (Vasilijević et al. 2015). The IUCN identifies a Transboundary Conservation Area (TCA) as a protected region that straddles international borders, further delineating them as Transboundary Protected Areas, Transboundary Conservation Landscapes and/or Seascapes, and Transboundary Migration Conservation Areas. Additionally, the IUCN designates Parks for Peace (formerly called peace parks) as those TCAs that are "dedicated to the promotion, celebration and/or commemoration of peace and cooperation" (Vasilijević et al. 2015, 8–13). According to the IUCN, peace parks may exist anywhere, but Parks for Peace are inherently transboundary in that they "celebrate the endurance of peace and the commemoration of peace in a region," like Waterton-Glacier International Peace Park; or, they "help to reinforce peace and cooperation" as in the Cordillera del Condor (see Chap. 8 in this book); or, they "could be used to promote peace at some point in future" (Vasilijević et al. 2015, 14). The example provided by the IUCN to illustrate this last type of Park for Peace is Korea's DMZ, which despite international support and local action, has remained a vision rather than a reality (Vasilijević et al. 2015, 14 and 42).

Large-scale military operations are among the greatest challenges to creating a Park for Peace in the DMZ. Such exercises have taken place within or near the DMZ and CCZ, on land and in the Yellow Sea, every year since the 1953 armistice. The ROK conducts its joint military training exercises with the United States every spring. Without fail the DPRK characterizes those activities as hostile to its security (despite its own annual winter training exercises) and threatens to retaliate with great shows of force (Cha 2016, 2017; Collins 2014). The North Korean threats, which have included warnings of nuclear attack, have yet to become direct action, with the possible exception of the *Cheonan* Incident of March 26, 2010, in which 46 ROK sailors died when their vessel, the *Cheonan*, sank in the Yellow Sea after being hit by what appears to have been a DPRK torpedo (the DPRK denies any involvement in the event; see Lendon 2010).

There are also more mundane obstacles to the possibility of a park in the DMZ. As Charles Chester notes in *Conservation Across Borders*, transboundary protection areas (TBPAs) raise the thorny issue of prioritizing wildlife and ecosystems over human health and subsistence (Chester 2006, 28–29). This is especially true when those living in or around a potential TBPA (or any proposed nature protection area) suffer from low standards of living, food insecurity, resource scarcity, or other deprivations. Such is the case on the Korean peninsula. While wealth and health indicators are relatively high in the ROK, they are among the lowest in the world in the DPRK. According to the CIA World Factbook (2017), South Korea and North Korea rank, respectively, 14 and 115 in comparative GDP and 45 and 211 in per capita GDP. While these data can be revealing, they do not tell the whole story. Between the mid-1990s and 2009, the DPRK suffered terrible famine, resulting in an estimated 330,000 deaths, due to a series of droughts and floods, general economic mismanagement, and a critical reduction in aid after the demise of the Soviet Union. Setting aside land for nature seems to fly in the face of humanitarian need.

Although many transboundary conservation areas have proven effective from both ecological and political standpoints (Ali 2007), transforming the DMZ from a warzone into an EcoPark poses a number of problems. One of these is, simply, where to begin? Ending the war and either establishing diplomatic relations between the two Koreas or reuniting them into one nation may seem the logical starting point; however, according to Arthur Westing, who has written extensively on transboundary parks and on the potential for one in the DMZ, setting aside the region in whole or part "would certainly be best accomplished before such time that a treaty is adopted" (Westing 2010, 20). Westing's concern is that without a prior agreement to protect the ecosystems and species of the DMZ, development—agricultural, industrial, commercial, and urban—would quickly encroach and likely eliminate the area's environmental health and value. Saleem Ali has made a similar point in his work on the Cordillera del Condor Corridor, arguing that the "'collateral values' of natural conservation that may be generated through conflict are vulnerable when peace is achieved without proper monitoring and enforcing mechanisms in place" (Chap. 8 of this book). Yet another obstacle to a peace park in the DMZ is a historical lack of commitment to nature preservation on the peninsula. Depending on the source and the variables included, protected areas comprise only 3–6% of the peninsula's entire territory (Westing 2010; UNEP 2003; IUCN 2009; World Bank 2017).[4]

The political, social, and economic divides across the DMZ may pose significant problems for conservation inside the DMZ, but they need not be insurmountable. As events in early 2018 demonstrated, anything is possible. The previous year had seen deepening tensions on the peninsula, including threats of nuclear war; in April 2018, the two Koreas had pledged to work toward a lasting peace (Choe 2017; McCurry 2017; Shorrock 2017; Choe 2018; DeThomas 2018). The thawing of inter-Korean relations opens the door for finding some common ground inside the DMZ. Within its limits lie sources for environmental and human wellbeing that are necessary for any society to thrive, regardless of ideological stance. Through its forests and rivers, the DMZ provides important ecosystem services (ES) that, if conserved, can benefit Koreans on either side of the border now and in the future. Conservation predicated on the ES model, when construed in purely economic terms, has garnered strong critique (Dempsey and Robertson 2012), but support for clean air and water and for stable hydrological and soil structures—key factors in a different interpretation of ES that directly connects human welfare to ecological health (MEA 2005, v)—may very well be the answer to debates over the future of the DMZ.

7.5 Out of War, Peace

Rife with uncertainty, uncontrollable in the extreme, war's execution and outcomes can rarely be predicted. The Korean Peninsula is no stranger to conflict. Sharing borders with China and Russia and separated at its closest point from Japan by only the 120-mile wide Korea Strait (or Tsushima Strait), Korea has been the object of expansionist empires for centuries. Competition between its larger neighboring

[4]Terrestrial protected areas are totally or partially protected areas of at least 1000 hectares that are designated by national authorities as scientific reserves with limited public access, national parks, natural monuments, nature reserves or wildlife sanctuaries, protected landscapes, and areas managed mainly for sustainable use. Marine areas, unclassified areas, littoral (intertidal) areas, and sites protected under local or provincial law are excluded. Source: United Nations Environment Programme and the World Conservation Monitoring Centre, as compiled by the World Resources Institute, based on data from national authorities, national legislation and international agreements.

powers made Korea the site of numerous conflicts, including the Imjin War (1592–1598) initiated by Japan, the Manchu (Jurchen or Qing) invasions in 1627 and 1636, and—after two hundred years of relative peace—the First Sino-Japanese War (1894–1895) and the Russo-Japanese War (1904–1905), each of which had control over Korea as an aim. These military confrontations had important consequences for Korea politically and economically, especially the later wars: China's loss in the First Sino-Japanese War paved the way for Japan to exert greater influence on the Korean peninsula and Russia's loss in the Russo-Japanese War led directly to Korea's formal colonization and annexation by Japan in the early twentieth century (Kang 2010; Lewis 2015; Totman 2004). All of these conflicts also had environmental consequences for the peninsula, resulting in widespread deforestation during the early wars due to increased logging to supply military needs and, later, under colonial rule, expanded exploitation of forest, mineral, and agricultural resources to support Japan's domestic and imperial growth (Totman 2004; Lee 2018; Fedman 2018).

The end of World War II and of Japan's colonial state in Korea promised self-determination on the peninsula. But deep divisions among the Korean populace eventually led to war. Both sides hoped to rule over a single nation, yet decades later the peninsula hosts two governments and two unique polities. For a people that had for centuries been one of the most culturally, ethnically, and linguistically homogenous nations on Earth, the middle of the twentieth century wrought enormous change. Since the July 1953 ceasefire, the Korean peninsula has been divided, with significant consequences for both human and natural systems. With the end of official combat, however, the physical manifestation of that rift has largely escaped the pressures of industrial development and economic competition.

From the standpoint of those who lived through the war that created the DMZ and for those who still have family on the other side of its borders, the region serves as a painful reminder of what was lost: communities divided by politics and war; farmland transformed into minefields; rivers, forests, and mountains made inaccessible for subsistence and recreation. But for some, it also represents hope for the future. Because its land and natural resources have not been developed, the DMZ has become a sort of storehouse for flora and fauna that have no viable habitat elsewhere on the peninsula. Those who advocate for its preservation do so with careful respect for those who wish to see a single Korea and who view the DMZ as the physical manifestation of the socio-political obstacles to that goal. Rather than dismiss those concerns out of hand, DMZ preservation activists view the cultural and ecological importance of the border as mutually constitutive. Protecting the DMZ's natural systems would honor long-held connections between the Korean people and their environment while providing a place to interpret and commemorate a tragic period in their recent history.

According to Seung-ho Lee, president of the DMZ Forum, natural and historic landscapes within the DMZ require protection: "Geological features in the DMZ such as a columnar joint located in the Hantan River and near the Imjin River and Cheorwon's lime rock cave have geologically high value for conservation. Within the DMZ, there are numerous historical and archeological treasures that have yet to be explored and preserved for posterity, including Gungye, an ancient capital city near Cheorwon. Important sites from the Choson period also exist in the DMZ, awaiting study and preservation. Many battlegrounds and other sites from the Korean War are located inside the DMZ. These, too, need to be examined and preserved for future cultural and historical study" (Lee 2010). In undertaking such protective action, Lee believes the political tensions in the region could be ameliorated, stating, "Collaborative efforts to turn Korea's Demilitarized Zone (DMZ) into a UNESCO World Heritage Site can serve as a trust-building measure" (Lee 2010). Thus, in both its material and symbolic forms, the DMZ reifies the discontinuities between north and south but, in very important ways, also serves as a green ribbon of hope, a place that commemorates the human sacrifices of military conflict and protects a variety of ecosystems and species that are found nowhere else on the peninsula.

Acknowledgements The author wrote this chapter during her fellowship at the Rachel Carson Center for Environment and Society (Munich, Germany). She would like to thank all the RCC staff and Fellows for their support. She also thanks Boise State University for supporting this research through a sabbatical award, as well as the volume editors for their keen insights, suggestions for improvement, and assistance in creating the maps.

References

Ali, S. H. (Ed.). (2007). *Peace parks: Conservation and conflict resolution*. Cambridge: MIT Press.

Bae, C., & Guyomarc'h, P. (2015). Potential contributions of Korean Pleistocene hominin fossils to palaeoanthropology: A view from Ryonggok Cave. *Asian Perspectives: The Journal of Archeology for Asia and the Pacific, 54*(1), 31–57.

Brady, L. (2008). Life in the DMZ: Turning a diplomatic failure into an environmental success. *Diplomatic History, 32*, 4.

Brady, L. (2012). Korea's green ribbon of hope: History, ecology, and activism in the DMZ. *Solutions, 3*:1. https://www.thesolutionsjournal.com/article/koreas-green-ribbon-of-hope-history-ecology-and-activism-in-the-dmz/. Accessed 1 May 2017.

Cha, V. (2016, March 7). How serious are North Korea's pre-emptive attack threats? *Center for Strategic and International Studies Korea Chair Snapshot*. https://www.csis.org/analysis/how-serious-are-north-korea%E2%80%99s-pre-emptive-attack-threats. Accessed 18 Mar 2017.

Cha, V. (2017, March 1). DPRK provocations likely during U.S.-ROK military exercises. *Center for Strategic and International Studies Korea Chair Snapshot*. https://www.csis.org/analysis/dprk-provocations-likely-during-us-rok-military-exercises. Accessed 18 Mar 2017.

Chester, C. (2006). *Conservation across borders: Biodiversity in an interdependent world*. Washington, DC: Island Press.

Choe, S.-H. (2016). South Korea to shut joint factory park, Kaesong, over nuclear test and rocket. *New York Times*. https://www.nytimes.com/2016/02/11/world/asia/north-south-korea-kaesong.html?_r=0. Accessed 17 Mar 2017.

Choe, S.-H. (2017, July 4). US confirms North Korea fired intercontinental ballistic missile. *New York Times*. https://www.nytimes.com/2017/07/04/world/asia/north-korea-missile-test-icbm.html. Accessed 27 July 2017.

Choe, S.-H. (2018). South Korea hands Kim Jong-un a path to prosperity on a USB drive. *New York Times*. https://www.nytimes.com/2018/05/10/world/asia/kim-jong-un-north-korea-south-usb-economy.html. Accessed 11 May 2018.

Chough, S. K. (2013). *Geology and sedimentology of the Korean peninsula*. London: Elsevier.

Collins, R. (2014). *A brief history of the US-ROK combined military exercises*. 38North.org. https://www.38north.org/2014/02/rcollins022714/. Accessed 11 May 2018.

Dempsey, J., & Robertson, M. M. (2012). Ecosystem services: Tensions, impurities, and points of engagement within neoliberalism. *Progress in Human Geography, 36*(6), 758–779. https://doi.org/10.1177/0309132512437076.

DeThomas, J. (2018). *Obituary: The maximum pressure policy*. 38North.org. https://www.38north.org/2018/05/jdethomas050918/. Accessed 11 May 2018.

DMZ Forum. *DMZ Forum mission and vision*. http://www.dmzforum.org/about/mission-and-vision. Accessed 30 Mar 2017.

Fedman, D. (2015). *The saw and the seed: Forestry and the politics of conservation in colonial Korea, 1895–1945*. Dissertation, Stanford University.

Fedman, D. (2018). Wartime forestry and the "low temperature lifestyle" in late colonial Korea, 1937–1945. *Journal of Asian Studies, 77*, 2.

Gelézeau, V., De Ceuster, K., & Delissen, A. (Eds.). (2013). *De-bordering Korea: Tangible and intangible legacies of the sunshine policy*. New York: Routledge.

Hoenig, B. (2014). Profoundly national yet transboundary: The Tatra National Parks. Environment & Society Portal, *Arcadia* no. 16. Rachel Carson Center for Environment and Society. http://www.environmentandsociety.org/node/6385. Accessed 17 Mar 2017.

International Crane Foundation. http://www.icf.org. Accessed 23 Mar 2017.

IUCN. (2009). *Korea's protected areas: Evaluating the effectiveness of South Korea's protected areas system*. International Union for the Conservation of Nature. https://portals.iucn.org/library/sites/library/files/documents/IUCN-2009-035.pdf. Accessed 3 Apr 2018.

Jeong, S.-J., Ho, C.-H., Choi, S.-D., Kim, J., Lee, E.-J., et al. (2013). Satellite data-based phenological evaluation of the nationwide reforestation of South Korea. *PLoS One, 8*(3), e58900. https://doi.org/10.1371/journal.pone.0058900.

Kang, D. C. (2010). *East Asia before the west: Five centuries of trade and tribute*. New York: Columbia University Press.

Kim, K. C. (1997). Preserving biodiversity in Korea's demilitarized zone. *Science, 278*, 242–243.

Kim, K. C. (2007). Preserving Korea's demilitarized corridor for conservation: A green approach to conflict resolution. In S. H. Ali (Ed.), *Peace parks* (pp. 239–260). Cambridge, MA: MIT Press.

Kim, K.-G. (2013). *The Demilitarized Zone (DMZ) of Korea: Protection, conservation and restoration of a unique ecosystem*. London/Berlin/Heidelberg: Springer.

Kim, M.-S. (2016). Trade with N. Korea falls to near zero. *Choson Ilbo*. http://english.chosun.com/site/data/html_dir/2016/05/13/2016051301098.html

Korea National Park Service (KNPS). (n.d.). *National parks of Korea*. http://english.knps.or.kr/Knp/AboutNP.aspx?MenuNum=1&Submenu=AboutNP. Accessed 27 July 2017.

Lee, S.-H. (2010). *Policy forum 10–046: A new paradigm for trust-building on the Korean peninsula – Turning Korea's DMZ into a UNESCO World Heritage Site*. http://nautilus.org/napsnet/napsnet-policy-forum/a-new-paradigm-for-trust-building-on-the-korean-peninsula-turning-koreas-dmz-into-a-unesco-world-heritage-site/. Accessed 1 May 2017.

Lee, J. (2018). Postwar pines: The military and the expansion of state forests in post-Imjin Korea, 1598–1684. *Journal of Asian Studies, 77*, 1–14.

Lendon, B. (2010). *S. Korea's final report affirms Cheonan was sunk by North Korean torpedo*. CNN.com. http://www.cnn.com/2010/WORLD/asiapcf/09/13/south.korea.cheonan.report/index.html. Accessed 11 May 2018.

Lewis, J. B. (2015). *The East Asian War, 1592–1598: International relations, violence, and memory*. London: Routledge.

McCurry, J. (2016). *Seoul shuts down joint North-South Korea Industrial Complex*. https://www.theguardian.com/world/2016/feb/10/seoul-shuts-down-joint-north-south-korea-industrial-complex-kaesong. Accessed 17 Mar 2017.

McCurry, J. (2017). Who is Moon Jae-In, South Korea's new president? *The Guardian*. https://www.theguardian.com/world/2017/may/09/moon-jae-in-the-south-korean-pragmatist-who-would-be-presidentc. Accessed 27 July 2017.

Millennium Ecosystem Assessment (MEA). (2005). *Ecosystems and human well-being: Current state and trends*. Washington, DC: World Resource Institute.

Republic of Korea (ROK). (2014). *The fifth national report to the convention on biological diversity*. https://www.cbd.int/doc/world/kr/kr-nr-05-en.pdf. Accessed 27 July 2017.

Shorrock, T. (2017). South Korean President Moon Jae-In prepares for a difficult meeting with Trump. *The Nation*. https://www.thenation.com/article/south-korean-president-moon-jae-in-prepares-for-a-difficult-meeting-with-trump/. Accessed 27 July 2017.

Taylor, A. (2018). North Korea's mountain mystery: Is Punggye-ri nuclear test site still functional? *Washington Post*. https://www.washingtonpost.com/news/worldviews/wp/2018/04/25/north-koreas-mountain-mystery-is-punggye-ri-nuclear-test-site-still-functional/?noredirect=on&utm_term=.89a2f9f39b15. Accessed 11 May 2018.

The World Factbook. (2017). Washington, DC: Central Intelligence Agency. https://www.cia.gov/library/publications/the-world-factbook/. Accessed 18 Mar 2017.

Totman, C. (2004). *Pre-industrial Korea and Japan in environmental perspective*. Leiden: Brill.

Trewartha, G. T., & Zelinsky, W. (1955). Population distribution and change in Korea 1925–1949. *Geographical Review, 45*, 1–26.

United Nations Environment Programme (UNEP). (2003). *DPR Korea: State of the environment*. Klong Luang/Pathumthani: United Nations Environment Programme.

Vasilijević, M., Zunckel, K., McKinney, M., Erg, B., Schoon, M., & Rosen Michel, T. (2015). *Transboundary conservation: A systematic and integrated approach* (Best Practice Protected Area Guidelines Series no 23). Gland: IUCN. https://portals.iucn.org/library/sites/library/files/documents/PAG-023.pdf.

Watts, J. (2008). South Korean tourist shot dead in North Korea. *The Guardian*. https://www.theguardian.com/world/2008/jul/11/korea. Accessed 15 Mar 2017

Westing, A. H. (2010). Towards environmental sustainability and reduced tensions on the Korean Peninsula. *Environment: Science and Policy for Sustainable Development, 52*, 20–23.

Woo, B.-M. (1982). Soil erosion control in South Korea. *Journal of Soil and Water Conservation, 37*(3), 149–150.

World Bank. (2017). *World Bank open data*. http://data.worldbank.org/indicator/ER.PTD.TOTL.ZS?view=chart. Accessed 19 Mar 2017.

Yonhap News Agency. (2014). Hyundai Asan faces 1 tln won loss on N. Korea tour suspension. http://english.yonhapnews.co.kr/northkorea/2014/11/18/73/0401000000AEN20141118002100320F.html. Accessed 17 Mar 2017.

Chapter 8
A Casualty of Peace? Lessons on De-militarizing Conservation in the Cordillera del Condor Corridor

Saleem H. Ali

Abstract The resolution in 1998 of the armed conflict between Peru and Ecuador through environmental peace-building negotiations, creating a transboundary conservation area, have been heralded as an exemplar of how ecological factors can foster collaboration between adversaries (Ali SH, Peace parks: conservation and conflict resolution. The MIT Press, Cambridge, MA, 2007; Simmons B, Territorial disputes and their resolution: the case of Ecuador and Peru. United States Institute of Peace, Washington, DC, 1999; Herz M, Nogueira JP, Ecuador vs. Peru: Peacemaking amid rivalry, International Peace Academy, occasional paper series. Rienner Publishers, Boulder: L, 2002). This was also a rare case of US mediation resolving a territorial conflict alongside Brazilian diplomacy. However, 20 years later the peace between the two countries has not reaped the conservation dividends expected as other extractive industry interests and drug gangs have found their way into this region. Some of the lessons from earlier peace settlements that went awry need to be considered (Hampson FO, Nurturing peace: why peace settlements succeed or fail. United States Institute of Peace Press, Washington, DC, 1996). This chapter explores the lessons of public-private partnerships for conservation in a post-conflict demilitarized hinterland. It argues that "collateral values" of natural conservation that may be generated through conflict are vulnerable when peace is achieved without proper monitoring and enforcement mechanisms in place. The importance of finding better ways of not just attaining peace but sustaining conservation after peace is addressed through interviews with key stakeholders in a detailed retrospective of this extraordinary case.

S. H. Ali (✉)
Department of Geography, University of Delaware, Newark, DE, USA
e-mail: Saleem@udel.edu

T. R. Lookingbill, P. D. Smallwood (eds.), *Collateral Values*, Landscape Series 25,
https://doi.org/10.1007/978-3-030-18991-4_8

8.1 Conflict Background

The territorial conflict between Ecuador and Peru goes back to the Spanish colonial period in the nineteenth century when Peru and Ecuador gained independence. In 1995, following several failed attempts at conflict resolution, an armed conflict broke out that lasted for about 3 weeks. A peace agreement signed in February of 1995 committed both countries to the withdrawal of forces "far" from the disputed zone. This plan was overseen by four guarantor countries: Argentina, Brazil, Chile, and the United States. In compliance with the plan, both nations organized the withdrawal of 5000 troops from the Cenepa Valley and supervised the demobilization of 140,000 troops on both sides. With this much accomplished, conservation groups became very active in trying to lobby for a peace park. It should be noted that Conservation International was actively involved in biodiversity fieldwork even before the resolution of the conflict; it had worked closely with the military when fieldwork on documenting the biodiversity of the region was conducted in 1993. Therefore they were gradually able to influence more "hawkish" army officers about the collective importance of conservation and its instrumental use for conflict resolution (Simmons 1999; Mittermeier et al. 2005).

In November 1997, the two nations agreed in the Declaration of Brasilia to address four areas: (1) a commerce and navigation treaty, (2) a border integration agreement that would stimulate much needed development in both countries, (3) a mutual security agreement designed to prevent future conflicts, and (4) a completion of demarcation of land borders. By February 1998, they were able to agree on the first three, but that left the most important one, the demarcation of land borders. Tensions arose again in August 1998 as 300 Ecuadorian soldiers spread out along an 11-km line, 3 km inside Peru and 20 km from the demilitarized zone.

To prevent further escalation, and with pressure from conservation groups, the presidents of Ecuador and Peru both met with President Clinton on October 9, 1998, and asked that the guarantor nations make a proposal to mark the border for them. With US satellite mapping they were able to arrive at an agreeable border demarcation. The terms of the peace agreement had some innovative features. The disputed stretch of border would be demarcated according to the Rio Protocol's line of division,[1] going back to the 1948 map, which was a major concession from the Ecuadorians. In return, Ecuador was given a square kilometer of private—but not sovereign—property across the Peruvian side of the border, extending to Tiwintza. Both countries would establish ecological parks on either side of the border, where unimpeded transit would be guaranteed and no military forces would be allowed.

[1] The "Rio Protocol" refers *The Protocol of Peace, Friendship, and Boundaries between Peru and Ecuador*, an international agreement signed in Rio de Janeiro, Brazil, on January 29, 1942, by the foreign ministers of Peru and Ecuador, with the participation of the United States, Brazil, Chile and Argentina as guarantors. The Protocol was intended resolve the long-running territorial dispute between the two countries, and brought about the official end of the Ecuadorian-Peruvian War of 1941–1942.

Ecuador was also granted non-sovereign navigation access to the Amazon and its tributaries in Peru and also allowed to establish two trading centers along the river.

Initially both countries declared national parks on their respective sides of the border (Fig. 8.1). However, in 2000, Conservation International and The International Tropical Timer Organization or ITTO (Duncan 2003) partnered with local conservation groups in Ecuador and Peru and with the indigenous communities (particularly the Schuar of Ecuador) to establish a bioregional management regime. The ITTO is an unusual intergovernmental organization which was set up in 1986 to promote sustainable forest management. The organization has its origins in the International Tropical Timber Trade Agreement of the United Nations Conference on Trade and Development. Currently the membership of ITTO comprises 35 timber-producing countries and 38 timber-consuming countries. The organization is headquartered in Yokohama, Japan and has been one of the leading supporters of transboundary conservation projects in forested regions.[2]

While the overall armed conflict has stopped, the implementation of various features of the peace agreement remain unfulfilled. This is largely been the case because the management of the peace park was highly dependent on external donor funds and the role of military after peace was not clearly articulated in the peace plan. Ideally, if both countries had made a commitment to reconfigure the role of the military to act as forest rangers or "green helmets," thereby maintaining environmental security, the long-term goals of the park could have been better achieved. Peace can create a power vacuum if the post-conflict process is not appropriately planned and nefarious players such as poachers, illegal miners or drug traffickers may take advantage of such an authority deficit.

The structural aspects of the peace treaty have also prevented the formation of a functional "peace park" where access to both sides of the border would be guaranteed. Instead of creating a shared zone, the peace treaty demarcated borders and established conservation areas as buffer zones. The operations of the Cordillera del Condor Transboundary Protected area (TBPA) could be vastly improved if these commitments were met. In addition, there is scant evidence of implementation of the Regulation of the Ecuadorian Peruvian Border Committees; The Programs of the Bi National Plan for border areas and development; The Comprehensive Peruvian Ecuadorian Agreement on Border Integration, Development and Neighborhood that includes as annexes the Rules of the Peruvian- Ecuadorian Neighborhood Commission.[3]

---

[2] Refer to the ITTO web site for further details www.itto.int

[3] USAID has funded numerous efforts for collaborative development along the Peru-Ecuador border. See for example, *Assessment of the USAID Peru– Ecuador Border Region Development Program.*

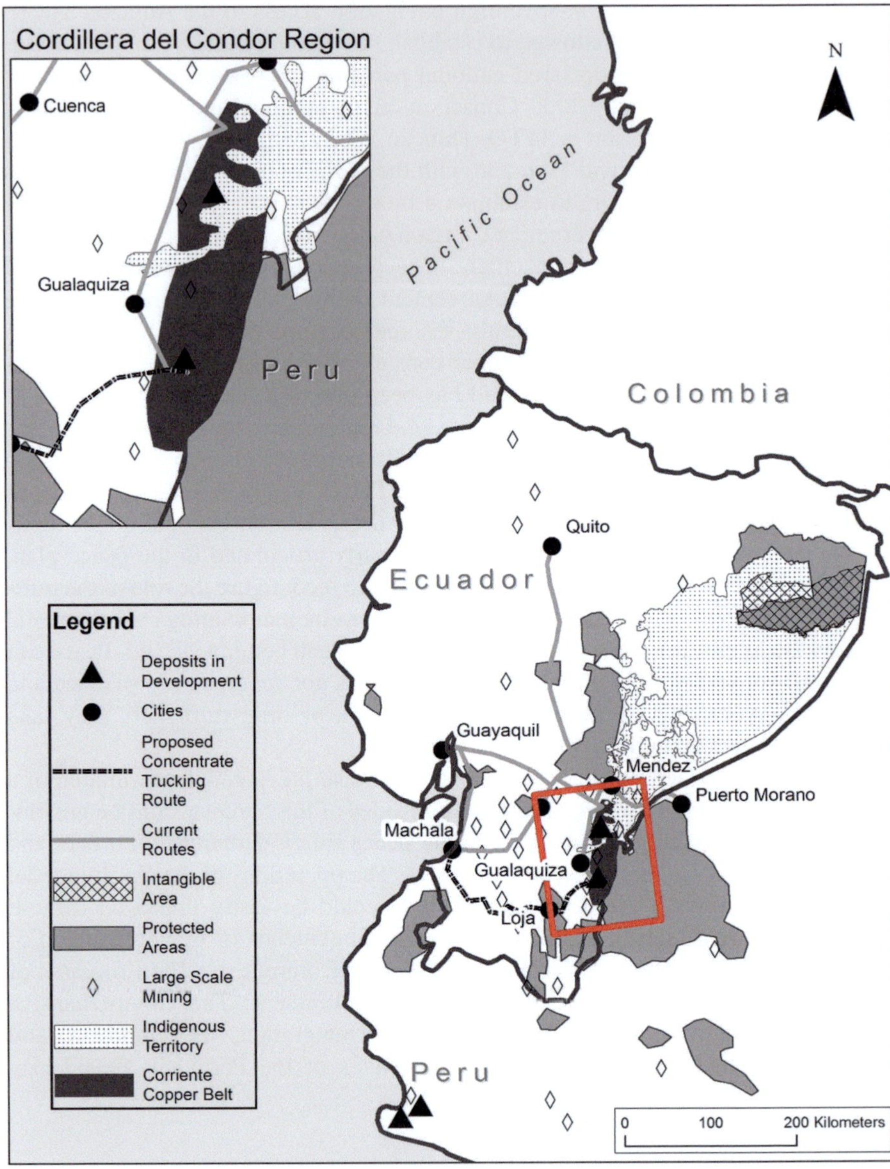

Fig. 8.1 Protected areas in the Cordillera Del Condor Region established as part of the resolution of the border dispute between Peru and Ecuador, and mining and other economic activities. (The "Intangible Area" in Ecuador is a zone where all extractive activity is banned)

8.2 The Key Stakeholders

The Condor region is sparsely inhabited but widely contested between the various indigenous communities that resident in its environs and have some collective attachment to the land. The ethnic denomination of the inhabitants of this region is Jivaroan – comprising a related group of languages and traditional beliefs. The most dominant indigenous population in this region are the Shuar tribe, who have a broad demographic range across Ecuador and Peru. They pride themselves as having remained independent of both Inca and Spanish colonial advances over the past millennium or so. Hence much of the Shuar narrative in the Condor region is around assertions of their sovereignty which was more formally operationalized in the formation of the Shuar Federation in 1964 in response to agrarian reform efforts by the Ecuadorian government (Perkins and Chumpi 2001), The Shuar have also had a long history of involvement in various ethnobotanical outreach from pharmaceutical researchers. There has been a recurring interest from American botanical gardens in working with the Shuar in this regard (Bennett et al. 2002). Thus within the community there is a tension between those who want to focus on more nature-based livelihoods through enterprises emanating from ethnobotany, cattle ranching and some tourism and those who wish to engage in extractive industries. However, the most prominent Shuar organization The Inter-provincial Federation of Shuar Centers (FICSH – Spanish Acronym that is commonly used for this organization), has been resolutely against mineral development in the region. The lowland tribes of this region are referred to as the Achuar but share many of the organizational affiliations with the upland Shuar and were among the last of the tribal groups to establish continued contact with Western populations in the 1970s.

On the Peruvian side of the border the dominant tribal group which shares ethnic similarities with the Shuar are the Awajun or Aguaruna. These groups also resisted Inca and Spanish conquest and have resisted mining ventures as well. In 2008, they attached a mining camp of exploration company Dorato resources. In addition, the Peruvian side hosts a very well-organized Jovaroan tribal group called the Wampis. In 2016, the Wampis were the first indigenous group to establish a formal internal government comprising a confederation of over 100 clans with a population of around 10,000 individuals. There is close cooperation on advocacy and activism between the Wampis and the Shuar in specific resource-related projects and concerns.

In addition to the tribal groups, there have been numerous international environmental organizations that have historically been involved in this region. Conservation International was originally the most prominent but has become less active in recent years and Amazon Watch has come more active in this region and often partner with local activist organizations and provide them logistical support. Furthermore, there is keen interest among environmental journalists to cover developments in this area with groups such as Mongabay and The Guardian sending correspondents frequently in the region to cover developments. Most recently, the arrest and subsequent protests of the FISCH leader Agustin Wachapa in early 2017 were widely covered by Mongabay. These groups are thus also important stakeholders in terms of the environmental narratives that are shaping geopolitics in the border region (Aguilar and Castagnino 2017).

The international funders who support particular indigenous organizations in the region are also significant stakeholders. For example, the Shuar community in Kumpiamis has received ongoing funding from the Global Green Grants Fund in Boulder, Colorado (Wald 2016). International government development donors have been more shy to contribute funds in this arena given the strong political ire this might draw from the Ecuadorian and Peruvian governments. However, there have been funds allocated to deal with illegal gold mining and resultant pollution in parts of Peru that are part of the Canadian government's program on extractive industries and development. Most of this work is occurring in other parts of the country but there is the spectre of artisanal miners making further inroads in the Condor region that could also lead to investment from the Canadian program in this regard.[4]

Ecuador and Peru both have government organizations which manage relationship between indigenous populations and the central government. These government agencies are also key stakeholders in the processes of negotiations and how the border areas conservation plans are governed. In addition, there are ministries of environment as well as ministries for natural resource extraction, which can often be competing for government resources and prioritization – a common occurrence in such contexts. These government agencies have different "clients" as well – the resource development arms of the government consider the private sector investors as their primary clients while the environment ministries consider the citizens at large as their clients. This dynamic is similar to any other challenge of conflicting government mandates worldwide. However, in the case of Ecuador and Peru the remoteness of the Condor region and its history of past conflict often leads to presidential level intervention on key matters. For example, the indigenous communities and the government on mining development has led to direct presidential orders for particular security actions to be undertake regarding eviction of particular communities in Ecuador as "paramilitary and criminal" (Watts 2017).

In terms of the extractive industries stakeholders, there are numerous exploration companies with interests in this region's rich mineral and timber resources but also the Chinese state-owned extractive sector has become increasingly prominent. This the Condor region which had largely been unaffected by external resource extraction interests due to its own internal conflicts has become a rallying point for a panoply of stakeholders.

8.3 From Macro-conflicts to Micro-conflicts

Although, the broader international conflict between Ecuador and Peru has abated as a result of the peace agreement, the space created by the absence of militaries has also allowed for micro-conflicts to arise internally on both sides of the border. Since

---

[4] The Canadian program is managed by the Artisanal Gold Council: http://www.artisanalgold.org/our-projects/peru/

2010, considerable conflicts between large and illegal medium scale mining industry are also happening, especially in Ecuador, including the presence of armed groups. Conflicts of small and informal miners from Ecuador can also pass to the zone of Peruvian concessions. Conflicts in the El Quimi reserve of Ecuador are also gaining strength because the Shuar want to contest the activities of the Ecuadorian Ministry of the Environment. Having clear mineral extraction zones as well as exploration guidance for artisanal miners could reduce the random impacts of mining moving forward.[5]

The Condor case highlights the importance of sustained engagement in peacebuilding following a treaty or conflict resolution process to ensure natural resource conflicts do not take other forms, particularly in remote inaccessible areas, far from governance oversight. However, the lack of consultation with the indigenous communities during the negotiations leading to the peace agreement are still manifest in resentment towards conservation efforts. In the words of one community leader from the Shuar:

> This was territory occupied by the Shuar Wampís and they were not consulted about the agreement. Therefore, it is illegal what has been done, declaring it property of the State of Peru, even if a reserve was built. And we are not against the conservation of the Natural Resources, but the name Park of the Peace is not considered appropriate. The conflicts that take place are by-products of the presence of transnational mining, oil, *cuíferas* and timber companies, that are yet to come. For who are we going to extract? To build more vehicles, computers, so they can sell it back to us at a more expensive price. Why the concentration of wealth happens? Why not allow the communities to participate? As for the timbering, mining, and oil extraction, they have to obey the policy of the native communities. In the process of decisions, they should participate with a policy describing how they want to do the extraction. About the compensation: if there is a proper participation of the native communities, then, compensation is not needed. However, if there is no participation, then the use of wood or other outside product shall not be accepted. The State is interested in generating energy through rivers inside the Shuar territories. The solution for this is that the international agencies will have to meet with the native communities and come up with a common decision.[6]

Despite the peace agreement, there is considerable difficulty in movement and border crossings. The ITTO research team that operated in the region till 2013 noted that the procedures for travel remain complicated. The community organizations have many ideas to improve cooperation through development efforts if some of these issues of access can be addressed. Another indigenous representative stated:

> Since the peace agreement was signed there should be freedom to come and go, but there has been a strict control and they want to suspend such freedom. La Federación de Comunidades Huambisa del Río Santiago (FECOHRSA) dreams about the creation of an environmental services fund, but there is no capacity for economic and technical exploration. It should be added that the projects proposed should be of Wampís authority and prepared by the communities themselves.[7]

[5] Respondent: Santiago Kingman, Coordinador del Proyecto ITTO PD 238/03 Respondiendo al cuestionario desde el punto de vista de las comunidades Macas, Ecuador. August 3, 2010.

[6] Respondent Ángel Nantip: Coordinador de Gestión Externa Comunicador del Pueblo Resident of Pueblo Shuar Arutan Fecha: 15 de julio de 2010.

[7] Kefren Graña, Presidente de FECOHRSA, Ruyer Chimpokat, Vice Presidente de FECOHRSA; Eliseo Chimshami, Tesorero de FECOHRSA, Comunidad de Kukuasa, Distrito del RSA FECOHRSA – Federación de las Comunidades Wampisas del Río Santiago.

Table 8.1 Implementation challenges of the Ecuador-Peru Peace Agreement

Commitment of Peace Agreement	Status as of September 2010 and in 2016
The agreement on the freedom of coming and going, vehicles, See Going and Fluvial Vessels and Aircrafts;	The freedom of coming and going is not guaranteed through the simple use of an identity card. People must spend several hours in the military posts of the two countries and leave their documents. There is no migration, except by the armed forces in the basin of Santiago
The Convention on Traffic in Persons, Vehicles, Maritime and Fluvial and Aircraft	The movement of people is not free, with the simple use of card. People can spend several hours by the military posts of the two countries and leave their documents. There is no migration but only armed forces in the basin of Rio Santiago
The Organizational Structure of the Bi National Plan for border areas and development;	The Bi-National Plan was created, but it didn't manage projects, nor did it foment them with local organizations of the Cordillera del Cóndor
Agreement of Acceleration and Deepening of Free Trade between Ecuador and Peru.	The free trade in the region has few prospects. The only thing that could develop is the sale of timber and non-timber forest resources
The Memorandum for Understanding on Electrical Interconnection	No observable implementation
The Ground Agreement for commissioning an improvement study for the Bi National Project for the transport of Hydrocarbons;	An increase of the oil activity in the region is expected. For now there is only invitation for bids and in the south of Ecuador does not yet exist hydrocarbon activity.
Memorandum of Understanding for Strengthening the Mutual Cooperation in Tourism	No observable implementation
Memorandum of Understanding in Educative Cooperation	No observable implementation

Table 8.1 provides a list of outcomes that were stipulated in the peace agreement and the paucity of progress in that vein, compiled by Santiago Kingman for ITTO in 2010. Subsequently, an effort was made by the author to get further update on the situation from field staff and there was no clear response. The table was sent to both Peruvia and Ecuadorian conservation organizations and ministry officials who did not deny that the situation has changed. However, there is no definitive information on the current situation given lack of access of data and an overall governance deficit in the area.

Ten years of support to Fundacion-Natura-Ecuador and Conservation International (CI) under grant support from the Moore Foundation, aimed at the Shuar is an example of positive engagement with the population. A staff member for CI-Peru, stated, however, that "the resources and situation is different and it is not expected that the experience will be replicated, but that it will at least serve as a lesson learned for the Wampis and Awajún and for the institutions that wish to develop projects in their territories."[8] Furthermore, "understanding that the native peoples have ties that surpass the borders and that maintain the same plans and traditions apply for the natural resources implies a modernization, a secular process involving social and political terms and the construction of a unity where there are only families."[9] While the conservation dimensions of the peace agreement are often downplayed in many policy analyses,[10] there is a general feeling among practitioners that ITTO's involvement in this effort has greatly helped to continue the momentum towards transboundary conservation.[11] Nevertheless, some of the critics of the mainstream transboundary protected areas movement such as Fall (2005), have noted that often the allure of peace can lead to a co-optation of grassroots dissent which needs to be more carefully considered by environmental organizations.

Fundacio Natura officials in Ecuador believe that "there is no doubt that biodiversity conservation has helped to create a climate of greater confidence between states and peoples, and to resolve some conflicts among peoples. However, other disputes regarding the extraction of non-renewable resources, such as mining, are still alive and resources are still insufficient to resolve mining conflicts.[12] As a pioneering effort to directly link conservation to conflict resolution in a violent border dispute, the Condor Case has become an important example for environmental peace-building (Conca and Dabelko 2002). However, the full potential for this agreement and the ITTO efforts at transboundary cooperation have yet to be realized. Addressing the concerns of indigenous communities, improving access across borders and regulating extractive industries will be key factors to ensure the efficacy of this project in reaching its goals of environmental peace-building (Busch 2008; Chester 2006).

8.4 The Mining Enclaves

On the Ecuadorian side of the border the indigenous Shuar community got deeply involved with the environmental and social assessment concerns pertaining to a large copper mining project which received approval from the Ecuadorian government in 2012. It may be argued that the development of this project was possible as a result of the peace that occurred as it was an explicitly stated goal of the peace agreement to allow for economic development in this region. Although the mine is not within the conservation zone itself, the Ecuadorian government and indeed the Chinese mining investment has been possible because there has a cessation of hostilities in the region. Since 2012, some prominent elements the Shuar indigenous

[8] Respondent: Bráulio Andrade Conservación Internacional Perú.

[9] Santiago Kingman, Coordinador del Proyecto ITTO PD 238/03 Macas, Ecuador, August 3, 2010.

[10] For example, a study of the peace agreement conducted by the U.S. Institute of peace bare mentions the central role of establishing the national parks as means of conflict resolution. Instead the focus is on the process of mediation. See Simmons 1999.

[11] Respondent: Luis Espinel, Director Ejecutivo, Conservación Internacional Perú.

[12] Respondent: Ruth Elena Ruiz, Fundacio Natura.

community has been actively involved in activism against the mine for environmental reasons but also due to forced evictions and resettlement. Ironically, there is also a history of indigenous opposition to environmental organizations for similar livelihood displacement that has occurred at times in Latin America due to establishment of protected areas (Chapin 2004; Dowie 2011; DeLeon 2011).

The establishment of the peace park and the involvement of prominent conservation organizations such as Conservation International was perceived originally as a buffer against high risk development ventures. Ecuador also has one of the most environmentally progressive constitutions where ecological rights are enshrined in congruence with human rights. However, the pressure to bring in foreign direct investment in order to augment national export earnings has been mounting on the Ecuadorian government. This has led to rapid approvals of development projects such as the Mirador mine. The portion of the Shuar community that is opposed to the mine, have accused the company of intimidation and have also called for the investigation of the suspicious drowning death of a prominent activist. Yet, the debut mine's production in 2017 are on target with little further recourse for community members who continue to periodically communicate their discontent through international activist groups. However, interviews conducted with local activist groups have revealed that the branch of the Shuar who reside within the protected area have limited contact with the group opposing the mine.[13] Thus the Mirador mine may be considered an incidental beneficiary of the peace in the region but the conservation corridor itself is somewhat tenuously connected to the project itself.

In the northern section of the corridor, there is another mining project which is also being developed by Lundin Gold Corporation, a Canadian company which has approached Conservation International for a biodiversity management plan as well as a voluntary compensation development plan. Such gestures by companies are aimed at assuaging risk of conflict that investors are becoming increasingly sensitive to regardless of regulatory requirements. Companies are also being asked to meet such requirements for listing in particular mutual funds or for certification efforts by stick exchanges. Unlike Colombia and other South American countries Ecuador does not have a compulsory compensation law for communities that are displaced by mining. This project known as "Fruta del Norte is located only 80 km form the city of Loja, the fourth largest city in the country. Again the existence of this project is likely possible due to peace in the region and responsibly executed mining could indeed bring prosperity to the region. However, the minimal presence of government in the region following the de-escalation of conflict makes the spectre of development highly dependent on the individual behaviour of corporate actors. Furthermore, the prevalence of illegal mining at small to medium scale remains a potent concern, particularly on the Peruvian side of the border. The Canadian government, through the Artisanal Gold Council, as well as the Canadian International Resources Development Institute, has been developing capacity to

---

[13] Two of the key groups that have been in contact with the Shuar are Accion Ecologica and The Indigenous Peoples and Community Conserved Territories and Areas Consortium www.iccaconsortium.org

address the formalization challenge of such mines since 2015. However, they have not engaged in any activities within the Condor corridor. Extractive industries development has thus become a very divisive issue in the erstwhile peace park and needs much greater attention by international development donors as well as the Peruvian and Ecuadorian governments alike.

8.5 Conclusion

The Cordillera del Condor was heralded as a true instrumental peace park when the establishment of the conservation zone helped to end a conflict between Ecuador and Peru in 1998. It was also a sterling example of a public-private partnership between Conservation International, the governments of the two countries and the International Tropical Timber Organization. Yet, the diminution of funding to sustain conservation efforts, as well as a lack of governance planning in the post-conflict period has led to a situation wherein environmental conflicts with indigenous communities have taken root. Lack of environmental enforcement remains a concern for both legal and illegal industrial activities in the area.

The military conflict of yesteryears had limited access to the region and there was a level of default conservation as a result. However, the military presence itself would have undoubtedly had an impact on the forest biomes as well. Perhaps the peace dividends of the Condor corridor rapproachment deserve to be considered at a broader level than just the park zone itself. The cessation of hostilities between Ecuador and Peru in this area has led to a concerted and comprehensive "Binational Plan," for development and security. This bi-national planning process has also included the establishment of a joint UNESCO biosphere reserves. The focus currently of this program has been on the Bosque Seco dryland forest biome which is much closer to population centers and could also be a greater tourism draw. Since the peace agreement, the Condor corridor itself has lost priority for the government due to its remoteness, but the process of cooperation set forth since the peace agreements has no doubt reaped many positive dividends in other parts of the two countries.

Thus we need to view demilitarization and the advent of peace in a more regional perspective rather than only focusing on the border zone of conflict itself. Military presence itself can have a huge ecological impact as well and it would be naïve to consider the mere presence of military forces as an antidote to destructive development. Nevertheless, what we need is more effective planning for demilitarization and post-conflict development. This can perhaps be achieved in transboundary areas with greater monitoring and enforcement of international agreements as well as using universal metrics of natural capital such as ecosystem services (National Research Council 1996; Lopez-Hoffman et al. 2010). At the end of the day, peace dividends of cases such as the Cordillera del Condor are a direct function of ensuring that there is no governance deficit following the end of hostilities.

Acknowledgements The author wishes to thank Taylor Holden and the Spatial Analysis Lab at the University of Richmond for preparing the map in Fig. 8.1.

References

Aguilar, D., & Castagnino, R. (2017, March 17). Shuar people ask for president's freedom amid mining dispute. *Mongabay*. https://news.mongabay.com/2017/03/shuar-people-ask-for-their-presidents-freedom-amid-mining-dispute/. Accessed 31 Mar 2017.

Ali, S. H. (Ed.). (2007). *Peace parks: Conservation and conflict resolution*. Cambridge, MA: The MIT Press.

Bennett, B. C., Baker, M. A., & Andrade, P. G. (2002). *Ethnobotany of the Shuar of eastern Ecuador*. Bronx: New York Botanical Garden.

Busch, J. (2008). Gains from configuration: The transboundary protected area as a conservation tool. *Ecological Economics, 67*, 394–404.

Chapin, M. (2004, November/December). A challenge to conservationists. *WorldWatch*. http://watha.org/in-depth/EP176A.pdf

Chester, C. C. (2006). *Conservation across borders: Biodiversity in an interdependent world*. Washington, DC: Island Press.

Conca, K., & Dabelko, G. D. (Eds.). (2002). *Environmental peacemaking*. Baltimore: The Johns Hopkins University Press.

DeLeon, C. P. (2011). *Citizenship rights and articulations of indigenous resistance in the Ecuadorian extractive frontier*. Masters thesis, York University, Toronto, Canada.

Dowie, M. (2011). *Conservation refugees: The hundred-year conflict between global conservation and native peoples*. Cambridge, MA/London: The MIT Press.

Duncan, P. (2003). *Changing landscapes: The development of the International Timber Organization and its influence on tropical forest management*. London: Earthscan Publications.

Fall, J. J. (2005). *Drawing the line: Nature, hybridity and politics in transboundary spaces*. Burlington: Ashgate Publishing.

Hampson, F. O. (1996). *Nurturing peace: Why peace settlements succeed or fail*. Washington, DC: United States Institute of Peace Press.

Herz, M., & Nogueira, J. P. (2002). *Ecuador vs. Peru: Peacemaking amid rivalry* (International Peace Academy, Occasional Paper Series). Boulder: L. Rienner Publishers.

Lopez-Hoffman, L., Varady, R. G., Flessa, K. W., & Balvanera, P. (2010). Ecosystem services across borders: A framework for transboundary conservation policy. *Frontiers in Ecology and Environment, 8*, 84–91.

Mittermeier, R. A., Kormos, C. F., Mittermeier, C. G., Gil, P. R., Sandwith, T., & Besancon, C. (2005). *Transboundary conservation: A new vision for protected areas*. Washington, DC: CEMEX/Conservation International.

National Research Council. (1996). *Biodiversity conservation in transboundary protected areas*. Washington, DC: National Academies Press.

Perkins, J., & Chumpi, S. M. S. I. (2001). *Spirit of the Shuar: Wisdom from the last unconquered people of the Amazon*. Rochester: Destiny Books.

Simmons, B. (1999). *Territorial disputes and their resolution: The case of Ecuador and Peru*. Washington, DC: United States Institute of Peace.

Wald, B. (2016). *Inside the struggle for Ecuador's Cordillera del Condor*. https://www.greengrants.org/2016/03/10/inside-the-struggle-for-ecuadors-cordillera-del-condor/. Accessed 31 Mar 2017.

Watts, J. (2017, March 19). Amazon land battle pits indigenous villagers against might of Ecuador state. *The Guardian*. https://www.theguardian.com/world/2017/mar/19/ecuador-indigenous-shuar-el-tink-mining-land-dispute. Accessed 31 Mar 2017.

Chapter 9
Legitimizing Militarization or Legitimate Conservation? Collateral Value and Landscapes of the Iron Curtain Borderlands

David G. Havlick

Abstract For most of the twentieth century, conservation efforts around the world were largely modeled after the pattern established in North America of protecting resource areas, such as forests and rangelands, or protecting wildlands that privileged rugged aesthetics and recreational opportunities. In recent decades, new forms of conservation have come into clearer focus, including the transition of militarized landscapes into new land uses dedicated to conservation. This chapter examines how changes along the Iron Curtain borderlands illustrate this type of conversion, as the region increasingly receives acclaim as the Green Belt of Europe. Examples here in central Europe, and others in North America and East Asia, challenge traditional notions of conservation in a variety of ways, but also contribute to new conservation strategies that may help reconnect people to places, even places long known for their contamination or danger. The mix of social and natural qualities at these militarized landscapes generates a diverse set of conservation practices that depend upon renegotiating ideas of public safety, beauty, restoration, and preservation. The recasting of such landscapes can be understood variably as a form of legitimating militarization or as a legitimate approach to conserving biodiversity. In either case, coming to terms with the particular contexts of politics, ecology, and history in these places proves essential if we are to adequately understand the collateral – and also conflicting – values generated by the relationship between conservation and militarization.

D. G. Havlick (✉)
Geography and Environmental Studies, University of Colorado-Colorado Springs, Colorado Springs, CO, USA
e-mail: dhavlick@uccs.edu

T. R. Lookingbill, P. D. Smallwood (eds.), *Collateral Values*, Landscape Series 25,
https://doi.org/10.1007/978-3-030-18991-4_9

9.1 Introduction

In protected areas across the United States, the year 2016 was marked by a variety of celebrations and commemorations for the centennial anniversary of the U.S. National Park Service. Beginning with the establishment of Yellowstone National Park in 1872, the U.S. has committed more than 84 million acres to national parks (U.S. National Park Service n.d.-a), and the system has been widely exported as an important model for conservation efforts worldwide. Signaling a commitment to the preservation-oriented approach to conserving lands championed by John Muir and other early national park advocates, the 1916 enabling legislation for the National Park Service emphasizes protection of "the scenery and the natural and historic objects and the wild life" of areas generally deemed to be "unimpaired" aesthetically and ecologically (U.S. National Park Service n.d.-b; see also Madron and Tilton, Chap. 2 of this book). Put more simply, for more than one hundred years the U.S. national parks have been designed to protect landscapes that fit a broadly-held notion of beautiful, natural landscapes.[1]

Since the late nineteenth century, conservationists and elected officials in the U.S. have promoted other designations that, over time, established tens of millions of hectares of protected lands in the form of National Forests, National Wildlife Refuges, Wilderness Areas, and other resource conservation areas. Each of these come with their own particular emphases, regulations, and unique histories, but most share a common focus of protecting valuable ecological, cultural, resource, or recreational areas from wanton exploitation. Added together, many of these places now constitute a cherished conservation legacy in the U.S. that dozens of other countries around the world have emulated in some form.

More recently, a rather different trend in land conservation has emerged in the U.S. and elsewhere, this time predicated not so much on the protection of pristine, "natural" areas, but focused instead on transforming heavily impacted lands to new purposes of conservation. Centered more on principles of ecological restoration than those of nature preservation, this alternative approach to conservation is becoming particularly common in the many militarized landscapes across the world that have emerged – particularly since the late 1980s – with new identities and new land use designations. This is perhaps nowhere more striking than along the Iron Curtain borderlands of central Europe, where recent decades have witnessed the transformation of the former death strip of this extensively militarized zone to a new reputation as the Green Belt of Europe (e.g. Pieck 2018; Coates 2014; Havlick 2014). Here, and in a number of other cases around the world, traditional notions of conservation are being challenged in important ways. These sites of military-to-wildlife repurposing may contribute to new conservation strategies that can help reconnect people

[1] This approach has also been subject to critique – for instantiating an improper separation of nature and society; for holding to an outdated notion of nature itself; and for privileging the protection of ostensibly pristine, "unimpaired" lands at the expense of indigenous or other populations that have made a home in these lands – but the national park ideal remains widely embraced as a means of protecting spectacular tracts of land. See, for example, Cronon 1996; Spence 1999; Jacoby 2001.

to places, even places long known for their contamination or danger. The mix of social and natural qualities at these militarized landscapes generates a diverse set of conservation practices that depend upon renegotiating ideas of public safety, beauty, restoration, and preservation (see Chap. 5 by Machado and Hupy, this book). The recasting of such landscapes can be understood variably as a legitimate approach to conserving biodiversity or as a form of legitimating militarization. In either case, coming to terms with the particular contexts of politics, ecology, and history in these places proves essential if we are to adequately understand the collateral – and also conflicting – values generated by the relationship between conservation and militarization.

9.2 From Death Strip to Green Belt

In 1946, when Winston Churchill first described "an iron curtain" descending across Europe, there was not yet a physical barrier extending across the region's borderlands. By the early 1960s, the political and ideological divide that Churchill described had hardened into a network of fencing, concrete walls, tank traps, concertina, minefields, patrol dogs, guard towers, and high voltage wires that effectively barricaded east from west and cost thousands of lives during the remainder of the Cold War. Throughout this period, from the Barents Sea in the north to the Black Sea in the south, the Iron Curtain served as the iconic feature of a divided central Europe (Figs. 9.1 and 9.2). Although the inner German border dividing East from West was the most commonly visited stretch of the Iron Curtain (see Eckert 2011), and likely remains today the area of a divided Europe that most often comes to mind, thousands of kilometers north and south of Germany were also strung with lethal fencing, guard posts, and barriers for much of the Cold War. The concrete wall dividing the city of Berlin was perhaps the most formidable of the Cold War barriers (though it also stood at least 135 kilometers east of the main line of the Iron Curtain), but many sections of the Iron Curtain borderlands featured similar walls, or combinations of walls, fencing, minefields, guards, or dogs. With human communities cleared from the borderlands along an extended swath, and casualties a regular occurrence, the Iron Curtain earned a fearsome reputation as a trans-European death strip (e.g. Harbutt 1988; Sheffer 2014).

Even as the Iron Curtain remained a central fixture of geopolitical attention during the Cold War, a number of related, collateral changes were occurring in the land uses and land cover of this extended militarized zone. With most existing land uses and residents prohibited and expelled from areas adjacent to the central European border, typically in a swath ranging from 500 m up to 10 km, cleared land steadily revegetated, cultivated areas grew feral, forests returned, and a broader, inadvertent process of rewilding took hold (Fig. 9.3). Similar processes are evident today in other militarized borders such as the Korean Peninsula's Demilitarized Zone (DMZ) (see Chap. 7 of this book by Brady) and the Green Line on the island of Cyprus. Ironically, even

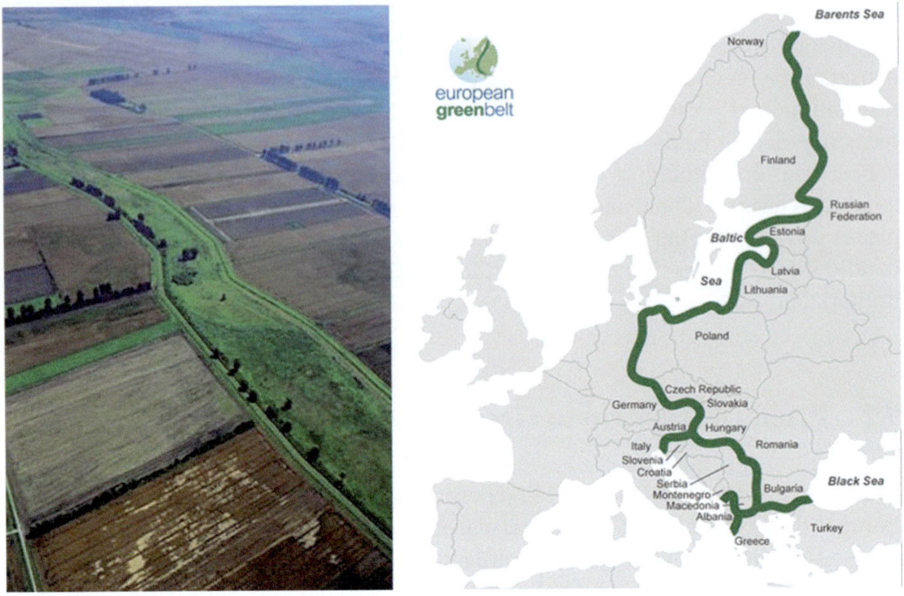

Fig. 9.1 The European Green Belt, tracing the path of the Cold War's Iron Curtain www.europeangreenbelt.org

Fig. 9.2 Iron Curtain Open Air Museum, Mödlareuth, Germany. (Photo by David Havlick)

Fig. 9.3 Rewilding forests along the Iron Curtain and its former patrol road. (Photo by David Havlick)

as central Europe's borderlands remained notorious as a "death strip," many areas within this swath were developing new, valuable ecological qualities.

Along the inner German border, these changes quietly attracted notice, and within weeks after the Berlin Wall fell in 1989 and the broader dismantling of Iron Curtain fortifications became likely, government officials and non-governmental organizations sought to turn the decades-long tragedy of the Iron Curtain into an opportunity for conservation and unification. In East Germany, at one of the government's final meetings in 1990, the council of ministers worked to establish large tracts of the Thuringian Rhone region as a biosphere reserve, a measure that was later worked into the Unification Treaty between East and West Germany. In the spring of 1991, the United Nations formally established 185,000 hectares of this area as a UNESCO reserve that now spills across these former inner German borderlands (Our Way Into the Future n.d.).

Today, the Iron Curtain borderlands include hundreds of protected areas and a series of national parks and reserves that are collectively known as "The Green Belt of Europe." In Germany alone, there are now more than 150 nature conservation areas along the former borderlands, with more than 120 additional conservation areas protected as "ribs" extending from the main spine of the former East-West border (Geidezis and Kreutz 2012; Fig. 9.4). In 2015, Germany announced that more than sixty additional former military bases would be converted to nature reserves, and images in the stories covering this news featured guard towers and

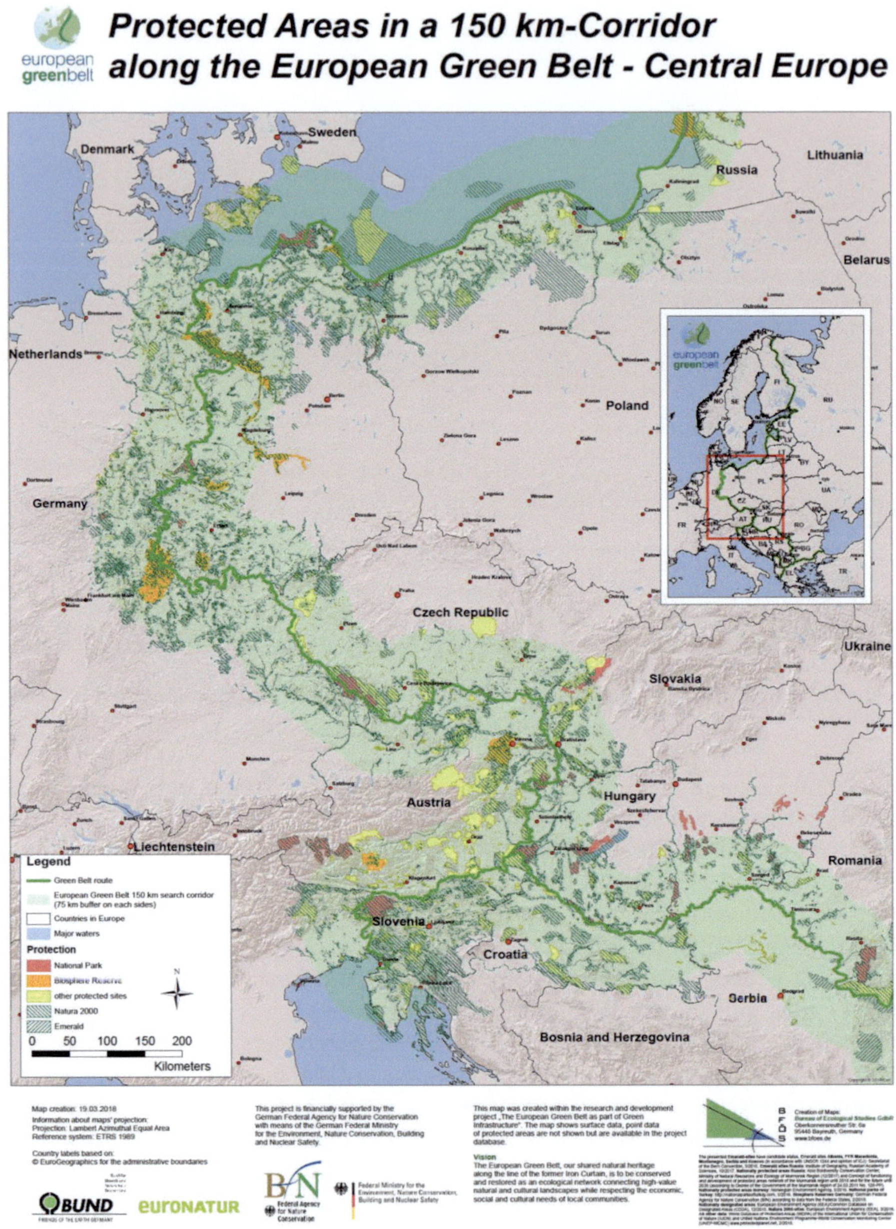

Fig. 9.4 Protected areas within 150-km corridor along the European Green Belt

scenes from the former Iron Curtain (e.g. Sola 2015). Although generally positive, the coverage of these military-to-wildlife conversions seemed invariably to highlight the irony of these transitions, positioning images of eagles against aging barbed wire and the derelict watchtowers of the new Green Belt. This, in fact, captures some of the ambiguity surrounding these changes, and notions of collateral ecological values more generally: should we be buoyed by the feel-good storyline of nature prevailing in these landscapes, after decades of militarization, isolation, and despair? Or should we more properly distrust the apparent happy ending and instead focus on the dislocation and death that in many cases preceded and, in a real sense, enabled these conservation moves? Better still, can we embrace this ambiguity and find ways that both appreciate the real conservation (and political) successes that the Death Strip-to-Green Belt changes signify, while also keeping the human cost and mixed histories of these land use changes clearly in view?

9.3 Military Environmentalism, or Ecological Militarization

Tourists who visit portions of the Iron Curtain borderlands today may well struggle to imagine the highly-militarized, lethal zone that existed here a just a few decades ago. Where electric fences and razor wire once stood, the former east-west borderlands are now marked by open fields, regenerating forests, and resurgent wildlife populations. Small towns dot the landscape, many bringing together traditional village features of the local pub, rathaus, or church, with more recent additions of solar panel-bedecked condominiums or modern resorts. It's easy, in other words, to encounter these lands as natural, leisure, or recreational landscapes that bear little explicit reminder of the 40 years of lethal Cold War fortifications.

To its credit, the European Union (EU) – along with more localized efforts – has worked to ensure that the history of these borderlands isn't entirely lost from view. In 2005, the EU formally designated an Iron Curtain Trail as one of Europe's longest bicycle routes. The vision for the trail goes beyond recreation and aspires to provide a means of "experiencing history," a model for sustainable tourism, and a route that fosters a broader sense of European identity (Cramer n.d.; The Iron Curtain Trail n.d.; Hammer 2009). Communities, non-governmental organizations, and EU member states increasingly value the lands of the former death strip for providing a variety of ecosystem and cultural services: as intact habitat and open space, a living memorial to the Cold War decades of a divided Europe, and as important areas of ecological revitalization, cultural meaning, and sustainable development (Cramer 2010, 2012).

For anyone who experienced the Iron Curtain during its Cold War period and has returned in the decades since the death strip's removal, the contrast could hardly be more striking. In the mid-1980s, I crossed from West Germany into Czechoslovakia, and later traveled through East Germany en route to Berlin. Almost 40 years later, I still recall my apprehension as we slowed for inspection by border guards, and the frightening search of my train car as armed guards pulled apart seat cushions, picked through my belongings, and examined my tourist visa and passport. Guard dogs patroled the train tracks, and the stark strip of cleared land discouraged any impulse to simply bolt and make a run for it (a book I carried by Milan Kundera, a Czech author whose works were banned at the time, made me anxious). Of course, there were many over the years who faced actual peril, rather than risks mostly just imagined, and did just that.

In fall 2013, I returned to these formerly imposing crossings to bicycle 1200 kilometers of the borderlands as part of a research sabbatical. Quiet roads and dirt paths crossed the inner German border, which in places was so inconspicuous that it past unnoticed. National parks and scenic footbridges filled crossings that for decades had been lethal, and it was easy to get swept away by the bucolic landscapes and many shades of green that now filled the once-fortified spaces. This, no doubt, is what excites boosters of the emergent Green Belt of Europe. The tangible sense of restoration, redemption, and resurrection is hard to shake. In a number of places, local communities have added their own touches, by preserving scraps of fenceline or border wall, by installing open air museums that vividly portray the Cold War array of barriers and lethal controls, and by introducing artwork that invites questions about the current and past uses of these borderlands. These serve as reminders, so the histories of specific places are not entirely lost, but in some ways they also serve to highlight the affirming contrast between past and present.

The triple border of Slovakia, Czech Republic, and Austria, for example, sits at the confluence of the Thaya and Morava Rivers. The nearest road crossing spans a quiet bridge between two small towns. To reach the actual confluence requires a three-kilometer detour along country paths and two-track ruts, which bear all the markings of a country picnic destination and no signs of militarized borderlands. At the river's edge, however, a small sculpture signals this Cold War past, presumably, with a pair of iron cast ankles shackled in chains (Fig. 9.5). There is, then, at least a trace mingling of past and present at this site, but overwhelmingly what the few visitors to this place encounter is a seemingly natural, or at least quietly rural landscape.

These experiences of a naturalized Iron Curtain borderland are now common in many places throughout central Europe, and surely contribute to a lasting impression of militarized borders greening successfully into conservation landscapes. In one sense, there is little reason to complain about this: ecologically, these lands truly are recovering, reforesting, or rewilding. In other words, collateral ecological values are tangible here. For this, conservationists likely have cause to celebrate as protected areas expand along the former Iron Curtain borders. At the same time, the deeper implications of this greening ought to be considered. What do these kinds of changes lead us to understand about these landscapes, and how do they nudge us toward certain views about nature and society? As we move to recognize or promote the merits of collateral values, shouldn't we also maintain an obligation to keep in view what social and political processes produced these conditions, and realize that the positive outcomes were in most respects accidental or subordinate to broader processes of militarization, the eviction of local communities, and widespread application of force? In this, it may be worth invoking not just the adjectival meaning of the term

Fig. 9.5 Sculpture on the banks of the Morava River at intersecting borders of Slovakia, Czech Republic, and Austria. (Photo by David Havlick)

collateral, which signifies the secondary benefits of habitat and open spaces that militarization has produced, but also to point to collateral as a noun, or the down payment in lives disrupted and lost that was required to achieve these gains.

9.4 Pulling Back the Curtain on Green Militarization

One of the more inspiring storylines to emerge from land use changes such as those found today along the Iron Curtain borderlands frames these transitions as evidence of ecological militarization or military environmentalism (for the use of this term, see Coates et al. 2011; Dudley 2012; Coates 2014; and more critically, Woodward 2004). Put this way, the reorientation of these lands from militarized borders to conservation reserves comes from a fundamental compatibility between military management and environmental protection. The explanation for these militarization-conservation affinities varies from a more passive restoration model, where nature simply filled the void created by militarized zones or lands made off-limits to other uses, to a casting of military activities as positively beneficial for conservation goals.

The European Green Belt initiative, which focuses on developing conservation networks along the former central European borderlands, offers a relatively passive view of how these changes have come to pass: "the border zone granted nature a pause. Unwittingly, the once-divided Europe encouraged the conservation and development of valuable habitats. The border area served as a retreat for many endangered species" (European Green Belt 2016). The European Green Belt efforts also point directly at the historical significance of the earlier Iron Curtain period of these borderlands, and work to ensure that the previous, militarized condition of the area remains evident and meaningful.

A more active view of military environmentalism is promoted in publications by the U.S. military and, in some cases, by non-governmental organizations who have cooperated with the U.S. Department of Defense (DOD) or similar agencies elsewhere in the world (e.g. Benton et al. 2008). Military training and environmental protection are cast as fundamentally compatible pursuits: "Fort Bragg not only is helping ensure the survival of this endangered bird [red-cockaded woodpecker] but also is enhancing the availability of realistic training for the nation's troops" (Stein 2008). Or, as a biologist for the Canadian Department of National Defense described the country's largest military base, it is "a veritable Serengeti… with over 1,100 documented species including over 25 species at risk, as well as massive herds of elk, deer, and pronghorn antelope" (Boyd 2014).

The Center for the Environmental Management of Military Lands (CEMML), based at Colorado State University (CSU) in Fort Collins, Colorado, takes a more systematic approach to documenting the environmental contributions of military lands. According to its website, CEMML consists of "a team of environmental professionals experienced in the conservation and sustainable management of natural and cultural resources on Department of Defense lands" (CEMML 2016). CEMML is supported largely by grants from the DOD, and contracts with nearly two hundred biologists and resource managers located either on the CSU campus or at more than forty military installations across the U.S. The center identifies explicitly with the ideas of military environmentalism, noting that, "CEMML recognizes that military land use and resource conservation are compatible goals that can be accomplished through the integration of sustainable land management practices" (CEMML 2016). Similar messages come through in articles that label such transitions, "From Bombs to Birds" and signage at the refuges themselves that point to the shift "From Weapons to Wildlife" (e.g. Weeks 2009, pp. 20–23).

In many cases, the emergence of ecological benefits from militarized landscapes is presented not merely as a sign of compatibility, but in more obligate terms. In this stronger version of military environmentalism – which can be described as *strong ecological militarization* (Havlick 2006) – the ecological qualities generated by military use are not just coincidental, but actually *depend* upon the military actions brought to these places. For example, at the U.S. Army's Jefferson Proving Ground, which is now the Big Oaks National Wildlife Refuge, grasslands and forest openings that currently provide valuable habitat for songbirds were created and maintained by munitions spotting and tests conducted during four decades following the Second World War. The U.S. Fish and Wildlife now prescribes fire to these same areas in an effort to replicate these military disturbances and stave off encroaching shrub and forest cover.

Similarly, at a number of sites in the southeastern United States, fires sparked by military training and testing are credited with maintaining fire-dependent longleaf pine ecosystems and related species such as the red-cockaded woodpecker. As an enthusiastic *National Geographic* account of the military's environmental stewardship at Eglin Air Force Base put it, "two tilt-rotor V-22 Osprey [aircraft] emerged above the treetops and arched down river and out of sight. These impressive metal birds symbolized not just national defense but natural defense..." (Ward 2015). The article's title captures the take-home message perfectly: "Bombing Range is National Example for Wildlife Conservation."

In 2013, the Obama Administration unveiled its "Sentinel Landscapes" initiative, which aimed to combine three key objectives: sustaining military readiness, restoring and protecting wildlife habitat, and preserving agricultural lands (McKalip and Jensen 2013). As of 2018, this joint program of the DOD, U.S. Department of Agriculture, and U.S. Department of the Interior included 7 sites across the United States, providing a vivid example of how defense and conservation interests are merging both in terms of messaging and through formal agreements (see www.Sentinellandscapes.org). The name of the program resonates in multiple directions, signaling the military duty of standing watch, but broadening the implications of this beyond national security to encompass conservation and agricultural lands as well. Maintaining lands for military readiness in this way can also be seen as providing for ecological and social well-being. Of course, military officials remain clear-eyed about the real purpose of the program; as the DOD's former Assistant Secretary for Energy, Installations and Environment pointed out, "Sentinel Landscapes will be a magnet for conservation activities, but the real motivation at DOD is creating the buffer we need to protect these critical [military] missions" (USDA 2015).

As environmental planners in the military and some conservation groups tend to point out, accounts of military lands providing ecological benefits are credible and in a number of instances can be backed up by empirical studies (some of which are funded by the DOD; e.g. Kitchen et al. 2000; see also Benton et al. 2008). They also perfectly fit the narrative of collateral ecological values: military activities are dedicated to a primary mission of national defense, but ancillary or subordinate benefits can come as a result of these actions. There are a number of reasons, however, to take a more critical view of military environmentalism. These range from reports that document the environmental *damage* caused by military activities (e.g. Quist et al. 2003) and broader processes of militarization, to concerns about historical erasure and the loss of cultural meaning that may come as new layers of land use obscure previous uses, and new names or reputations for militarized landscapes take hold.

9.5 Critiquing Military Environmentalism

In Colorado, the *Colorado Springs Gazette* newspaper, much like the city's politics, is notoriously pro-military. For the past five decades, Colorado Springs has staked a claim to being a military town, home of the U.S. Air Force Academy, the Army's Fort Carson, Shriever and Peterson Air Force bases, and the headquarters for the North American Aerospace Defense Command (NORAD) that is burrowed into a 9500-foot high granite peak on the edge of town. Lately, though, the news about the military's effect on Colorado Springs hasn't been quite so rosy.

In June 2016, Colorado newspapers broke a story that the drinking water for 80,000 residents in communities downstream from Peterson Air Force Base was contaminated with toxic perfluorinated chemicals (PFCs). The chemicals, which can persist in human bodies for years, had been used for decades at Peterson as a fire-fighting foam (Finley 2016; see also NIH 2016). News coverage over the next several months shifted from concern to outrage, and in late October 2016 an investigative report by the *Gazette* documented that, "the Air Force ignored decades of warnings from its own researchers in continuing to use a chemical-laden firefighting foam that is a leading cause of contaminated drinking water for at least 6 million Americans, including thousands of people south of Colorado Springs" (Roeder and Rodgers 2016).

Even as the controversy crested into national news, with coverage in the *New York Times* (Turkewitz 2016), the Air Force continued to use the toxic foam and resist public notification of known spills. In mid-October 2016, Peterson Air Force base inadvertently dumped 150,000 gallons of PFC-contaminated water into the Colorado Springs wastewater system, which in turn delivered the polluted water into the area's principal stream, Fountain Creek. Air Force environmental officials waited 6 days before notifying the public of the release and, when pressed, simply responded that they were not required by law to alert downstream users about this, "non-regulated substance" (Roeder and Rodgers 2016). Defense officials have now acknowledged at least 2000 sites – mostly Air Force bases – contaminated by PFCs, and some experts expect that defense-related contamination from the endocrine-disrupting chemicals will eventually be documented in every U.S. state (Turkewitz 2016).

The PFC contamination is but one of many examples of continued and long-lasting environmental and public health damage associated with military base operations domestically. Barnett (2001) reported that the U.S. military generates more toxins than the top five U.S. chemical corporations combined (p. 95). By the end of the twentieth century, approximately 25,000 U.S. Army, Navy, and Air Force sites required some degree of cleanup to meet basic environmental regulations, with an anticipated cost of remediation exceeding $80 billion (adjusted to 2016 dollars) (Barnett 2001; Durant 2007).[2] Just in terms of energy consumption, the U.S. military also creates a massive ecological footprint: the DOD is responsible for 2–3% of all the energy consumed in the United States, roughly one-fourth of all jet fuel

[2] Durant, p. 78, cites a 1989 DOD estimate of $42.5 billion.

consumed worldwide, and generates more than 10% of U.S. CO2 emissions (see Renner 1991; Shulman 1992; Barnett 2001, p. 95; and Woodward 2004, p. 73).

By many accounts, the scope of environmental neglect is even more acute at the hundreds of miltary installations the U.S. maintains internationally, where in many cases environmental oversight can be scant and environmental regulations non-existent (see Gerson and Birchard 1991; Lutz and Enloe 2009; Vine 2015). At the Indian Ocean base of Diego Garcia, for example, four incidents on the small island over a 15-year period spilled more than 1.3 million gallons of jet fuel, which contaiminated groundwater and soils (Vine 2015, p. 139). At Johnston Atoll, in the Pacific Ocean, U.S. military operations left behind 16,000 tons of soil laced with dioxin and 7000 tons of soil tainted with PCBs. The atoll, which is now managed as a national wildlife refuge, was also contaminated with plutonium when nuclear-armed missile tests failed catastrophically on multiple occasions in 1962 (TenBruggencate 2003; Cleaning Up 2005).

Military training and testing activities are, of course, simply forms of preparation for the United States' sharper point of supporting a large military: the ability to assert lethal force broadly across the planet. This fundamental war-fighting mission of the military clearly brings its own acute forms of social and environmental impacts, both inadvertently and as a direct objective. The litany of these impacts is too great to attempt to list comprehensively here, but include human casualties, dislocations and social upheaval, the destruction of infrastructure, hazards created from munitions and explosive hazards, chemical contamination, soil disturbance, water and air pollution, loss of biological diversity, and lasting political instability (see, for example, Sanders 2009). To overlook these widespread and persistent consequences of military action, or to obscure these impacts by highlighting conservation success stories from various training facilities or recovering sites of militarization, is to disregard the fullest accounting of the role the military plays in the world.

To bring the focus back to military environmentalism, however, and the condition of militarized landscapes as these relate to conservation outcomes, it is worth considering DOD lands across the U.S. more broadly. A majority of the country's most severely contaminated sites – included on the National Priorities List for Superfund designation – are found on military training and testing lands (e.g. Nazaryan 2014; see also Vine 2015). Somewhat paradoxically, DOD lands are also considered the most biologically diverse of any federal lands in the U.S., with a greater concentration of Threatened and Endangered species than lands such as national parks and national forests that are more commonly associated with conservation and habitat protection (e.g. Benton et al. 2008). Though these conflicting qualities can seem challenging to reconcile in some cases, at the very least they highlight the heterogeneous character of military impacts on the environment. To take either the environmental abuse wrought by militarization or the environmental amenities found in militarized landscapes as the singular story is clearly too simplistic. *Both* are features of the military-environment relationship, and both ought to be kept in view as we move forward to develop policies that seek to protect positive environmental qualities while also repairing and holding the military accountable for the considerable damage it causes.

9.6 Collateral Values and Militarized Landscapes

In militarized landscapes transitioning explicitly to new purposes of conservation, the risk of losing sight of prior histories, and the social and environmental damage wrought by military activities, is particularly acute (e.g. Hourdequin and Havlick 2016; Havlick 2011; Davis 2005, 2007, 2015). As the earlier example from Florida's Eglin Air Force Base points out, there are, however, also lands that remain more actively militarized but still gain recognition for the ecological and conservation amenities they provide.

The Green Line dividing Cyprus is the site of the United Nations' longest-running peacekeeping mission, where since 1974 UN patrols have maintained a buffer zone put in place after decades of violent conflict that ultimately split the island's Greek and Turkish populations (Cassia 1999; Coates 2014; Chan 2016). The borderland of the Green Line remains heavily militarized and largely off-limits, even as Cypriots on both sides of the line have cooperated to address a variety of environmental, social, and cultural concerns (Grichting 2014, p. 430), and since 2003 have endured relatively few travel restrictions across the line (Chan 2016).

Within the Green Line's buffer zone, which ranges from 3.5 to 5 km in width, most all buildings and infrastructure have been isolated and left to slow ruin during the course of more than three decades. As the built environment gradually disintegrates, however, the natural environment seems to have steadily flourished. Biological inventories of the Green Line conducted since 2007 have documented rare plants and birds, small mammals, amphibians and reptiles (Grichting 2014). The population of Cyprus mouflon (*Ovis orientalis ophion*), an endemic species of wild sheep once on the brink of extinction,[3] now numbers in the thousands, and the endangered Mediterranean monk seal has been sighted along the Green Line seacoast (Coates 2014; Grichting 2014).

A continent away, on the Korean peninsula, the demilitarized zone (DMZ) separating North from South has also attracted attention for its conservation prospects amid its longstanding militarized condition (see Brady, Chap. 7 of this book). By many accounts, the DMZ is the most heavily militarized border found anywhere in the world (e.g. Lah and Kwon 2015; Moss 2014), a title previously applied to Europe's Iron Curtain. Former U.S. President Bill Clinton famously called the DMZ's 240-kilometer long border, "the scariest place on earth" (Havely 2003). The DMZ has drawn attention in recent decades, however, as a de facto wildlife reserve (Higuchi et al. 1996; Kim 1997; Turner 2005; Thomas 2009; see also Weisman 2007; Card 2008; Brady 2008). In the four-kilometer swath "preserved" between the two Korean states by watchtowers, fences, armed patrols, and nearly two million land mines, an ecosystem has managed to thrive: the DMZ now provides important habitat for migratory birds along the East Asia/Australasia flyway, and serves as a rare terrestrial sanctuary for resident species including Asiatic black bear, musk

[3] Mouflon in Cyprus are considered a distinct subspecies by the IUCN, but have a complicated genealogy and history of geographic of distribution; see for example, Pedrosa et al. 2005.

deer, Amur leopard cats, raccoon dogs, Amur gorals, and possibly Siberian tigers and Amur leopards (Card 2008; Platt 2011; see also Bird Life International n.d.).

The various representations of the social and natural qualities found in the DMZ illustrate how militarized landscapes can press us to renegotiate ideas of public safety, beauty, restoration, and preservation. South Korea increasingly presents the DMZ as a tourist attraction, and a number of tourist websites and government efforts rebrand the zone in markedly non-militarized terms. In recent years, South Korea has sought to rename the DMZ and a southern buffer zone of agricultural lands the "Peace and Life Zone" or PLZ (see DMZ Tour Course Guidebook 2009; Cain 2014). The Korea Tourism Organization offers PLZ tours and its website explains, "The name 'Peace and Life Zone' pays reference to the unpolluted natural environment and the people's general hope for the arrival of a new peaceful era to both sides of the border" (Demilitarized Zone Tours 2015). Although the website acknowledges broad outlines of the DMZ's history, and clearly still identifies the DMZ as a militarized borderland, it also casts the militarization of the zone very much as *historical*: "The DMZ and its surroundings were once the site of fierce battles during the Korean War, but has recovered from its wounds over the last half-century to become a quiet lush green area inhabited by diverse living creatures" (Demilitarized Zone Tours 2015). Another DMZ tour company website encourages prospective visitors to "Explore the Excitement of Silence" (DMZ Tours 2014). The more detailed text on the site explicitly acknowledges certain aspects of the zone's militarization, but emphasizes qualities of naturalization and the *peacefulness* of the place.

The DMZ in this way is presented as a site where, absent human activity, nature is thriving. Of course, the area might also – rather differently – be seen as land sacrificed to the security ambitions of a divided Korea, or the lingering outcome of intrapeninsular hostilities and years of violent conflict, but when its ecology is singled out, the DMZ instead can be valorized – and commodified – as territory affirmatively providing environmental and ecotourist amenities to the region. Tourists from around the world now come to the DMZ to pose for pictures in faux-North Korean classrooms, complete with framed portraits of Dear Leader Kim Jong-Il, scurry through tunnels ostensibly dug by the North in preparation for a broad military assault, buy DMZ-oriented trinkets, and enjoy a theme park named "Peace Land" (Pearson and Ingrassia 2013).

These examples from the DMZ highlight how politicized the greening of militarized space can become, even as these transitions may play out in popular media more simply as examples of nature's resilience. As the disparate representations of the DMZ suggest, casting militarized landscapes as ecological havens can create its own set of problems, with still-dangerous landscapes airbrushed into seemingly benign attractions. Elsewhere, the militarization of conservation – or what Lunstrum (2014) labels *green militarization* – remains more visibly problematic, as conservation objectives are leveraged to justify the use of deadly force against local human populations (see also Bocarejo and Ojeda 2016). At these sites and others, it remains important to find ways to highlight the military-environment relationship without diminishing the violence that occurs, but also without entirely disregarding the conservation opportunities some of these places provide (or the real threats that poaching can pose). It seems essential, then, to ask: how might we preserve the memory, meaning, and continuing toll of these cultural impacts while also engaging with the conservation merits of these lands?[4]

9.7 Memory, Meaning, and Conservation

Dangerous military sites remain alarmingly common globally, but in the United States a number of environmental regulations and regulatory agencies exist to try to address the most egregious cases of physical contamination. As noted earlier, the U.S. National Priorities List includes many sites of military training and testing now designated for Superfund cleanups. At the federal level, the U.S. Environmental Protection Agency (EPA) oversees remediation efforts at many of these locations, and state and county public health agencies are often in the regulatory mix as well. At the Rocky Mountain Arsenal National Wildlife Refuge just outside Denver, Colorado, for example, cleanup of this former chemical weapons manufacturing site had to clear both EPA and State standards thanks to a series of lawsuits that held the U.S. Army and a private lessee of the site, Shell Chemical Corporation (now Shell Oil), accountable. Although the remediation and consolidation of contaminants at this and similar sites remain controversial, there are at least specific legal standards that pertain to the environmental degradation that took place (Nazaryan 2013; Redeker 2002; see also Iversen 2012; Draper 2014). At the Rocky Mountain Arsenal, these requirements led to a multi-billion dollar restoration effort designed to make the site safe for wildlife refuge workers and visitors (though it's worth noting this standard is weaker than what would be required of residential or commercial uses). The wildlife inhabiting the area have also been tested repeatedly for signs of higher-than-normal chemical loads.

It likely comes as no surprise that cleanup of contaminants, and financial and legal accountability, drive remediation efforts at military-to-wildlife sites such as the Rocky Mountain Arsenal. What often gets left behind in these efforts is the cultural attributes and lasting meaning of these places. As one official at the Rocky Mountain Arsenal wildlife refuge pointed out, "The number one priority is to restore as many disturbed acres as we possibly can… for the benefit of wildlife and land conservation" (author interview, 2012). This approach fits the overarching mission of the U.S. National Wildlife Refuge System: "to administer a national network of lands and waters for the conservation, management, and where appropriate, restoration of the fish, wildlife, and plant resources and their habitats within the United States for the benefit of present and future generations" (U.S. Fish and Wildlife Service 2015). Although people are ultimately presented as the beneficiaries of the conservation services provided by the wildlife refuge system, the human histories and cultural meaning from refuge sites are not directly considered. Refuge managers

[4] On a related question, see Smallwood 2014, p. 302.

often point to their agency's "wildlife first" mission (and funding shortages) as a key reason why they're not able to attend to cultural concerns at their refuges, even as many of these same managers acknowledge that the cultural layers of their refuges merit attention.

The views of citizen volunteers and organizations from communities adjacent to military-to-wildlife refuges often support this interest in historic preservation and cultural memory. When interviewed, many of the restoration volunteers at the Rocky Mountain Arsenal indicated their concerns about the ecological character of the site and their volunteer efforts to remove invasive plants, but they also highlighted the importance of learning about the site's history. As one volunteer explained, "I think people should know the history behind the Arsenal. For a family to come and enjoy it, they have to know why they turned this place into a wildlife refuge and they have to know why it is this and not another type of place" (see Havlick et al. 2014). Framed this way, the history of the site and the changes that have taken place can actually *add* value and meaning to the emerging ecologies that the processes of militarization, ecological restoration, and conservation have produced. In nearby Commerce City, Colorado, and in other communities located near transitioning military installations, local citizens have banded together to form historical associations explicitly dedicated to keeping the cultural attributes – and in many cases, the sacrifices made by local communities – visible.

The idea that history and ecology *both* ought to be represented in militarized landscapes that are now recognized for their ecological values is an important, recurring theme for those working on the Green Belt of Europe project as well. As the main EU Parliament sponsor of the Iron Curtain Trail told me in a 2013 interview, "We can't only look to nature, that would be crazy. Culture, politics, nature, and history all need to be considered together" (Cramer 2013). The broader European Green Belt initiative similarly foregrounds both the ecological promise of the changing central European borderlands and their cultural significance: "Besides its extraordinary ecological importance, the initiative is a living example of how Europe and its diverse cultures can truly grow together. From the European Green Belt, we can learn that biological diversity goes hand in hand with cultural diversity. It is a symbol for transboundary cooperation and a Europe's [sic] shared natural and cultural heritage" (European Green Belt 2016).

The comparison between military-to-wildlife refuges in the U.S. and the changes along the former Iron Curtain is revealing, at least in part, for the structural differences in policy that underpin the respective efforts. U.S. Fish and Wildlife Service officials tasked with managing refuges that come from previous military uses are limited by their legal mandate ("wildlife first"), and when faced with ever-limited budgets often need simply to focus on conservation priorities fairly exclusively. The mandate for Europe's Green Belt, on the other hand, points to "high-value natural and cultural landscapes" (European Green Belt 2016). Given these structural differences in how the U.S. and European lands are to be managed, it's also worth examining how experiences at the ground level may influence what meaning visitors take away from these transitioning landscapes of collateral conservation.

9.8 Curation of Cultural Landscapes

Access to military-to-wildlife refuges in the U.S. varies considerably, from wide open to completely off-limits, but many of these sites provide some form of public use.[5] At sites that are open to the public, visitors often engage with these places at particular locations managed specifically for a visitor interface: visitor centers, trailheads, or kiosks installed at parking lots. Although many U.S. national wildlife refuges include thoughtful interpretive signs or exhibits highlighting cultural features, visitor surveys at several of these sites suggest that people often fail to register these efforts at curation, the cultural attributes of the refuges generally, or the more specific military histories and impacts at these sites (on this and what immediately follows, see Havlick 2016). When asked an open-ended question about what words they would use to describe the Rocky Mountain Arsenal National Wildlife Refuge, visitors' most common cluster of words were "peaceful" or "quiet," "wild" or "natural," and "beautiful." Each of these appeared in at least 25% of visitor responses, whereas only 4% of the responses made any explicit mention of remediation efforts or the military and chemical production that characterized the site for decades. When visitors at a different former military installation, now the Assabet River National Wildlife Refuge, were asked the same open-ended question, 48% responded "beautiful," 36% said "peaceful" or "quiet," and 23% suggested "wild" or "natural." Just 3% of respondents commented on the history of the site or its 50 years of military use as an ammunition storage facility.

It's worth emphasizing that the Fish and Wildlife Service does not try to hide the military history of either site. To the contrary, exhibits in new visitor centers at both refuges feature a mix of ecological and cultural displays. The landscapes themselves also contain evidence of their military use. Portions of the Rocky Mountain Arsenal remain off-limits to public use and are marked, if visitors look carefully, as Army-owned landfills for contaminated military and chemical waste. At Assabet River, dozens of large concrete igloos (or "bunkers") are scattered across the refuge, and local historians and refuge volunteers periodically offer popular "bunker tours."

These findings raise the question of what could be done differently to impress upon visitors that these refuges are not just sites of ecological flourishing, but rather exist as examples of collateral values that come from mixed cultural and ecological processes. Even with explicit curation that points out the military histories at these sites, most visitors seem to pay more notice to the live roaming bison, for instance, at the Rocky Mountain Arsenal, and not the bomb casing or hazmat-suited mannequin mounted as visitor center exhibits. What's perhaps most striking about the refuge is exactly what visitors commented on in the surveys: it appears *natural*. Where the South Plants chemical manufacturing operations once sprawled across the interior of the Rocky Mountain Arsenal site, visitors today encounter a resurgent

[5] U.S. National Wildlife Refuges in general are typically open only from dawn to dusk, and may have designated entrance locations and signed boundaries, but most are free to the public, have few if any entrance locations staffed by agency employees, and many are large enough that boundaries are generally unmonitored and porous.

shortgrass prairie and extensive prairie dog towns. Basin F, which just a few decades ago was described as the single most toxic square mile on the planet, now sports prairie grasses against a backdrop of the refuge's growing herd of bison. Little wonder then, that despite materials that depict and describe the Arsenal's history of chemical production, for most visitors the takeaway message from their time at this refuge is simply that plants and wildlife are thriving on this valuable scrap of habitat tucked between Denver, Commerce City, and Denver International Airport.

Perhaps there's no harm in this ecological flattening of the Arsenal's history, but what happened previously on this site matters, and the sacrifices demanded of the land, the neighboring communities, and the persistent contamination of groundwater and soils should be part of the public understanding of this place. To keep this in view would not require gutting the ecological commitment the U.S. Fish and Wildlife Service is required to make here, nor would it need to detract from visitors' enjoyment of their time at the refuge. The cultural layers here could be represented simply and directly, with auto tour stops that provide before and after images of the chemical facilities-turned-prairie, interpretive signs that mark the location of former schoolhouses and homesteads that predated the Army takeover of the site, and maps that indicate where munitions and napalm and rocket fuels manufactured here were used, transported, or stored. Curation of this site could also be more oblique, perhaps signaled by art installations, murals and mosaics, or even fields dotted with neon army figures (see Drenthen 2016). Each of these could serve to disrupt the tempting reception of this site as *simply a wildlife refuge*, and instead spur visitors to question the meaning of the place and the images or figures stationed upon it.

This is, in fact, the approach taken at a number of locations along the Iron Curtain borderlands. In addition to open air museums and reconstructions of the fortifications that once characterized this stretch of land, today's Green Belt of Europe is dotted with reminders that this hasn't always been a landscape known for its ecological features. Most every road crossing along the former inner German border is marked with a large brown sign that depicts a map of divided Germany and lists the date and hour that the barriers of the Iron Curtain lifted at that spot.

Elsewhere, border locations in Slovakia, the Czech Republic, Austria, and Germany are marked by metal sculptures, lists of those who died at a particular crossing, neglected border control stations, or kilometer after kilometer of overgrown patrol roads. Guard towers have intentionally been left standing, many abandoned and left to ruin, others refurbished as unique structures for lodging, restaurants, or sight-seeing. Scraps of the original walls and fences have also been left standing in a few places, but more common than all of these are the crosses. Some are elaborate, with rococo flourishes or detailed inscriptions, others stand stark in their simplicity. One, standing more than five meters high, was made of refabricated fencing torn down from the inner German border that it had long barricaded. Each of these, in their own ways, serves to disrupt the tidy acceptance of these borderlands as simply natural or naturalizing landscapes, and instead challenges those who pass through or live along these areas to question and reflect.

This highlights the lasting challenge of collateral values more generally: that we keep in view the fact that the environmental qualities emerging from such places –

and perhaps militarized lands most dramatically – have mostly not come by design, but coincidentally or by ancillary convenience. We should not deceive ourselves into thinking that military priorities have necessarily softened or yielded to a new ecological ethic. "Mission first" remains the underlying creed of military institutions, and that mission retains as its foundation the application of far-flung lethal force. As militarization and conservation emerge in various contexts as compatible objectives, it will remain important not to confuse one with the other.

References

Barnett, J. (2001). *The meaning of environmental security: Ecological politics and policy in the new security era*. London: Zed Books.
Benton, N., Ripley, J. D., & Powledge, F. (Eds.). (2008). *Conserving biodiversity on military lands: A guide for natural resource managers*. Arlington: NatureServe.
Bird Life International. (n.d.) *East Asia/Australasia flyway*. http://www.birdlife.org/datazone/userfiles/file/sowb/flyways/8_EastAsia_Australasia_NEW.pdf. Accessed 30 Oct 2016.
Bocarejo, D., & Ojeda, D. (2016). Violence and conservation: Beyond unintended consequences and unfortunate coincidences. *Geoforum, 69*, 176–183.
Boyd, D. (2014, November 14). *Bombs and biodiversity: Why military lands make great conservation areas*. PRI (Public Radio International). http://www.pri.org/stories/2014-11-18/bombs-and-biodiversity-why-military-lands-make-great-conservation-areas. Accessed 21 Oct 2016.
Brady, L. M. (2008). Life in the DMZ: Turning a diplomatic failure into an environmental success. *Diplomatic History, 32*, 585–611.
Cain, G. (2014, October 30). South Korea tries to re-brand dmz as rare animal sanctuary. *NBC News*. http://www.nbcnews.com/news/asian-america/south-korea-tries-re-brand-dmz-rare-animal-sanctuary-n231641. Accessed Aug 18 2015
Card, J. (2008). Paradise in no man's land. *Earth Island Journal (Spring)*. http://www.earthisland.org/journal/index.php/eij/article/paradise_in_no_mans_land/. Accessed 18 Aug 2015.
Cassia, S. P. (1999). Martyrdom and witnessing: Violence, terror and recollection in Cyprus. *Terrorism and Political Violence, 11*, 22–54.
CEMML. (2016). *Center for the environmental management of military lands*. https://www.cemml.colostate.edu/cultural/09476/cemmliraqenl.html. Accessed 26 June 2016.
Chan, S. (2016, November 7). Cyprus: Why one of the world's most intractable conflicts continues. *New York Times*. https://www.nytimes.com/2016/11/08/world/europe/cyprus-reunification-talks.html. Accessed 14 July 2017.
Cleaning up Johnston Atoll. (2005, November 25). *APSNet special reports*. http://nautilus.org/apsnet/cleaning-up-johnston-atoll/. Accessed 22 Feb 2016.
Coates, P. (2014). Borderland, no-man's land, nature's wonderland: Trouble humanity and untroubled earth. *Environment and History, 20*, 499–516.
Coates, P., Cole, T., Dudley, M., & Pearson, C. (2011). Defending nation, defending nature? Militarized landscapes and military environmentalism in Britain, France, and the United States. *Environmental History, 16*, 456–491.
Cramer, M. (2010). *Iron Curtain Trail, part 3*. Rodingersdorf: Verlag Esterbauer.
Cramer, M. (2012). *Iron curtain trail, part 2*. Rodingersdorf: Verlag Esterbauer.
Cramer, M. (2013, September 19). Interview with author, Literaturhaus Berlin, Berlin, Germany.
Cramer, M. (n.d.). *Iron Curtain Trail brochure*. The Greens/EFA in the European Parliament. http://www2.ironcurtaintrail.eu/uploads/brochure_iron_curtain_trail.pdf. Accessed 31 Oct 2016.
Cronon, W. (1996). The trouble with wilderness; or, getting back to the wrong nature. In W. Cronon (Ed.), *Uncommon ground: Rethinking the human place in nature* (pp. 69–90). New York: W.W. Norton.

Davis, J. S. (2005). Representing place: 'Deserted isles' and the reproduction of Bikini Atoll. *Annals of the Association of American Geographers, 95*, 607–625.
Davis, J. S. (2007). Military natures: Militarism and the environment. *GeoJournal, 69*, 131–134.
Davis, S. (2015). *The empire's edge: Militarization, resistance, and transcending hegemony in the Pacific*. Athens: University of Georgia Press.
Demilitarized Zone Tours. (2015). From DMZ to PLZ. 2015. Visit Korea. http://english.visitkorea.or.kr/enu/SI/SI_EN_3_4_1.jsp. Accessed 18 Aug 2015.
DMZ Tour Course Guidebook. (2009). *The road to peace and life, 545 km*. Seoul: Korea Tourism Organization.
DMZ Tours. (2014). *Explore the excitement of silence*. http://www.dmztours.com. Accessed 29 Aug 2014.
Draper, E. (2014, February 8). Former rocky flats site stirs concerns for some neighbors. *The Denver Post*. http://www.denverpost.com/2014/02/08/former-rocky-flats-site-stirs-concerns-for-some-neighbors/. Accessed 30 Oct 2016.
Drenthen, M. (2016). Layered landscapes, conflicting narratives, and environmental art. In M. Hourdequin & D. G. Havlick (Eds.), *Restoring layered landscapes: History, ecology, and culture* (pp. 239–262). New York: Oxford University Press.
Dudley, M. (2012). *An environmental history of the UK defence estate, 1945 to the present*. London: Bloomsbury.
Durant, R. F. (2007). *The greening of the U.S. military: Environmental policy, national security, and organizational change*. Washington, DC: Georgetown University Press.
Eckert, A. M. (2011). Greetings from the zonal border': Tourism to the Iron Curtain in West Germany. *Zeithistorische Forschungen/Studies in Contemporary History, 8*, 9–36.
European Green Belt. (2016). *9th Pan-European Green Belt Conference*. http://www.european-greenbelt.org/. Accessed 24 Oct 2016.
Finley, B. (2016, June 15). Drinking water in three Colorado cities contaminated with toxic chemicals above EPA limits. *Denver Post*. https://www.denverpost.com/2016/06/15/colorado-widefield-fountain-security-water-chemicals-toxic-epa/
Geidezis, L., & Kreutz, M. (2012). Green Belt Europe – Structure of the initiative and significance for a Pan European ecological network. In I. Marshall, M. Muller, & G. Mathias (Eds.), *Proceedings of the 1st GreenNet conference: The Green Belt as a European ecological network – Strengths and gaps. No. 10* (pp. 12–21).
Gerson, J., & Birchard, B. (Eds.). (1991). *The sun never sets: Confronting the network of foreign U.S. military bases*. Philadelphia: American Friends Service Committee.
Grichting, A. (2014). Cyprus: Greening in the Dead Zone. In K. G. Tidball & M. E. Krasny (Eds.), *Greening in the Red Zone: Disaster, resilience and community greening* (pp. 429–443). Berlin: Springer.
Hammer, J. (2009, July 26). A barrier gone, but not forgotten. *New York Times*. TR1.
Harbutt, F. J. (1988). *The Iron curtain: Churchill, America, and the origins of the Cold War*. Oxford: Oxford University Press.
Havely, J. (2003). *Korea's DMZ: 'Scariest Place on Earth'*, CNN.com World. http://edition.cnn.com/2003/WORLD/asiapcf/east/04/22/koreas.dmz. Accessed 28 Oct 2016.
Havlick, D. G. (2006). *Bombs away: New geographies of military-to-wildlife conversions in the United States*. Doctoral dissertation, The University of North Carolina at Chapel Hill.
Havlick, D. G. (2011). Disarming nature: Converting military lands to wildlife refuges. *Geographical Review, 101*, 183–200.
Havlick, D. G. (2014). The Iron Curtain Trail's landscapes of memory, meaning, and recovery. *Focus on Geography, 57*, 126–133.
Havlick, D. G. (2016). Restoration, history, and values at transitioning military sites in the United States. In M. Hourdequin & D. G. Havlick (Eds.), *Restoring layered landscapes: History, ecology, and culture* (pp. 160–180). New York: Oxford University Press.
Havlick, D. G., Hourdequin, M., & John, M. (2014). Examining restoration goals at a former military site: Rocky Mountain Arsenal, Colorado. *Nature and Culture, 9*, 288–315.

Higuchi, H., Ozaki, K., Fujita, G., Minton, J., Ueta, M., Soma, M., & Mita, N. (1996). Satellite tracking of white-naped crane migration and the importance of the Korean Demilitarized Zone. *Conservation Biology, 10*, 806–812.

Hourdequin, M., & Havlick, D. G. (Eds.). (2016). *Restoring layered landscapes: History, ecology, culture*. New York: Oxford University Press.

Iversen, K. (2012). *Full body burden: Growing up in the nuclear shadow of Rocky Flats*. New York: Random House.

Jacoby, K. (2001). *Crimes against nature: Squatters, poachers, thieves, and the hidden history of American conservation*. Berkeley: University of California Press.

Kim, K. C. (1997). Preserving biodiversity in Korea's Demilitarized Zone. *Science, 278*, 242–243.

Kitchen, A. M., Gese, E. M., & Schauster, E. R. (2000). *Changes in coyote activity patterns due to reduced exposure to human persecution*. (USDA National Wildlife Research Center – Staff Publications, paper 658). Fort Collins: USDA, Animal and Plant Health Inspection Service.

Lah, K., & Kwon, K. J. (2015). *Tired of conflict: Life near the Korean Demilitarized Zone*. CNN World. http://www.cnn.com/2015/08/31/asia/south-korea-life-near-dmz/. Accessed 28 Oct 2016.

Lunstrum, E. (2014). Green militarization: Anti-poaching efforts and the spatial contours of Kruger National Park. *Annals of the Association of American Geographers, 104*, 816–832.

Lutz, C., & Enloe, C. (2009). *Bases of empire: The global struggle against U.S. military posts*. New York: NYU Press.

McKalip, D., & Jensen, J. (2013). Supporting military readiness and training through environmental conservation. *White House blog*. https://www.whitehouse.gov/blog/2013/07/17/supporting-military-readiness-and-training-through-environmental-conservation. Accessed 21 Oct 2016.

Moss, L. (2014, September 18). *Wildlife thrives in 'the most dangerous place on earth'*. MNN.com. http://www.mnn.com/lifestyle/eco-tourism/stories/wildlife-thrives-in-the-most-dangerous-place-on-earth. Accessed 28 Oct 2016.

Nazaryan, A. (2013, October 11). Tiptoeing through America's sarin stash. *Newsweek*. http://www.newsweek.com/2013/10/11/tiptoeing-through-americas-sarin-stash-238108. Accessed 30 Oct 2016.

Nazaryan, A. (2014, July 17). The US Department of defense is one of the world's biggest polluters. *Newsweek*. http://www.newsweek.com/2014/07/25/us-department-defence-one-worlds-biggest-polluters-259456.html. Accessed 31 Oct 2016.

NIH. (2016). Perfluorinated chemicals (PFCs) National Institute of Environmental Health Sciences (flyer). https://www.niehs.nih.gov/health/materials/perflourinated_chemicals_508.pdf. Accessed 4 Nov 2016.

Our Way Into the Future. (n.d.). *Rhon Biosphere Reserve brochure*. UNESCO-Rhon Biosphere Reserve, 24 pp.

Pearson, J., & Ingrassia, P. (2013, October 25). Tank traps meet tourist traps at Korea's Demilitarized Zone. *Reuters*. http://www.reuters.com/article/us-korea-dmz-tourism-idUSBRE99O0SP20131025. Accessed 22 June 2016.

Pedrosa, S., Uzun, M., Arranz, J. J., Gutiérrez-Gil, B., San Primitivo, F., & Bayón, Y. (2005). Evidence of three maternal lineages in Near Eastern sheep supporting multiple domestication events. *Proceedings of the Royal Society of London B: Biological Sciences, 272*, 2211–2217.

Pieck, S. K. (2018). What stories should a 'National Nature Monument' tell? Lessons from the German Green Belt. Cultural Geographies, 26(2), 195–210.

Platt, J. R. (2011, September 27). South Korea seeks to protect endangered species in Demilitarized Zone. *Scientific American*. http://blogs.scientificamerican.com/extinction-countdown/south-korea-seeks-rotect-endangered-species-demilitarized-zone/. Accessed 18 Aug 2015.

Quist, M. C., Fay, P. A., Guy, C. S., Knapp, A. K., & Rubenstein, B. N. (2003). Military training effects on terrestrial and aquatic communities on a grassland military installation. *Ecological Applications, 13*, 432–442.

Redeker, B. (2002, December). Toxic site to serve as wildlife refuge. *ABC News*. http://abcnews.go.com/WNT/story?id=129958&page=1. Accessed 30 Oct 2016.

Renner, M. (1991). Assessing the military's war on the environment. In L. Brown (Ed.), *State of the World 1991* (pp. 132–152). New York: Norton.

Roeder, T., & Rodgers. J. (2016, October 23). *Toxic legacy: Air Force studies going back decades show danger of foam that contaminated Colorado Springs-area water*. Colorado Springs Gazette, A1.

Sanders, B. (2009). *The Green Zone: The environmental costs of militarism*. Oakland: AK Press.

Sheffer, E. (2014). *Burned bridge: How East and West Germans made the Iron Curtain*. Oxford: Oxford University Press.

Shulman, S. (1992). *The threat at home: Confronting the toxic legacy of the U.S. military*. Boston: Beacon Press.

Smallwood, P. D. (2014). Creating Afghanistan's first national park in the midst of conflict. In K. G. Tidball & M. E. Krasny (Eds.), *Greening in the Red Zone: Disaster, resilience and community greening* (pp. 297–303). Berlin: Springer.

Sola, K. (2015). Germany to turn 62 military bases into nature sanctuaries for birds, beetles, and bats. *The World Post*. http://www.huffingtonpost.com/2015/06/19/german-military-bases-nature-reserves_n_7623882.html. Accessed 10 Oct 2016.

Spence, M. D. (1999). *Dispossessing the Wilderness: Indian removal and the making of the national parks*. New York: Oxford University Press.

Stein, B. A. (2008). Biodiversity and the military mission. In N. Benton, J. D. Riple, & F. Powledge (Eds.), *Conserving biodiversity on military lands: A guide for natural resource managers* (pp. 2–33). Arlington: NatureServe.

TenBruggencate, J. (2003, November 23). Atoll still harbors toxic dangers. *Honolulu Advertiser*. http://the.honoluluadvertiser.com/article/2003/Nov/23/ln/ln12a.html. Accessed 4 Nov 2016.

The Iron Curtain Trail. (n.d.). *The Iron Curtain Trail: Experiencing the history of Europe's division*. http://www.ironcurtaintrail.eu/en/der_iron_curtain_trail/index.html. Accessed 3 Dec 2013.

Thomas, J. A. (2009). The Exquisite corpses of nature and history: The case of the Korean DMZ. Asia-Pacific. *Journal, 7*, 1–13.

Turkewitz, J. (2016, July 26). Tainted water near Colorado bases hints at wider worries. *New York Times*. A17.

Turner, T. (2005, November 18). Turn DMZ into a peace park. *USA Today*. http://usatoday30.usatoday.com/news/world/2005-11-18-turnerdmz_x.htm. Accessed 23 June 2016.

U.S. Fish and Wildlife Service. (2015). *National Wildlife Refuge System – About: Mission*. https://www.fws.gov/refuges/about/mission.html. Accessed 31 Oct 2016.

U.S. National Park Service. (n.d.-a). *Frequently asked questions*. https://www.nps.gov/aboutus/faqs.htm. Accessed 26 Sept 2016.

U.S. National Park Service. (n.d.-b). About us. https://www.nps.gov/aboutus/history.htm. Accessed 26 Sept 2016.

USDA. (2015, April 8). *Sentinel landscape partnership announced at Fort Huachuca and NAS Patuxent River-Atlantic test ranges*. U.S. Department of Agriculture News Release 0090.15. http://www.usda.gov/wps/portal/usda/usdamediafb?contentid=2015/04/0090.xml&pr table=true&contentidonly=true. Accessed 21 Oct 2016.

Vine, D. (2015). *Base Nation: How U.S. military bases abroad harm America and the world*. New York: Henry Holt.

Ward, C. (2015). *Bombing range is national example for wildlife conservation*. National Geographic Florida Wildlife Corridor (blog), http://voices.nationalgeographic.com/2015/04/01/bombg-range-is-national-example-for-wildlife-conservation/. Accessed 21 Oct 2016.

Weeks, J. (2009, Winter). From bombs to birds. *Defenders Magazine*, pp. 20–23.

Weisman, A. (2007). *The world without us*. New York: St. Martin's Press.

Woodward, R. (2004). *Military geographies*. New York: Wiley.

Chapter 10
From Buffer Zone to National Park: Afghanistan's Wakhan National Park

Peter D. Smallwood and Chris C. Shank

Abstract On March 30th, 2014, Afghanistan declared the Wakhan Corridor as its second national park. At over 10,000 km^2, the park is larger than Yellowstone National Park in the USA. It is high country, ranging from 2500 meters at its west end, to a mountain pass to China at 5000 meters in the east, and peaks of 7000 meters along its southern border. Despite its elevation, the Wakhan National Park is home to iconic wildlife species such as Marco Polo sheep and the snow leopards. It is also home to some 17,000 people. The Wakhan has had a long journey from geopolitical buffer zone to national park, a journey that is not yet complete. It became defined as a specific region during The Great Game of the nineteenth century between the two great empires of the age: Tsarist Russia, and the British Raj in India. The great powers wanted a buffer zone between them, an effort to keep their competition from accidentally spilling over into war. Neither the British, the Russians, nor the Afghan Emir could have known that in the twenty-first century, this buffer zone would come to be valued for its natural capital. While there were ceremonies to declare the park in 2014, it is not yet clear how the park will be managed. The park faces many challenges, but has great potential to preserve rare mountain habitats for the people who live there, and the world beyond its borders.

10.1 Introduction

One of the more prominent geographical features of Afghanistan is the long finger of territory reaching out from its northeast corner, called the Wakhan Corridor (Fig. 10.1). It is bordered by Tajikistan to the north, and Pakistan to the south with the extreme northeastern end touching a similar, though much smaller stretch from

P. D. Smallwood (✉)
Department of Biology, University of Richmond, Richmond, VA, USA
e-mail: psmallwo@richmond.edu

C. C. Shank
Cochrane, AB, Canada
e-mail: ccshank@gmail.com

T. R. Lookingbill, P. D. Smallwood (eds.), *Collateral Values*, Landscape Series 25,
https://doi.org/10.1007/978-3-030-18991-4_10

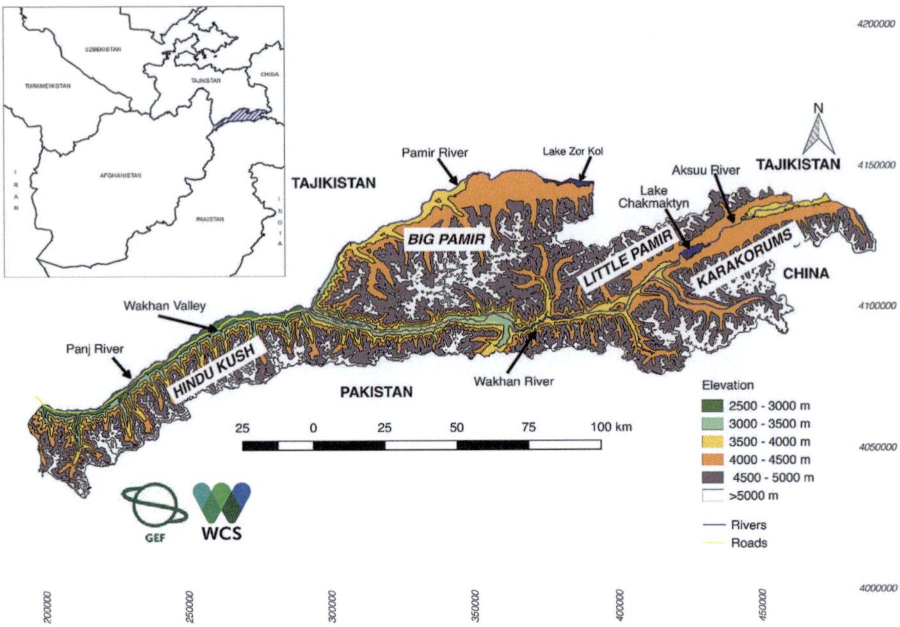

Fig. 10.1 Major geographic features and elevations of the Wakhan National Park

China. It became part of Afghanistan partially as a result of the Great Game between the British Empire in India and the Russian Empire in the 1800s. Both empires were expanding into Central Asia and each regarded the other as a dangerous competitor. The contest between them was not unlike the Cold War between the USSR and the USA in the second half of the twentieth century; two great powers, struggling for dominance and allies, fighting many small proxy wars, but carefully avoiding direct conflict. If Russia were to encroach on British territory in India, the most likely route was through the Wakhan. Thus, the British were keen to keep Afghanistan as a buffer and at the same time, to secure British India's border with Afghanistan. The resulting boundary delimitation resulted in the Wakhan Corridor, a narrow strip of land separating what was then Russia from British India. More than 100 years later, it became Wakhan National Park (WNP), Afghanistan's second national park. Here, we tell the story of how the Wakhan Corridor came to be part of Afghanistan in the nineteenth century, and briefly review its development to the present. We then describe the current conditions of the national park, and its potential to preserve ecosystem services, guide sustainable development, and protect mountain ecosystems.

The Wakhan Corridor comprises 10,950 km^2, larger than Yellowstone National Park in the USA. It is over 200 km in length and 17–63 km wide, lying between Tajikistan, Pakistan, and China (Fig. 10.1). There are four significant rivers (the Wakhan, Panj, Pamir and Aksuu Rivers) in the Wakhan, all of which ultimately flow into the Amu Darya (the Oxus River in older literature). The ecology of

Wakhan is determined largely by its high elevation and the very large changes in elevation (approximately 2512–7492 m = difference of 4980 m). The lowest-lying areas of the Wakhan are along the Panj River on the eastern part of the Corridor. The highest point is Mount Noshaq (7492 m) located in the extreme western corner of the Park. The Wakhan is occupied by two ethnic communities, the Wakhi and the Kyrgyz.

10.2 How the Wakhan Corridor Came to Be

To understand the origins of the current political boundaries of the Wakhan Corridor, we begin in the Great Game of the nineteenth century. In the early part of the century, successive Tsars of Russia expanded their domain from a small Duchy of Moscow into a vast empire, stretching from eastern Europe to the Pacific, and had begun to conquer territories to the south (Kollmann 2016). India was ruled by a private, for-profit company, the East India Company, based in London (albeit by this time, the British government largely controlled the company). British rule in India stretched from Burma in the east, to undefined borders with Afghanistan. The Company was expanding its dominion westward into those undefined territories (Lawson 2014). The British government regarded India as the most important of its colonial possessions, and was increasingly concerned that Russia intended to continue its expansion into India (MacMunn 1919).

In the first half of the nineteenth century, the British recognized Afghanistan as a separate nation ruled by an Emir, Dost Mohammad Khan. The British wanted to keep Afghanistan as a buffer between its Indian colony and the expanding Russian Empire. However, successive British governments vacillated between two strategies for how to accomplish this. The "forward" strategy advocated taking a very strong hand in the affairs of Afghanistan, while the "stationary" strategy was more restrained. In 1838, the Governor General of India became frustrated with the Emir over his reluctance to establish closer relations with the British and renounce contacts with Russia, and the forward strategy temporarily prevailed. The result was that Britain invaded Afghanistan in 1839, deposed Dost Mohammed and installed the unpopular Shah Shuja as a puppet ruler. It soon became clear that the Shah Shuja could not hold the throne without the continued occupation of British troops. The occupation would prove disastrous.

Afghans grew resentful of the occupation. In November of 1841, a rebellion began in Kabul with riots and the murder of a celebrated British diplomat. The rebellion grew over the next 2 months, forcing the British Army to surrender in January 1842, give up their weapons, and agree to evacuate the city. During the retreat from Kabul to Jalalabad, the Army was completely destroyed with the loss of virtually all 4500 troops and as many as 12,000 camp followers. This catastrophe strongly influenced British policy toward Afghanistan for the rest of the Great Game, including the establishment of her borders.

Following vengeful reprisals by the British, Afghanistan was returned to its own rule, albeit with agreements that its foreign affairs would be managed by the British, and particular warnings against contact with the Russians. Very little was accomplished in the first Anglo-Afghan war, and Afghanistan mostly returned to the status quo ante. In fact, the deposed Dost Mohammad Khan regained the throne and ruled Afghanistan for another 20 years (see Hopkirk 1990; Ewans 2002, and Dalrymple 2013 for engaging accounts of this Anglo-Afghan war).

During his second reign, Dost Mohammad re-established control from Herat in the west to Jalalabad in the east, but was unable to retake Peshawar, which aligned with and eventually became a part of British India. To the southwest, he reclaimed Kandahar, but could not take Quetta and eventually gave up on his ambition to bring the kingdoms of Baluchistan back under Afghan rule. As with Peshawar, they became aligned with British India as Princely States, leaving Afghanistan as a landlocked country (see Fig. 10.3 for location of these cities). While not yet mapped or precisely fixed, the southern and eastern border of Afghanistan began to take shape: an arc from the western border with the Persian empire, slowly curving northward to cut between Kandahar and Quetta, and between Jalalabad and Peshawar (Barfield 2010). However, the far northeast corner, including the Wakhan, remained completely undefined.

While Dost Mohammad consolidated his kingdom of Afghanistan, the Tsars of Russia continued their march to the south, approaching British India. They conquered Islamic Khanates in what is now Turkmenistan, Uzbekistan and Tajikistan, approaching Afghanistan and the far northwest corner of China (Sergeev 2013; Meyer and Brysac 1999). Meanwhile, the British tightened their grip on India. Princely States effectively came under the British Raj, and eventually Queen Victoria claimed the title Empress of India, to show the Russians that the United Kingdom was thoroughly committed to keeping her most profitable colony (MacMunn 1919; Lawson 2014). As the empires drew nearer, the Great Game became more dangerous, with Afghanistan caught in the middle (Fig. 10.2).

Tensions appeared to ease a bit in 1872, when Russia and Britain exchanged diplomatic letters recognizing Afghanistan as a neutral zone separating the two empires (Ewans 2002). Subsequent letters resulted in the Anglo-Russian Agreement of 1873, which established the border between Afghanistan and Russia in Wakhan as the Panj River and the Pamir River as far east as its source in what is now called Lake Zor Köl (Office of the Geographer of the U.S. Department of State 1983). This recognized at least part of the Wakhan Corridor as belonging to Afghanistan, however poorly mapped the region might be (Hopkirk 1990). Of course, these letters were between the great powers of Russia and the British Raj; the Emir of Afghanistan (now Sher Ali, son of Dost Mohammad) was not consulted.

Keen to accurately map the extent of this new buffer zone between themselves and the Russians, the British sent an expedition to the Pamirs in 1874 (Gordon 1876). In addition to now being Afghan territory, these mountains were rumored to be too high and rugged to permit much of an Army to pass from Russia to British Indian territory of Kashmir. The expedition made two distressing discoveries: first, they

Fig. 10.2 Punch was a weekly magazine of humor and satire, published in London during the Great Game era. Here, the Afghan Emir is trapped between the Russian bear and the British lion. (Public Domain)

found passes through these mountains that a determined army could march over, even hauling cannon with them. Second, there was a gap between the established border at Lake Zor Kül and the western border of Kashgaria (then a renegade province from China). Much of the “Pamir Gap,” variously described as 50 or 60 miles wide, was unoccupied and undefended (Skrine and Nightingale 1973; Shahrani 2002; Kreutzmann 2017), and authority over the area to the south as far as Chitral and Hunza was not established. If Russia claimed this territory, it would own passes through which it could attack Britain's most valuable “jewel in the crown,” India (Hopkirk 1990).

The Great Game continued for the next 20 years with exploration, intrigue, and diplomacy which has been adroitly captured by Hopkirk (1990) and Meyer and Brysac (1999), among others. British policy toward Afghanistan continued to flip between forward and stationary strategies, resulting in a second Anglo-Afghan war in 1878–1880. It was practically a replay of the first: “forward” with an invasion to install a more cooperative emir; occupation; the murder of a celebrated diplomat at the start of a rebellion against the occupation; embarrassing military defeat followed by punishing reprisals; and finally the British return to the “stationary” strategy. The army withdrew, leaving Afghans to rule themselves (albeit with the British in charge of Afghanistan's foreign affairs). Through it all, Britain remained anxious that Russia had designs on British India (Rowe et al. 2010).

That anxiety was piqued again in 1890, when a British agent named Francis Younghusband toured the area. He too found a "Pamir Gap" with strategic passes. He met with the Chinese governor of the region at Kashgar (the renegade province now back in Chinese control) and attempted to persuade him to extend his control westward to the Taghdumbash Pamir and areas further north, which would at least partially close the gap. Returning to the Pamirs to check on a rumor, Younghusband encountered a platoon of Cossacks on horseback in the Little Pamir, commanded by a Russian officer. While quite courteous, the Russian claimed *all* of the Afghan Pamirs now belonged to Russia, and Younghusband must leave. Eventually the Tsar's government claimed it was all a mistake and the overeager Russian officer had far overstepped his authority. But the event greatly heightened the tension between the empires (Skrine and Nightingale 1973; Hopkirk 1990; Shahrani 2002; Kreutzmann 2017).

On top of all that, the Emir of Afghanistan, Abdur Rahman Khan, was stirring up the frontier tribes in the Princely States in British India. All parties, British, Afghan, and Russian, decided that it was in their best interests to formally define the boundaries of their respective spheres of influence. Sir Mortimer Durand was dispatched to Kabul and reached an agreement with Rahman Khan on a roughly sketched line separating India from Afghanistan (as well as the short Pamir Gap between Russia and Afghanistan). The Durand Line, as it came to be known, stretched 2430 m from the Wakhan Corridor to the Persian border. The current border between Afghanistan and Pakistan roughly follows this line (Fig. 10.3).

However, this agreement would not be effective until the line could be surveyed on the ground, boundary markers established, and more detailed, accurate maps redrawn. Three boundary commissions were appointed to undertake on-the ground surveys south of the Wakhan (McMahon 1909). The northeastern boundary of the Wakhan portion proved especially difficult to survey, so a joint British/Russian boundary commission delineated the Russian/Afghan border in the "Pamir Gap" area east of Lake Zor Köl (Gerard 1897) to the Taghdumbash Pamir. Along the southern border, the height of land was defined as the boundary. Although this border separated Afghanistan from both British India and China, neither Afghan nor Chinese officials were involved (Rowe et al. 2010). When he was informed of his new possessions in the Wakhan, the Afghan Emir declared that he did not want these territories, but the combination of pressure and additional British subsidies convinced him to sign the agreement (Ewans 2002; Rowe et al. 2010). These borders, and indeed the entire Durand Line, remain a great source of controversy and tension between Afghanistan and Britain's successor, Pakistan, largely because of disagreement as to whether the Durand Line was intended to be a firm border between sovereign states or a more informal delineation of spheres of influence (Omrani 2009; Omrani and Ledwidge 2009; Leake and Haines 2017), and because it divided the Pashtun tribes between Afghanistan and Pakistan (Fig. 10.3).

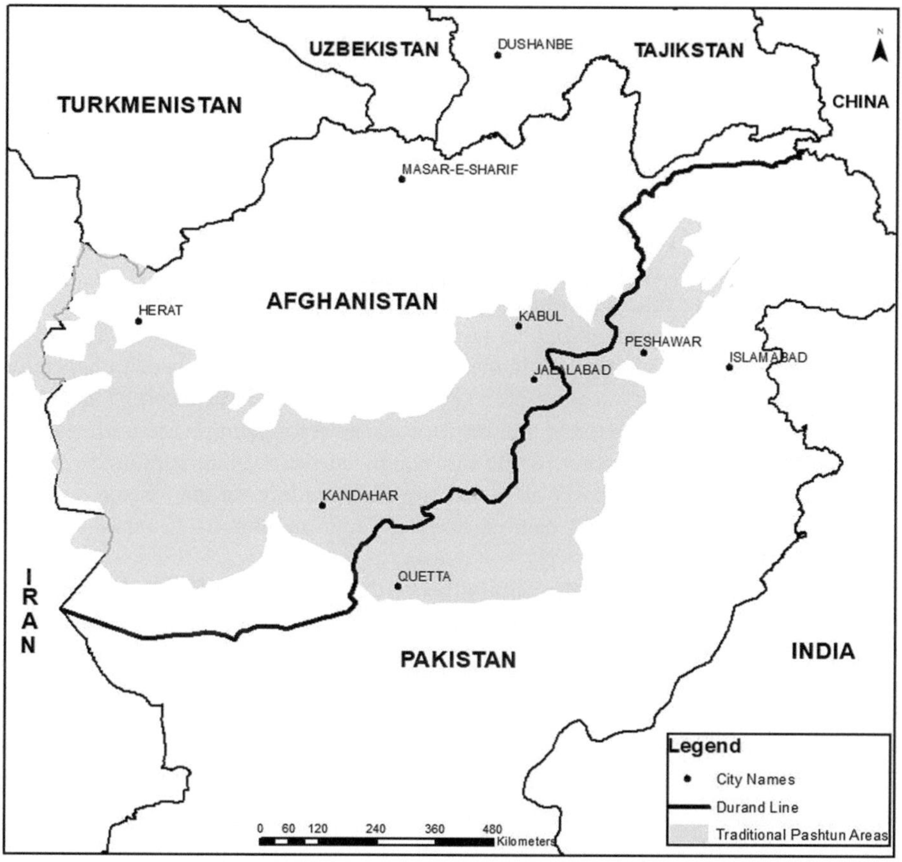

Fig. 10.3 Traditional Pashtun areas of Afghanistan and neighboring countries. Afghanistan's southern border (shown here as a heavy line) roughly follows the Durand Line, established in the 1890s. It passes through majority Pashtun areas, dividing majority Pashtun areas between Afghanistan and Pakistan

10.3 From Buffer Zone to National Park

The first half of the twentieth century reshaped the world. The "Great War" in Europe resulted in new boundaries being drawn across Europe, North Africa, and the Middle East. It also helped fuel the Bolshevik revolution in Russia. In 1917, the Bolsheviks toppled their government, and established the Union of Soviet Socialist Republics (USSR) in its place. The USSR reclaimed the former Russian Republics. Although many resisted, by the end of the 1920s, all of the republics along Afghanistan's northern border were subdued, including Tajikistan along the northern border of the Wakhan Corridor. The British Raj in India ended in 1947, resulting in two new independent states: India and Pakistan. Thus, the Durand line now separated Afghanistan from Pakistan all the way from Iran to its short border with China.[1] And in 1949, Mao Zedong led the communist forces to victory in China.

Despite these dramatic events all around the Wakhan, the shape and location of its borders remained essentially the same (Office of the Geographer of the U.S. Department of State 1969, 1983). However, the revolutions in Afghanistan's neighbors greatly changed the character of the borders. Initially, these borders were porous, and the Wakhi and Kyrgyz people travelled freely into neighboring countries. The seasonally nomadic Kyrgyz continued their traditional migrations into the USSR and the Wakhi continued trading in India and China. However, in the 1930s, the Soviets began to limit movement across the northern border. The Kyrgyz were perceived as anti-Soviet, resulting in punitive cross-border raids by the Soviets in which several Kyrgyz were killed. In the mid-1950s, the Chinese closed their border with Afghanistan, thereby sealing off a significant branch of the ancient Silk Road. The Soviets sponsored economic and institutional development on the north side of Panj River in Soviet Tajikistan, while the Wakhi on the Afghan south side of the river remained impoverished. Wakhi and the once widely-ranging Kyrgyz effectively became prisoners in a narrow strip of high mountains (Shahrani 2002; Callahan 2013).

While the politics of the Great Game attached the Wakhan Corridor to Afghanistan and drew its borders, the hardening of those borders in the twentieth century may have been a critical factor in preserving its mountain ecosystems. The short growing season in this high country could not have sustained a large ecological footprint. However, the isolation severely limited economic opportunities for the people of the Wakhan, and made it very difficult for outsiders to even visit, much less exploit the Wakhan's natural resources. Some of the terrible consequences of these limits on the Kyrgyz and Wakhi people are presented later in this paper. However, the isolation also limited their ecological impact on what is, after all, a relatively fragile environment. Thus, the ecosystems of the Wakhan remain relatively intact; in some places, almost pristine. These ecosystems and a few examples of the wildlife they support are also described below. We hope these ecosystems may now provide sustainable economic opportunities for the people, via the WNP.

It is easy to forget that for much of the twentieth century, Afghanistan was relatively stable and peaceful, albeit very poorly developed. A young King Zaher Shah took the throne in 1933, and held it for 40 years, developing a constitutional monarchy. A modest tourist industry grew, as stops in Afghanistan became part of "the hippie trail" (Gemie 2017). Photographs of Afghan cities from the 1950s and 60s contrast starkly with current impressions of the country (Taylor 2013; Ingersoll 2013). Even the very isolated Wakhan had some visitors: adventurous big-game hunters visited the Wakhan in those days, along with occasional trekkers and biologists (Habibi and Petocz 2011). Those days came to a sharp end with a bloody coup in 1978 (Ewans 2002).

---

[1] The easternmost part of the Wakhan Corridor borders the disputed region of Gilgit-Baltistan (formerly Northern Areas), claimed by both India and Pakistan but administered by Pakistan.

There were efforts toward conservation in the Wakhan during Zaher's reign. In the 1950s, the King ordered the upper Istimoch Valley in the western Big Pamir to be declared a Royal Hunting Reserve. In 1968, he granted use of the reserve to the Afghan Tourist Organization (ATO) to run a tourist-hunting program that continued until 1979. In 1978, four valleys comprising 67,938 ha were formally gazetted by the Afghan government as a Wildlife Reserve (Petocz 1978; Sayer and van der Zon 1981; Habibi and Petocz 2011). Throughout a period of 25 ± years, no livestock grazing was allowed in the Reserve. However, protection did not survive the decades of turmoil following 1979 and hunting by local people and grazing in the former Reserve resumed.

The Wakhan's iconic wildlife species are wide-ranging and do not respect political boundaries (Schaller et al. 1987). Consequently, the noted conservationist George Schaller resurrected the idea of a transboundary protected area, apparently first proposed in 1914 by a Russian named Tienshaski (Xie et al. 2007). In 2006, a meeting was held in Urumqi, China to develop a strategy to implement a transboundary Peace Park with participants from the governments of Afghanistan, Tajikistan, Pakistan, and China, as well as from non-governmental and donor organizations. A draft protected area map was created and agreed upon, delineating the borders of a four-country transboundary reserve (Fuller 2007; Schaller 2007; Xie et al. 2007). Unfortunately, political events in the region over the ensuing years blocked forward momentum for the plan (Schaller 2012).

Nevertheless, interest in the concept remains. Recently, the International Centre for Integrated Mountain Development (ICIMOD) has proposed a regional program called the Karakoram-Pamir-Wakhan Landscape Conservation and Development Initiative encompassing six protected areas in four countries. The intent is to provide a platform for policy makers, conservation managers, scientists, and local communities to discuss transboundary issues and exchange information (ICIMOD 2015). Other transboundary projects in the region that could either be linked to these initiatives or from which lessons could be learned are described in Rosen and Zahler (2016).

Afghanistan declared Band-e-Amir in central Afghanistan its first national park in 2009 (Smallwood et al. 2011). In 2014, it declared the entire Wakhan Corridor as its second national park.

10.4 The People of Wakhan National Park

The official government population estimate for the entire Wakhan District in 2012–2013 was 14,500 persons (Central Statistics Organization of Afghanistan n.d.). Other estimates put the Wakhi population alone at closer to 15,000 (Ostrowski and Rajabi 2017). The Kyrgyz are far fewer: 1100–1200 (Callahan 2013; Callahan 2016). By any estimate, the Wakhan is very sparsely populated, with less than 1.5 people per km^2.

The Wakhi are Nazari Ismaili Muslims located in permanent settlements in the Panj and Wakhan River Valleys. The Wakhi language is an old variant of Persian (Shahrani 2002). They are agropastoralists farming the bottomlands and lower areas of the Pamirs while grazing higher elevations in the Hindu Kush and Big and Little Pamirs between May and October. The primary crops grown are barley, wheat, flax, millet, lentils, beans, and grass pea. Most of their livestock is sheep and goats, with some cattle and yaks as well.

The Kyrgyz are a Turkic-speaking group of Sunni Muslims that first began to utilize the Afghan Pamir as summer range during the eighteenth century (Callahan 2013). There are two distinct populations of Kyrgyz in WNP; one in the eastern Big Pamir and one in the Little Pamir. Once nomadic, the Kyrgyz of the Little Pamir now undertake short seasonal migrations from the south side of the Wakhan in summer to the north side in winter, the "closed frontiers nomadism" of Shahrani (2002). The harsh climate of their high-altitude homelands precludes crop agriculture. Kyrgyz are completely dependent on livestock, trading livestock for all other needs (e.g., flour, salt, wood for framing their yurts).

In 1972–1973 there were about 1380 Kyrgyz living in the Little Pamir (Shahrani 2002). Following the bloody coup in Kabul in April 1978, about 1300 of them fled from the Little Pamir to Pakistan. The majority of these refugees in Pakistan emigrated to Turkey in 1982, although about 300 later returned to the Little Pamir (Callahan 2013). By 2007–2008, there were fewer than half the number of people occupying the Little Pamir compared to the early 1970s (Callahan 2013). By contrast, the Kyrgyz of the Big Pamir never left the area.

Because of the closed borders and rugged terrain, there is only a single, rough road extending only half the way up the Wakhan Corridor, leaving the people largely cut off from the rest of the world. Because of their isolation, they have received little assistance from the government or international aid organizations. Consequently, it is perhaps the poorest and least educated area in Afghanistan. For example, in the Big Pamir, Duncan and Duncan (2006) estimated Kyrgyz maternal mortality at 4000 per 100,000 live births, relative to a global rate of 216, and more than half of infants dying before their 5th birthday; far higher than the global average of 4.3% (UNICEF 2015).

10.5 Current Environment

Following Curzon (1896), a "pamir" is a large, plain-like mountain valley that has been filled with glacial debris and alluvium (although colloquially, the names Big Pamir and Little Pamir often refer to the mountain massifs as well). There are 7–8 pamirs located in Central Asia of which two are located in Afghanistan. The Big or Great Pamir is the flat, high-elevation area straddling the Tajikistan-Afghanistan border around Lake Zor Köl and the Pamir River (Fig. 10.1). The Little Pamir is the broad, 100 km-long valley of the Aksuu River flowing northeast and the Wakhan River flowing southwest. The Little Pamir is bordered on the south by Hindu Kush

Fig. 10.4 Meadow/Grassland habitat in the Wakhan Valley. (Photo by Ayub Alavi, ©Wildlife Conservation Society)

and Karakorum Ranges (Fig. 10.1). The proper noun "Pamir" also refers to the entire mountain range lying largely to the north of Afghanistan. Much of the landscape is visually striking (Fig. 10.4).

The Wakhan Corridor is far from moderating oceanic temperatures and moisture sources, resulting in a typical continental climate with cold winters, short, hot summers, and limited precipitation. The nearly 5000 m of vertical relief of the WNP has a large effect on both precipitation and temperature.

Meltwater from glaciers is a major source of water used by Wakhan communities for irrigation and domestic use as well as feeding the rivers that support downstream communities. Haritashya et al. (2009) evaluated satellite data for the Big Pamir and determined an average retreat rate of 36 m per year. They found many new glacial lakes to have formed recently and that existing ones have increased in size. Glacier meltwater lakes tend to be unstable and can cause downstream devastation from outburst floods.

Nearly 59% of the Wakhan is barren ground and over 14% is glacier. In total, less than 25% is vegetated with the most common vegetation classes being *Artemisia* (e.g., sagebrush, wormwood) types (11.9%) and alpine grass and forbs (6.5% of area). There are only 4.9 km^2 of forest/shrub mix in the entire area (Bedunah 2009). The consensus is that the western Big Pamir is heavily overgrazed and ecologically degraded and has been for many years. The Little Pamir has experienced less grazing pressure in recent years and the range is not as degraded as in the Big Pamir (Petocz et al. 1978; Bedunah 2008, 2009; Winnie and Harris 2007; Winnie 2009).

Teggermansu, at the extreme northeastern end of the Wakhan, has rarely been grazed in about 30 years and range conditions are essentially pristine (Winnie 2009).

According to the World Wildlife Fund's (WWF) Global Terrestrial Ecoregion classification, WNP is characterized by three separate global ecoregions other than rock and ice (Olson et al. 2001). The Pamir Alpine Desert and Tundra Ecoregion (PA1014), Karakoram-West Tibetan Plateau Alpine Steppe (PA1006), and Parapamisus Xeric Woodland (PA1322) are all classified as globally Vulnerable. None of WNP's ecosystems appear on IUCN's Red List of Ecosystems (http://iucnrle.org). Two areas are listed globally as International Bird Areas (IBAs) and are therefore considered as interim Key Biodiversity Areas (KBAs) (BirdLife International 2018). The Critical Ecosystem Partnership Fund (CEPF) is currently assessing the entirety of WNP according to the new KBA criteria as part of the larger Mountains of Central Asia biodiversity hotspot profiling (Critical Ecosystems Partnership Fund 2017).

10.6 Wildlife

The Wakhan does not host a wide diversity of species. There are currently 25 species of mammals known from the Wakhan, about 250 species of birds, four species of reptiles, possibly two amphibians, and only three species of fish. There have as yet been no systematic, comprehensive surveys for insects or plants.

Perhaps the most iconic species of the Wakhan is the snow leopard, *Panthera uncia* (Fig. 10.5). Snow leopards are arguably the world's most elusive and charismatic of the large cat species, and the subject of immense international interest by nature lovers and conservationists. More than any other single factor, snow leopards are likely to draw worldwide attention to WNP.

IUCN acknowledges the difficulties in estimating their population size with the most recent estimates for a global population of snow leopard adults at 7000–8000

Fig. 10.5 Wildlife of Afghanistan. Left, camera trap photo of radio-collared Snow Leopard. Right, bachelor herd of Marco Polo Sheep. Left, by WCS Snow Leopard Team; Right, by Ali Madad Rajabi. Both, ©Wildlife Conservation Society

(McCarthy et al. 2016). Based on these data, snow leopards were recently down-listed to Vulnerable on the IUCN Red List. Snow leopards are listed as nationally Endangered by the Government of Afghanistan, and are listed by the Convention on the International Trade in Endangered Species (CITES) as an Appendix I species, thereby prohibiting international trade in snow leopards and their products.

WNP represents the core population of snow leopards in Afghanistan, although there have been confirmed sightings elsewhere in Badakhshan Province and in neighboring Nuristan Province. The Wildlife Conservation Society began studying snow leopards in 2009, and currently estimates the Wakhan population at somewhere between 50 and 100 individuals (Moheb and Paley 2016). It is certain that Wakhan snow leopards' ranges span international borders.

Marco Polo sheep (*Ovis ammon polii*) is also an iconic Wakhan species (Fig. 10.5). Marco Polo sheep are generally considered to be a subspecies of argali (*Ovis ammon*), although recent analysis (Groves and Grubb 2011) considers them as a full species – *Ovis polii*. They are the longest-horned sheep in the world, having a horn length of up to 190 cm (Castelló 2016). Marco Polo sheep are long-legged relative to other argali subspecies and are adapted to move quickly through the high-altitude, rolling topography typified by the Pamirs.

Marco Polo sheep are particularly important as a "flagship species" for the entire Pamir ecosystem. Effective protection of Marco Polo sheep "…can help all species, and the local peoples, in this unique ecosystem to survive and prosper" (Schaller and Kang 2008). The gvernment of Afghanistan lists Marco Polo sheep as protected, which means that Marco Polo sheep cannot be legally harvested in Afghanistan. IUCN lists all argali subspecies as "near threatened" globally (Harris and Reading 2008).

Petocz et al. (1978) and Luikart et al. (2011) consider there to be three populations of Marco Polo sheep in the WNP; the Big Pamir, Little Pamir and Teggermansu populations. Luikart et al. (2011) demonstrated that there is considerable gene flow between Afghan and Tajik/Chinese populations of Marco Polo sheep. However, migration between the three Afghan populations (Big Pamir, Little Pamir and Teggermansu) is small enough that these populations can be considered as separate for management purposes.

10.7 Development of Wakhan National Park

The Environment Law, passed by the Afghan Parliament in 2007, gives authority over protected areas to the National Environmental Protection Agency (NEPA). Subsequently, NEPA delegated on-the-ground management responsibility and authority to the Ministry of Agriculture, Irrigation, and Livestock (MAIL) while retaining a policy and oversight role. The Wildlife Conservation Society (WCS) has been cooperating with the government of Afghanistan on research and management of the country's protected areas since 2005 (https://afghanistan.wcs.org).

In 2010, 577 km^2 of the Big Pamir was proposed as a Wildlife Reserve by NEPA, with a management plan approved in 2016. The Teggermansu Valley, at the extreme northeast end of the Wakhan Corridor, was proposed as a Wildlife Reserve in 2012, with a management plan approved in 2017. In 2014, NEPA declared the entire Wakhan District as a national park. A series of community meetings was held in 2015 in the Wakhan Valley to explain the initiative. Full consultations with all Wakhi clusters and the Kyrgyz of Little Pamir were carried out in 2016. These consultations included full participation by national and regional governments and local parties. A draft management plan was developed from those consultations, which would supersede the management plans for the two wildlife reserves. It is still under review at this writing (mid-2018).

The draft plan calls for a variety of zones in the park: Strict Protection zones where resource use would be prohibited; Conservation Zones where some grazing, fodder collection, and other uses of the land may be permitted when compatible with the conservation goals of the park; and a Settlement Zone in which private infrastructure is permitted.

Despite being called a "national park," the park will be managed as an IUCN Category VI, termed a "Protected Area with Sustainable Use of Natural Resources" (Dudley 2008). This type of protected area is intended to protect natural places, together with associated cultural values and traditional natural resource management systems. IUCN Category VI protected areas are generally large with most of the area in a natural condition with a portion of the area under sustainable, non-industrial use of natural resources. This is compatible with the category "managed resource protected area," defined in Afghanistan's Environment Law as "an area containing predominantly unmodified natural systems managed mainly for the sustainable use of natural ecosystems and to ensure long-term protection and maintenance of biological diversity, while providing at the same time a sustainable flow of natural products and services to meet community needs." It is also consistent with Afghanistan's National Protected Area System Plan and moves Afghanistan further toward meeting its commitments under the Convention on Biological Diversity (Johnson et al. 2012).

10.8 Benefits of Wakhan National Park to Local People

Although WNP has not been officially gazetted, protection of the Wakhan has been on-going since 2008 with the technical and financial assistance of the WCS. WNP is currently patrolled by "community rangers," contracted by a local organization, with funding support from WCS, and by "government rangers" hired as MAIL employees. All are Wakhi or Kyrgyz locals. In 2016, there were 43 community rangers (eight in Little Pamir, five in Big Pamir, 26 in the Wakhan Valley/Hindu Kush, four dedicated to snow leopard research) and 16 government rangers (M. Alavi, pers. comm., March 2017). WCS has equipped rangers with uniforms and basic equipment, but these will require upgrading and continual replacement.

Rangers have received some training, but lack sufficient supervision as a result of the large distances and poor communications.

Many other programs have been conducted; programs that may seem tangential to nature conservation but are deemed vital by improving the living standards of local people. A few examples follow: Local tinsmiths were trained on the construction of fuel-efficient stoves, and local women were trained in their installation and use. The stoves not only reduce the health consequences of smoke inhalation; they reduce the amount of woody brush each family needs to collect to prepare meals. Veterinarians and paravets were trained in administering vaccines to livestock, how to take samples from animals for necropsy, and in disease recognition for disease surveillance. Diseases pass both ways between livestock and wild herbivores; these efforts may improve health in both. Predator-proof corrals to protect livestock from wolves and snow leopards were developed and deployed, which reduces human-wildlife conflicts. Environmental Education Programs have been developed and local teachers trained in their delivery.

The potential for tourism in WNP may promote a virtuous cycle: tourists come to see beautiful landscapes, not denuded, eroded lands. They come hoping to see wildlife as well as people living a traditional lifestyle. They spend money to hire local guides and stay with or near local people, which in turn incentivizes local people for conservation. There are international companies offering guided tours into the Wakhan. Even though the Wakhan has to date been free from attacks by terrorists or insurgent groups, the overall security situation in Afghanistan is likely to limit the growth of tourism in the near term.

WCS has undertaken considerable research in the Wakhan including recording wildlife observations by rangers, mammal, bird, and fish surveys, snow leopard radio-collaring, and monitoring of rangeland condition, glacial recession, livestock numbers, distribution, and health, and human-predator conflict. WCS and several other NGOs have provided support of many aspects of community development (WCS Afghanistan 2018).

Another dimension to the collateral values that come with conservation are the developments of good governance practices for a civil society (Smallwood et al. 2011; Zahler and Schaller 2014). Afghanistan's first national park, Band-e-Amir, set a precedent for how Afghanistan's protected areas may be managed. The Band-e-Amir Protected Area Committee was established to guide management decisions for the park. It is composed of representatives from the local villages and government officials, including MAIL and NEPA. Each of the villages elected its representatives (Mohibbi and Cochard 2014). Villagers told us that this was their first experience with democracy, and with having their voice heard by the provincial and national governments. The draft management plan for WNP calls for a similar arrangement. However, the large distances and poor transportation and communications infrastructure will be a challenge for such a committee in the Wakhan.

This all seems promising for the people of Wakhan, most of whom live in very poor economic conditions and poor health. The park could be managed in a way that improves their grazing practices and other land uses, provides economic opportunities, and generally improves their livelihoods. It could lead to better governance locally and at the federal level. The development of the WNP could promote further development of these collateral values from this former buffer zone between nineteenth century great powers. However, the challenges are formidable.

10.9 Challenges

To put it simply, the insurgency in Afghanistan is still active, violent, and growing. The US-led military surge from 2010–2012 may have increased the government's control of the provinces for a time, but it did not last. By 2015, not only were attacks by Taliban and affiliated groups on the rise, but an affiliate of the Islamic State (ISIS) was operating in Afghanistan (McFate et al. 2015). Estimates of the proportion of Afghanistan threatened or influenced by the Taliban or other rebel groups vary widely, but most agree that it is growing (Sharifi and Adamou 2018; Mashal and Sha 2018; Cooper 2018). In April and May 2017, the Taliban attacked the Border Police at Zebak and Ishkashim, just outside the western boundary of WNP, but were repelled (McKay 2017). Afghanistan's economy has stalled; after years of rapid growth, GDP has been essentially flat since 2012 (World Bank 2018). This is due in part to the loss of economic opportunities associated with servicing the military surge and declining funding for development NGOs, and in part due to the reluctance of businesses to invest in a country with declining security (Economist 2017). As a consequence, Afghanistan is rated among the most fragile states in the world (Fund for Peace 2018).

These conditions would be a great challenge even to long-established, highly popular governments. Afghanistan's government is neither. A new government with a new constitution was established in 2004. The most recent presidential election in 2014 was bitterly disputed and an extra-constitutional office of Chief Executive was invented to prevent the dispute from devolving into a civil war: the power-sharing arrangement has not worked well (Fair 2018). President Ghani's legitimacy and authority has been openly questioned by Atta Muhammad Noor, who was the country's most powerful provincial governor and remains both a political and military force (DesMarais and Forrest 2018; D'Souza 2018).

As a consequence, the government is not functioning quickly or well. Areas of natural resource management and protected areas are not exempted from these problems. Thus, although the park was officially declared by NEPA and MAIL in 2014, at this writing in the summer of 2018, they have not approved a management plan. In fact, legal status of both WNP and Afghanistan's first national park, Band-e-Amir remain tenuous, as neither Afghanistan's Parliament nor President have approved them as parks, and they have not been gazetted (published as law) as called for in Afghanistan's Environment Law.

Even if the central government of Afghanistan allows local conservation efforts to continue in WNP, it faces other daunting challenges from outside Afghanistan. Chief among them is climate change. According to an analysis by the University of

Notre Dame Global Adaptation Index (ND-GAIN 2018) Afghanistan ranked in the top 10 countries as being most vulnerable to climate change and least prepared to deal with climate change. Projections suggest that median annual temperatures in the Wakhan will increase, relative to 1970–1999 conditions, by 1.8 °C, 3.5 °C, and 5.7 °C in the 2030s, 2060s, and 2090s respectively (McSweeney et al. 2012). This will increase glacial melt and evaporation, leading to less soil moisture and stream flow. These changes will shift Wakhan ecosystems in unpredictable ways and create more challenging conditions for crops, livestock, and wildlife.

For many years there has been talk of China building a road through the Wakhan linking China with Pakistan and/or Afghanistan. They have already built a road in western China reaching to within 10 km of the Afghan-Chinese border (Malik 2014). Construction of a road through the Little Pamir would open WNP to development pressures detrimental to Park values. Recently, there has been a Chinese military presence in Wakhan, presumably with the tacit agreement of the Afghan government, and persistent rumors that the Chinese intend to build a military base in the area (Toktomushev 2018). If WNP were deemed to be a legally sanctioned protected area under provisions of the Environment Law, such developments would require an Environmental Impact Assessment.

And yet, despite these challenges, we remain hopeful. Progress in conservation and human development continues to be made, however slow and fitful that progress may be. Our hope is tempered by a realistic assessment of the challenges facing the WNP, but hope persists. This narrow strip of land, which came into existence as a buffer zone between two competing empires in the nineteenth century, isolated in the twentieth century, home to two impoverished peoples and to a rich biota, may be able to develop collateral values of conservation.

Acknowledgements The authors would like to thank the Wildlife Conservation Society for providing us with opportunities to work for conservation in Afghanistan, and the entire WCS Afghan Team for their dedication to protecting the Afghan environment. We would especially like to thank Stephane Ostrowski for his comments on an earlier draft and Qais Sahar for assistance in locating suitable photos. Finally, we thank Holden Taylor and Amanda Waggoner of the University of Richmond's Spatial Analysis Lab for assistance in preparing the maps in this chapter, and Samantha Guss, University of Richmond Librarian for finding the political cartoon in Fig. 10.2.

References

Barfield, T. (2010). *Afghanistan*. Princeton: Princeton University Press.

Bedunah, D. (2008). *Rangeland assessment of the Wakhan Corridor study areas*. Unpubl Rpt, WILDLIFE CONSERVATION SOCIETY, Kabul, Afghanistan. https://afghanistan.wcs.org/Publications.aspx Accessed 19 July 2018.

Bedunah, D. (2009). *Description of Wakhan Corridor vegetation land classes delineated in the supervised land classification*. Unpubl Rpt, WILDLIFE CONSERVATION SOCIETY, Kabul, Afghanistan. https://afghanistan.wcs.org/Publications.aspx. Accessed 19 July 2018.

BirdLife International. (2018). http://www.keybiodiversityareas.org/site/factsheet/46714. Accessed 19 July 2018.

Callahan, E. M. (2013). *To the roof of the world: Power and patronage in Afghan Kyrgyz society*. PhD dissertation, Boston University Graduate School of Arts and Sciences.

Callahan, T. (2016). Khans, kings, communists, warlords and presidents: Afghan Kirghiz socio-economic strategies for extorting and extracting from the state. In H. Kreutzmann & W. Teiji (Eds.), *Mapping transition in the Pamirs* (pp. 79–93). Springer International.

Castelló, J. R. (2016). *Bovids of the world: Antelopes, gazelles, cattle, goats, sheep and relatives*. Princeton: Princeton University Press.

Central Statistics Organization of Afghanistan. (n.d.). *Settled population of Badakshan Province… 2012–13*. http://www.cso.gov.af/Content/files/Badakhshan(1).pdf. Accessed 18 July 2018.

Cooper, H. (2018, March 8). U.S. braces for return of terrorist safe havens to Afghanistan. *New York Times*, p. A8. https://www.nytimes.com/2018/03/12/world/middleeast/military-safe-havens-afghanistan.html. Accessed 19 July 2018.

Critical Ecosystems Partnership Fund. (2017). https://www.cepf.net/sites/default/files/mountains-central-asia-ecosystem-profile-eng.pdf. Accessed 19 July 2018.

Curzon, G. N. (1896). *The Pamirs and the source of the Oxus*. London: The Royal Geographical Society.

D'Souza, S. M. (2018). *Afghanistan: Battle of narratives, power contestations & cycles of violence* (ISPSW strategy series: Focus on defense and international security, issue no. 540). http://www.ispsw.com/wp-content/uploads/2018/03/540_DSouza.pdf. Accessed 19 July 2018.

Dalrymple, W. (2013). *Return of a king: The battle for Afghanistan* (pp. 1939–1942). New York: Alfred A. Knopf.

DesMarais, S., & Forrest, C. (2018, February 3). *Afghanistan political showdown between Ashraf Ghani and Mohammad Atta Noor*. Institute for the Study of War, Report. http://www.understandingwar.org/backgrounder/afghanistan-political-showdown-between-ashraf-ghani-and-mohammad-atta-noor. Accessed 19 July 2018.

Dudley, N. (Ed.). (2008). *Guidelines for applying protected area management categories*. Gland: IUCN. x + 86pp.

Duncan, A., & Duncan, E, (2006). *Report of survey of the Kirghiz community in the Great Pamir*, Wakhan District, Badakhshan Province, Afghanistan. Report prepared for Orphans, Refugees and Aid International (Afghanistan).

Economist. (2017, Januray 7). *Why states fail and how to rebuild them: Lessons from Afghanistan and South Sudan*. https://www.economist.com/international/2017/01/07/why-states-fail-and-how-to-rebuild-them. Accessed July 2018.

Ewans, M. (2002). *Afghanistan: A short history of its people and politics*. New York: Perennial.

Fair, C. C. (2018). Afghanistan in 2017: Another year of running in place. *Asian Survey, 58*(1), 110–119.

Fuller, S. (2007). Linking Afghanistan with its neighbors through peace parks: Challenges and prospects. In S. H. Ali (Ed.), *Peace parks: Conservation and conflict resolution* (pp. 291–312). Cambridge, MA: MIT Press.

Fund for Peace. (2018). *Fragile states index*. http://fundforpeace.org/fsi/. Accessed 17 July 2018.

Gemie, S. (2017). Cross-cultural communication and the hippy trail 1957–78. *Interventions, 19*(5), 666–676.

Gerard, M. G. (1897). *Report on the proceedings of the Pamir Boundary Commission*. Calcutta: Office of the Superintendent of Government Printing.

Gordon, T. E. (1876). *The roof of the world: Being a narrative of a journey over the high plateau of Tibet to the Russian frontier and the Oxus sources on Pamir*. Edinburgh: Edmonston and Douglas.

Groves, C., & Grubb, P. (2011). *Ungulate taxonomy*. Baltimore: Johns Hopkins University Press.

Habibi, K., & Petocz, R. (2011). *Journeys in the hinterlands of Afghanistan: The Golden years of nature conservation* (pp. 1971–1978). Los Gatos: Smashwords, Inc.

Haritashya, U. K., Bishop, M., Shroder, J. F., Bush, A. B. G., & Bulley, H. N. N. (2009). Space-based assessment of glacier fluctuations in the Wakhan Pamir, Afghanistan. *Climatic Change, 94*, 5–18.

Harris, R. B., & Reading, R. (2008). *Ovis ammon*. The IUCN Red List of Threatened Species 2008. e.T15733A5074694. https://doi.org/10.2305/IUCN.UK.2008.RLTS.T15733A5074694.en. Accessed 17 July 2018.

Hopkirk, P. (1990). *The great game: On secret service in high Asia*. London: John Murray, Publishers.

ICIMOD. (2015). *Karakoram-Pamir-Wakhan landscape conservation and development initiative*. http://lib.icimod.org/record/31172/files/KPWL_15.pdf. Accessed 25 July 2018.

Ingersoll, G. (2013, February 8). Afghanistan before the war. *Business Insider*. http://www.businessinsider.com/astonishing-photos-of-prewar-afghanistan-show-everyday-life-in-peaceful-kabul-2013-2. Accessed 17 July 2018.

Johnson, M. F., Kanderian, N., Shank, C. C., Rahmani, H., Lawson, D., & Smallwood, P. D. (2012). Setting priorities for protected area planning in a conflict zone – Afghanistan's National Protected Area System Plan. *Biological Conservation, 148*(1), 146–155.

Kollmann, N. S. (2016). *The Russian Empire 1450–1801*. Oxford: Oxford University Press.

Kreutzmann, H. (2017). *Wakhan Quadrangle: Exploration and espionage during and after the Great Game*. Wiesbaden: Harrassowitz Verlag.

Lawson, P. (2014). *The East India company: A history*. New York: Routledge.

Leake, E., & Haines, D. (2017). Lines of (in)convenience: Sovereignty and border-making in post-colonial South Asia, 1947–1965. *The Journal of Asian Studies, 76*(4), 963–985.

Luikart, G., Amish, S. J., Beja-Pereira, A., Godinho, R., Allendorf, F. W., & Harris, R. B. (2011). High connectivity among argali sheep from Afghanistan and adjacent countries: Inferences from neutral and candidate gene microsatellites. *Conservation Genetics, 12*, 921–931.

MacMunn, G. F. (1919). *Afghanistan: From Darius to Amanullah*. (1977 printing by Gosha-E-Adab, Quetta, Pakistan).

Malik, H. Y. (2014). Geo-political significance of the Wakhan Corridor for China. *Fudan Journal of the Humanities and Social Sciences, 7*, 307–323.

Mashal, M., & Shah, T. (2018, March 13). Taliban briefly take Afghan district as security worsens. *New York Times*, p. A8. https://www.nytimes.com/2018/03/12/world/a/sia/afghanistan-taliban-farah.html. Accessed 19 July 2018.

McCarthy, T., Mallon, D., Sanderson, E. W., Zahler, P., & Fisher, K. (2016). What is a Snow Leopard? Biogeography and status overview. In T. McCarthy & D. Mallon (Eds.), *Snow leopards* (pp. 23–42). New York: Elsevier.

McFate, J. L., Denaburg, R., & Forrest, C. (2015, December 10). Afghanistan threat assessment: The Taliban and ISIS. Institute for the Study of War. *Backgrounder*. http://www.understandingwar.org/sites/default/files/Afghanistan%20Threat%20Assessment_The%20Taliban%20and%20ISIS_3.pdf

McKay, H. (2017, May 11). Afghan soldiers claim small but important victory in Badakhshan. *Fox News*. http://www.foxnews.com/world/2017/05/11/afghan-soldiers-claim-small-but-important-victory-in-badakhshan.html. Accessed 28 July 2018.

McMahon, A. H. (1909). *Letters on the Baluch-afghan boundary commission 1896*. Calcutta: Baptist Mission Press.

McSweeney, C., New, M., & Lizcano, G. (2012). *UNDP climate change country profiles: Afghanistan*. http://www.geog.ox.ac.uk/research/climate/projects/undpcp/index.html?country=Afghanistan&d1=Reports. Accessed 20 July 2018.

Meyer, K. E., & Brysac, S. B. (1999). *Tournament of shadows. The Great Game and the race for empire in Central Asia*. Washington, DC: Counterpoint.

Moheb, Z., & Paley, R. (2016). Snow leopard status and conservation in Afghanistan. In T. McCarthy & D. Mallon (Eds.), *Snow leopards* (pp. 409–417). New York: Elsevier.

Mohibbi, A. A., & Cochard, R. (2014). Residents' resource uses and nature conservation in Band-e-Amir National Park, Afghanistan. *Environmental Development, 11*, 141–161.

ND-GAIN. (2018). Notre dame global adaptation index, country index. https://gain.nd.edu/our-work/country-index/rankings/. Accessed 19 July 2018.

Office of the Geographer of the U.S. Department of State. (1969, May 1). International Boundary Study: Afghanistan – China boundary. No 89.
Office of the Geographer of the U.S. Department of State. (1983 September 15). International Boundary Study: Afghanistan – USSR. No. 26, *REVISED.*
Olson, D. M., Dinerstein, E., Wikramanayake, E. D., Burgess, N. D., Powell, G. V. N., Underwood, E. C., D'Amico, J. A., Itoua, I., Strand, H. E., Morrison, J. C., Loucks, C. J., Allnutt, T. F., Ricketts, T. H., Kura, Y., Lamoreux, J. F., Wettengel, W. W., Hedao, P., & Kassem, K. R. (2001). Terrestrial ecoregions of the world: A new map of life on Earth. *Bioscience, 51*(11), 933–938.
Omrani, B. (2009). The Durand Line: History and problems of the Afghan Pakistan border. *Asian Affairs, 40*(2), 177–195.
Omrani, B., & Ledwidge, F. (2009). Rethinking the Durand line: The legality of the Afghan-Pakistani frontier. *The RUSI Journal, 154*(5), 48–56.
Ostrowski, S., & Rajabi, A. M. (2017). One health in Wakhan: a summary of health investigations carried out by WCS in the Wakhan District, Afghanistan (2006–2014). Unpubl Rpt, WILDLIFE CONSERVATION SOCIETY, Kabul, Afghanistan. http://stephane.ostrowski.free.fr/pdf/3-2017-002.pdf. Accessed 19 July 2018.
Petocz, R. G. (1978). Report on the Afghan Pamir, part 3: Management plan for the Big Pamir Wildlife Reserve. FAO, FO:DP/AFG/74/016 Field Document No 7.
Petocz, R. G., Habibi, K., Jamil, A., & Wassey, A. (1978). Report on the Afghan Pamir. Part 2: Biology of Marco Polo argali (Ovis ammon polii). FAO:DP/AFG/74/016 Field Document No 6.
Rosen, T., & Zahler, P. (2016). Transboundary initiatives and snow leopard conservation. In T. McCarthy & D. Mallon (Eds.), *Snow leopards* (pp. 267–276). New York: Elsevier.
Rowe, W. C., Diener, A. C., & Hagen, J. (2010). The Wakhan Corridor: Endgame of the great game. In A. C. Diener & J. Hagen (Eds.), *Borderlines and borderlands: Political oddities at the edge of the nation-state* (pp. 53–68). New York: Rowman and Littlefield Publishers, Inc.
Sayer, J., & van der Zon, A. F. W. (1981). *National parks and wildlife management in Afghanistan: a contribution to conservation strategy*. FAO Technical Report. FAO, Rome.
Schaller, G. B. (2007). A proposal for a Pamir international peace park. *USDA Forest Service Proceedings RMRS, 49*, 227–231.
Schaller, G. B. (2012). *Tibet wild: A naturalist's journeys on the roof of the world.* Washington: Island Press.
Schaller, G. B., & Kang, A. (2008). Status of Marco Polo sheep *Ovis ammon polii* in China and adjacent countries: Conservation of a vulnerable subspecies. *Oryx, 42*, 100–106.
Schaller, G. B., Hong, L., Hua, L., Junrang, R., Mingjiang, Q., & Haibin, W. (1987). Status of large mammals in the Taxkorgan Reserve, Xinjiang, China. *Biological Conservation, 42*(1), 53–71.
Sergeev, E. I. (2013). *The Great Game, 1856–1907: Russo-British relations in Central and East Asia*. Washington, DC: Woodrow Wilson Center Press.
Shahrani, M. N. (2002). *The Kirghiz and Wakhi of Afghanistan: Adaptation to closed frontiers and war*. Seattle: University of Washington Press.
Sharifi, S., & Adamou, L. (2018, Januray 31). Taliban threatens 70% of the country, *BBC finds*. BBC World Service. https://www.bbc.com/news/world-asia-42863116. Accessed 18 July 2018.
Skrine, C. P., & Nightingale, P. (1973). *Macartney at Kashgar: New light on British, Chinese and Russian activities in Sinkiang, 1890–1918*. London: Routledge.
Smallwood, P. D., Shank, C. C., Dehgan, A. O., & Zahler, P. (2011). Wildlife conservation in Afghanistan? Conservation projects multitask in conflict zones, blending development and conservation goals. *Bioscience, 61*, 506–511.
Taylor, A. (2013, July 2). Afghanistan in the 1950s and '60s. *The Atlantic*. https://www.theatlantic.com/photo/2013/07/afghanistan-in-the-1950s-and-60s/100544/. Accessed 19 July 2018.
Toktomushev, K. (2018, January, 17). China's military base in Afghanistan. *The Diplomat*. https://thediplomat.com/2018/01/chinas-military-base-in-afghanistan/. Accessed 19 July 2018.

UNICEF. (2015). *Levels & trends in child mortality*; Report 2015. https://data.unicef.org/wp-content/uploads/2015/12/IGME-report-2015-child-mortality-final_236.pdf. Accessed 19 July 2018.

WCS Afghanistan. (2018). Initiatives https://afghanistan.wcs.org/Initiatives.aspx and Publications. https://afghanistan.wcs.org/Publications.aspx. Accessed 23 July 2018.

Winnie, Jr. J. (2009). *Results of the Argali survey in the Big Pamir and Wakhjir Valley by the Marco Polo sheep survey team in 2008* (Unpublished Report). Kabul: WILDLIFE CONSERVATION SOCIETY. https://afghanistan.wcs.org/Publications.aspx. Accessed 19 July 2018.

Winnie, Jr. J., & Harris, R. (2007). *2007 Marco Polo argali research in the Big Pamir of Afghanistan* (Unpublished Report). Kabul: WILDLIFE CONSERVATION SOCIETY. https://afghanistan.wcs.org/Publications.aspx. Accessed 19 July 2018.

World Bank. (2018). *The World Bank in Afghanistan*. https://www.worldbank.org/en/country/afghanistan/overview. Accessed 19 July 2018.

Xie, Y., Kang, A., Wingard, J., & Zahler, P. (2007). *The Pamirs Transboundary Protected Area*. WCS Report on the 2006 International Workshop on Wildlife and Habitat Conservation in the Pamirs. Urumqi, China.

Zahler P, Schaller G (2014, February) Saving more than just snow leopards. *The New York Times*, p. SR4.

Chapter 11
Guantánamo 2.0: Transforming Gitmo into a Peace Park and Ecological Research Center

Joe Roman

Conservation is a basis for permanent peace.
– Franklin Delano Roosevelt, 1944

Abstract Cuba has a long history of environmental protection, with a network of more than 250 national parks and protected areas, and relatively high levels of fish biomass and marine biodiversity in marine parks that are unparalleled in the Caribbean. There is concern that the normalization of relations between the United States and Cuba might reverse the country's advances in ecological conservation. In this chapter, I propose an approach to protect Cuba's coastal ecosystems and enhance conservation and ecological research throughout the Caribbean. After helping Cuba fight for independence from Spain, the United States occupied the island in 1898. As part of the Cuban-American Treaty, which granted Cuba independence in 1902, the new country was required to rent Guantánamo Bay to the United States as a coaling and naval station, a perpetual lease that could be broken only by mutual consent. The present U.S. policy is that withdrawal from the base is not an option. Cuba insists on an unconditional return of the land as part of normalization. There is a third path that would benefit Cuba, the United States, and the rest of the world. Once the military prison at U.S. Naval Station Guantánamo Bay is closed, the entire base should be repurposed into a state-of-the-art research institution and peace park, a conservation zone to help resolve conflicts between the two countries. A first step in returning the land to Cuba, this model could unite both nations in joint management, rather than serve as a wedge between them. By bringing together Cuban, U.S., and international scientists, artists, and scholars, Guantánamo could help all countries meet the challenges of climate change, mass extinction, and declining coral reefs.

J. Roman (✉)
Gund Institute for Environment, Rubenstein School of Environment and Natural Resources, University of Vermont, Burlington, VT, USA

T. R. Lookingbill, P. D. Smallwood (eds.), *Collateral Values*, Landscape Series 25,
https://doi.org/10.1007/978-3-030-18991-4_11

11.1 Introduction

Cuba has approximately 5000 km of coastline, including coral reefs, mangroves, seagrass beds, and tropical wet forests. Long stretches of coast remain undeveloped, with relatively high levels of fish biomass and marine biodiversity that is unparalleled in the Caribbean (Whittle and Rey Santos 2006; Newman et al. 2006). At the same time, Cuba has transformed agricultural production: since the 1990s, when support from Russia ended along with the dissolution of the Soviet Block, it turned to low-input production methods, such as organic farming and urban agriculture.

But things are changing quickly. Just a few of years ago, few would have foreseen the 2014 reopening of the U.S. embassy in Havana or Obama's 2016 visit to Cuba. These changes are enticing, laying the groundwork for the end of 50 years of U.S. economic sanctions (the embargo or *bloqueo*) against the Caribbean nation. They have the potential to increase economic opportunities in Cuba, but conservation efforts and sustainable agriculture are both potentially threatened by the opening of U.S. tourism and business. Will pressure from U.S. companies such as the agricultural giant Cargill, Starwood Hotels, and Carnival Cruise Line catalyze large-scale changes on the Cuban islands, or will environmental legislation protect the nation's coastal ecosystems? All too often, biodiversity — native animals, plants, and fungi — becomes collateral damage on the road to economic development.

There are hopeful signs that the United States and Cuba can work together to conserve nature. An agreement between the two countries created sister sanctuaries in November 2015, including the Florida Keys National Marine Sanctuary and Guanahacabibes National Park on the Cuban west coast. The U.S.-based Environmental Defense Fund has been working with Cuba to help implement its first National Plan of Action for Sharks and Rays. *The Bulletin of Marine Science*, based at the University of Miami, recently published a special issue on Cuban marine ecology and conservation, after years of rejecting manuscripts by Cuban scientists without review under U.S. sanctions (Roman 2018). After decades of being cut off from the United States, the Cuban people will make environmental policy decisions, with the anticipated pressures of normalization. But there is one 116.5-square-km block of land and water on Cuba's coast where United States will play a direct role in coastal management: Guantánamo Bay.

The Guantánamo Bay Naval Base is unique among U.S. overseas naval stations (Fig. 11.1). A treaty signed by the United States and Cuba in 1903, after the war for Cuban independence from Spain, established the United States' right to maintain naval stations in Cuba. The U.S.A. continues to assert the validity of this treaty. Cuba claims that the treaty, signed when it was under occupation, is illegal and the area should be returned. The United States has made it clear that the future of Guantánamo Bay is not on the table in negotiations with Cuba (Holpuch 2015). At the same time, former Cuban president Raúl Castro demanded the return of the base, accepting nothing short of complete and unconditional withdrawal by the United States. Many former U.S. officials have acknowledged that the return of the

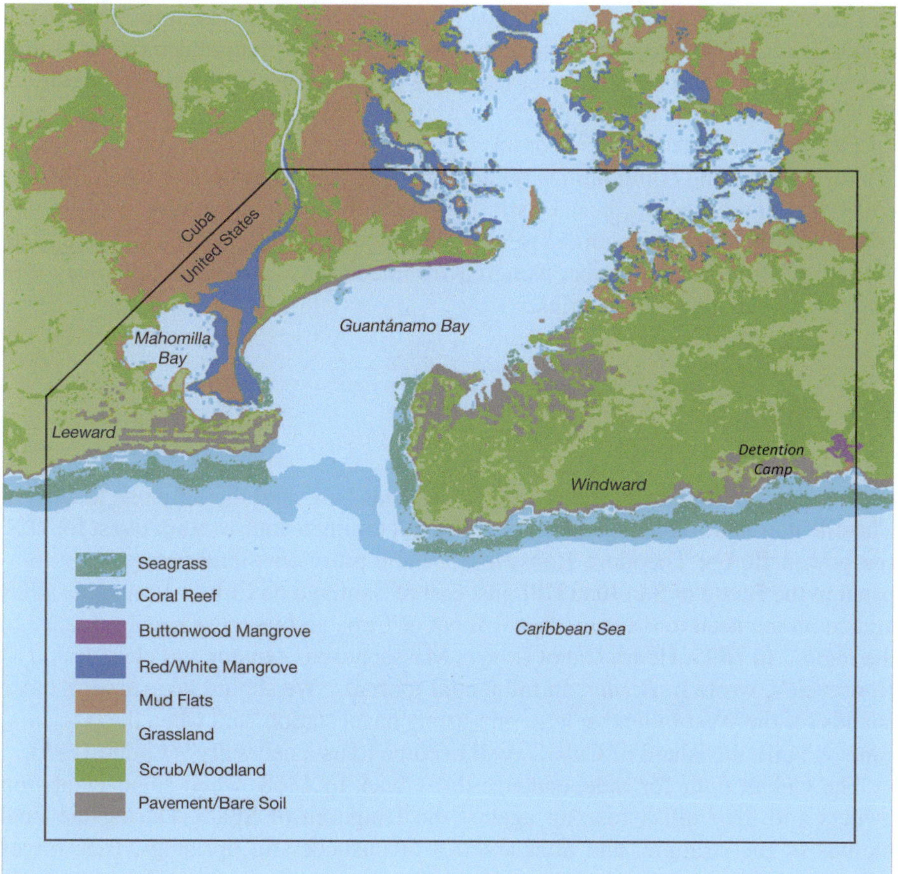

Fig. 11.1 Natural wetland and marine habitats on the Guantánamo Naval Base

base will occur at some point, though progress is unlikely until the remaining Guantánamo Bay detainees, foreign terror suspects captured after the attacks of September 11, 2001, are transferred, tried, or released. Despite the campaign rhetoric of Donald Trump, at this writing, 18 months into the new administration, there have been no new detainees sent to Guantánamo and one release (New York Times 2018). "It's probably inevitable that we'll have to give it back to Cuba," Admiral James Stavridis, the Supreme Commander of NATO between 2009 and 2013, has acknowledged, "but it would take a lot of diplomatic heavy-lifting" (Miroff 2015). He later added, "The odds of the U.S. needing the base for combat operations are essentially nil" (Stavridis 2015).

In this chapter, I present a third path for the future of Gitmo: converting the base into an international peace park and research station dedicated to international conservation and the enhancement of diplomatic relations between the two countries. The restoration of the Guantánamo Naval Base will require imagination, research, and good stewardship, an approach that recognizes the historical links between the countries, acknowledges the uses and tensions in the region, and provides a protected landscape and seascape that can serve as a hub for Caribbean conservation. As a first step, the restricted waters surrounding the base could be set aside as a no-take marine-protected area, conserving endangered species and acting as a nursery for surrounding communities reliant on fishing. A second phase would involve converting the base to a peace park, dedicated to protecting native wildlife and converting the military facilities to a research station. These moves would initiate the return of the area to Cuba, using environmental and scientific diplomacy as a bridge between the two countries.

11.2 History

Guantánamo is the United States' oldest overseas base, and the only one located in a hostile nation. Its roots are embedded in the late-nineteenth-century quest for U.S. sea power. Before Theodore Roosevelt became president—and before he participated in the Battle of San Juan Hill, just east of Santiago de Cuba—he and his allies agreed on the need to drive the Spanish out of Cuba and increase naval presence in the region. In 1895, Henry Cabot Lodge, Massachusetts senator and close friend of Roosevelt's, wrote in *Forum*, an influential journal, "We should have among those islands [of the West Indies] at least one strong naval station, and when the [Panama] canal is built, the island of Cuba… will become to us a necessity" (Lodge 1895).

The Cuban fight for independence dates back to 1868, when sugar-plantation owners and their allies rose up against the Kingdom of Spain. During the final months of the struggle, and after a series of unsuccessful uprisings, U.S. forces joined the Cubans, occupying Guantánamo Bay for use as a naval coaling station. In 1898, the United States and Spain signed the Treaty of Paris: Spain relinquished sovereignty over Cuba, and a U.S. Army officer was appointed as the provisional governor. As the Cubans gathered for their Constitutional Convention in 1900, the administration of President William McKinley pressed to establish naval bases on the island. The Americans were concerned about German plans—perhaps exaggerated in the interest of expanding U.S. influence in the region—to seize a base in the Caribbean, for a possible attack on the U.S. East Coast (Schwab 2009). After American gunboats surveyed the Cuban coastline, one commander claimed that the "best site in the Caribbean," was Guantánamo Bay, with a hundred miles of waterfront, a deep-water channel, and 25 square miles of open water.

In 1901, with American forces still in Cuba, the U.S. Senate passed the Platt Amendment, which included provisions that allowed the United States to access land for naval bases and interfere freely in Cuban affairs. Delegates at the Constitutional Convention were told that the United States would not withdraw its troops until the amendment was approved in its entirety. The amendment passed on a vote of 15–14, with a few minor changes and plenty of protests. (The United

Fig. 11.2 U.S. fleet in Guantánamo Bay, 1917 (Naval Historical Foundation)

States held firm to the exact words as passed by the Senate; the amendment was later passed as written.) Two years after the convention, a U.S.-Cuba survey team set the boundaries, and the U.S. Navy took control of Guantánamo Bay for a coaling and naval station, "and for no other purpose," according to the amendment. The rent was $2000 a year, to be paid in gold, and there was no cutoff date. A Cuban diplomat wrote about the U.S. fleet's occupation of the bay: "Now that they have seen Guantanamo, they will never renounce their control over it."[1]

Why Guantánamo, as opposed to one of the many bays and harbors around Cuba? The capacious bay opened onto one of the busiest sea lanes in the Western Hemisphere: the Windward Passage connects the Eastern Seaboard, Gulf of Mexico, and Central and South America. And the lanes were about to get even busier; the Panama Canal would open in 1914. It was also fairly uninhabited: the parched terrain had attracted few settlers since Christopher Columbus sailed into the bay in 1494, naming it Puerto Grande. (The name Guantánamo comes from the Taíno who occupied the region, and now refers to the base, city, and province.) In subsequent years, the remote location became a haven for pirates and runaway slaves. Soon after the U.S. Atlantic Fleet moved in, it began conducting gunnery and war-simulation exercises in the harbor (Fig. 11.2).

In 1934, U.S. president Franklin D. Roosevelt initiated his Good Neighbor Policy toward Latin America, which included renouncing the unilateral right to intervene in the internal affairs of other countries and abrogating the Platt Amendment in

[1] For more information, see Paul Kramer's article "A useful corner of the world: Guantánamo," *The New Yorker*, July 30, 2013; the Guantánamo Public Memory Project (http://gitmomemory.org/), and Stephen Irving Max Schwab's *Guantánamo, USA: The Untold History of America's Cuba Outpost* (University Press of Kansas, 2009).

Cuba. Although many Cubans and an American commission of foreign-policy experts called for a complete withdrawal of U.S. naval forces from Gitmo, the Roosevelt administration resisted. The Panama Canal and Caribbean were considered pivotal to the U.S. fleet in the buildup to World War II amidst growing concerns of a fascist invasion of the Western Hemisphere. The new treaty, signed by Cuba, allowed the U.S. to continue using Gitmo as a naval station, with Washington maintaining jurisdiction over the area. During the war, the base became a hub in the U.S. Caribbean convoy system; it was the second busiest port in the Western Hemisphere after New York—with more 17,768 ships in convoy during 1943 and 1944—and one of the safest, with extensive air patrols (Schwab 2009).

In the postwar years, the base was well funded, housing bowling alleys, ice-cream shops, and affordable servants. Historian Paul Kramer dubbed it, "Mayberry with iguanas." Perhaps, but it was a short drive to Cuba's biggest brothel: the nearby town of Caimanera offered prostitution, alcohol, and other trade that attracted $23 million a year from the base (Rodríguez Milán 2016). The relative tranquility would last until 1959, when Fulgencio Batista, a U.S. ally considered so corrupt even American officials predicted revolution as early as 1944, was driven from the country. The base was a source of friction between the United States and the new Castro regime—Fidel called Gitmo "a dagger plunged into Cuban soil"—and American officials wondered how the U.S. would respond if Cuba attacked the base. These concerns became escalated after the U.S. supported an invasion of Cuba in 1961. The invading force was beaten in 3 days at the Bay of Pigs, strengthening Castro's leadership, his antipathy toward the United States, and his ties to the Soviet Union. During the Cuban missile crisis, the Soviets positioned three ballistic missile launchers about 15 miles from the base in 1962. Pointed at Guantánamo, each of the missiles had a warhead with the explosive capacity comparable to the bomb dropped on Hiroshima. Secret negotiations between the United States and U.S.S.R resulted in the removal of all Soviet missiles and a public promise from the United States to respect Cuba's territorial sovereignty. No more invasions.

Tensions were still high in 1964, when Castro shut off the flow of potable water to Gitmo. Soon after, President Lyndon Johnson terminated the services of 1504 Cuban workers commuting to the base, two-thirds of the work force, reducing the flow of money into Cuba by $3 million. In February 1964, the Navy contracted Westinghouse International to construct three water desalination facilities to replace the water that had been piped in from nearby Yateras. By summer, water use returned to normal. The swimming pools on the base were filled by July 30 (Schwab 2009). When Castro doubted the veracity of these claims, Vice Admiral John Bulkeley, the base's commanding officer, invited reporters to join him at the Northeast Gate, where he cut the pipe connecting Gitmo to the Cuban water system. It was bone dry. The U.S. began jetting in food, and many Cuban laborers were replaced with workers from Jamaica and the Philippines.

In 1977, top-ranking Navy commanders and State department officials proposed using Guantánamo as a political bargaining chip to help normalize U.S. relations with Havana (Binder 1977). By the end of the Cold War, Guantánamo was on a target list of bases considered for closure (Moore 1990). But then it took on a new role. After Haitian president Jean-Bertrand Aristide was overthrown in a brutal coup d'état in 1991, hundreds of thousands of refugees fled the country. The U.S. Coast Guard rescued more 38,000 Haitians in international waters. Many were brought to Guantánamo for processing, where they were guarded by almost 2000 soldiers in prisonlike conditions and far from legal aid (Cushman Jr. 1994). The Justice Department chose the base, in part, because it contended that the rights and privileges granted by the U.S. Constitution did not apply to an overseas base (Schwab 2009). Ten thousand, seven hundred forty-seven Haitians were given refugee status, but many others were returned to Haiti. After a federal judge condemned the conditions at the camp—and the denial of Haitians' rights—the detention center was shut down. (It was reopened a year later, after lawyers for the detainees reached a deal with the Clinton administration to speed up their clients' release.) A new use for the base had been discovered: the storage of people (Kramer 2013).

On September 9, 2001, there were fewer than a thousand military personnel stationed on the base. The roads were in such bad shape that Humvee accidents were the major concern for the troops. The base was transformed after Al Qaeda's terrorist attacks on New York and the Pentagon, and the U.S.-led invasion of Afghanistan. In 2002, detainees were flown in, 20–30 at a time in shackles and black-out goggles, in an effort to eliminate any threats they might pose and to provide intelligence. As photographs with men in orange jumpsuits at Camp X-Ray—an outdoor prison that resembled a kennel complex—began to circulate, Gitmo became a lightning rod for the Taliban and terrorist organizations. Interrogation techniques were devised on the base by a resident psychologist and psychiatrist, who had been sent there to care for American troops; they included escalating tactics such as extended isolation, 20-hour interrogations, sleep deprivation, yelling, and hooding (Fink 2016). Interrogators used strobe lights, loud music, threats against family members, painful shackling, and sleep deprivation on some detainees. At least one detainee was sexually assaulted by female interrogators and beaten. Mohammed Al-Qahtani, suspected of being an intended hijacker on September 11, was menaced with military dogs, injected with intravenous fluids to make him urinate himself, and interrogated for 18–20 h at least 48 times. The Central Intelligence Agency employed the controversial use of waterboarding, a method of torture that involves water being poured into the nose and mouth of a victim lying on his back, with feet inclined above the head, on many prisoners. In 2005, the Amnesty International secretary general said, "Guantánamo has become the gulag of our times, entrenching the notion that people can be detained without any recourse to the law." Camp X-Ray was closed, but the United States held firm at Gitmo. Military contractors completed Camp Delta, a long-term facility constructed of shipping containers, at the southeastern corner of the base.

Since 2001, Guantánamo has held as many as 780 prisoners. The Military Commissions Act, signed by President George W. Bush in 2006, asserted that the base is beyond the jurisdiction of the U.S. legal system, including the Supreme Court. The policy attracted international and domestic criticism, becoming a high-profile topic during Barack Obama's 2008 presidential campaign. On his third day in office, Obama signed an executive order aimed at closing the detention center. In November 2009, Attorney General Eric Holder announced a plan to transfer Khalid Sheikh Mohammed and four co-defendants charged with planning and organizing the September 11 attacks, to face prosecution in federal court in New York. New York politicians and the public resisted, and the Obama rescinded the plan in 2011. His later attempts to move some detainees to federal prisons in the United States were blocked by Congress, which barred the use of federal funds to hold them on U.S. soil. It was easier to send detainees to their home countries or to third countries such as Spain and Bulgaria. Forty-eight men were placed under indefinite detention: even if they couldn't be prosecuted for past crimes, they were considered a threat to the security of the United States. In a 2013 speech, Obama acknowledged the toll that the prison had taken on the United States: "Gitmo has become a symbol around the world for an America that flouts the rule of law. Our allies won't cooperate with us if they think a terrorist will end up in Gitmo… There is no justification beyond politics for Congress to prevent us from closing a facility that should never have been opened."

Guantánamo has been costly, in money and manpower, and to human rights and American values. Many of those involved with the detention center at the base believe it should be shut down. Major General Michael Lehnert arrived at the base 3 months after the 9/11 attacks, as the commander charged with constructing and operating a short-term detention facility to hold Taliban and al Qaeda detainees. (His initial assignment was for 60 days.) Since retiring, he has become one of the facility's most strident critics, calling it "a blight on our history," one of more than 60 retired generals and admirals who have called for Guantánamo to be closed (Lehnert 2015).

By May 2018, there were 40 detainees held at GTMO (New York Times 2018). Previous estimates suggested that there were 33 guards and staff per detainee, at a cost of $445 million a year (Bruck 2016; Rosenberg 2016). The word Guantánamo had become synonymous with terror and incarceration. When I entered the word into Google, the two top images were the barbed wire entrance to Camp Delta and ten men in orange jumpsuits on their knees (Fig. 11.3).

Beyond Camp Delta, the primary missions of Guantánamo Bay are to serve as a strategic logistics base for the U.S. Navy's Atlantic Fleet, support counterdrug operations in the Caribbean, and process undocumented aliens for U.S. refugee status or repatriation to their home countries (Schwab 2009). Much of this mission could be met by Naval Air Station Key West, less than 900 kilometers away. Once the prison is closed, the most compelling reason for the Pentagon to possess the base will disappear. Guantánamo's mission could be transformed from its roots, in a nineteenth-century demand to expand the presence of warships, into an international response to the degradation of the oceans and the environment.

Fig. 11.3 Top two images for a Google search of the word "Guantanamo" in October 2016 (https://en.wikipedia.org/wiki/Guantanamo_Bay_detention_camp)

11.3 Natural Assessment

The Department of Defense is one of the largest land stewards in the United States—managing more than 10 million hectares, with more endangered species per hectare than any federal agency—and one of the biggest landowners worldwide (Cohn 1996). In managing Gitmo, the Defense Department must follow regulatory laws such as the Endangered Species Act, but protecting natural environments is also an important part of the military's mission, in that they provide important, realistic training grounds (Snider 2011).

What are the collateral values that could be protected and restored in transforming the base into a peace park and research facility? Because of its isolation from the rest of Cuba, the naval base at Guantánamo Bay supports a variety of threatened and endangered species, including reptiles, mammals, corals, and other marine life (Fig. 11.1). A rapid ecological assessment of the station conducted by ProAmbiente and The Nature Conservancy in the 1990s has helped inform the natural history of the area and my understanding of the base (Sedaghatkish and Roca 1999). The bay, which covers an area of about 36 km^2, divides the base into two parts. The eastern windward side is rugged with steep-sided hills and low coral plateaus. The leeward side is dominated by a flat floodplain, with extensive mud and salt flats surrounding the Guantánamo River. The shallow and deep oceanic waters to south of the base are also restricted, extending about two nautical miles from the shoreline. Much of the land and waters on the base are unoccupied, providing habitat for many native marine and terrestrial species.

11.3.1 Plants and Habitats

The U.S. Navy is the steward of an important array of natural communities on the base, including rare and endemic species (Table 11.1). The rapid ecological assessment found relatively pristine areas of several plant communities, including thorn scrub, coastal cactus scrub, and palm (*Coccothrinax*) forest (Sedaghatkish and Roca 1999). These Caribbean dry tropical forests are among the most endangered ecosystems on Earth, and they are relatively rare in Cuba, too. Of the 193 plant species that were identified at Gitmo, 5 are endemic to Guantánamo and the dry forest around the station, 48 are endemic to Cuba, and 19 are endemic to the Caribbean islands (Sedaghatkish and Roca 1999). Of particular interest are several cactus species that are endemic to the region, including *Dendrocereus nudifloris* and *Pereskia zinniiflora*. Coastal limestone terraces are typically sparsely vegetated, dominated by sand-fly bush (*Rachicallis americana*) and *Caribea littoralis*, a rare plant endemic to a few localities of southeastern Cuba.

The station also hosts important Caribbean coastal habitats, such as coral reefs, mangroves, sandy beaches, and seagrass beds. Coral reefs, seagrass beds, and mangroves are among the most important habitats for marine life at Gitmo, with coral reefs covering about 873 hectares, seagrass beds 434 hectares, and mangroves 557 hectares (Department of the Navy 2006). The rapid assessment of the reefs conducted in 1996 reported that coral cover was well above values reported for other Caribbean reefs, even though herbivores were in low abundance; *Acropora cervicornus*, the staghorn coral, was common and disease rare (Chiappone et al. 2001). In addition to these well-preserved areas, many other natural communities have been damaged by the construction of roads and facilities.

11.3.2 Reptiles

Guantánamo harbors a high number of amphibians and reptiles, many of which play important ecological roles on the base. More than 31 species have been recorded at Gitmo, a fifth of the diversity in Cuba (Sedaghatkish and Roca 1999). In this section, I focus on the base's reptiles, which have been more extensively studied than the amphibians.

Table 11.1 Threatened, endangered, and rare endemic species of Guantánamo Bay Naval Base

Species	Status
Caribea littoralis	Rare plant, endemic to southeastern Cuba
Dendrocereus nudiflorus	Endemic Cuban cactus found in five identified sites; endangered, IUCN Red List
Pereskia zinniiflora	Endemic Cuban cactus found in four identified sites; vulnerable, IUCN Red List
Cuban ground iguana (*Cyclura nubila nubila*)	Threatened, U.S. Endangered Species Act; vulnerable, IUCN Red List
Cuban boa (*Epicrates angulifer*)	Near threatened, IUCN Red List
Green turtle (*Chelonia mydas*)	Threatened, U.S. Endangered Species Act; endangered, IUCN Red List
Hawksbill turtle (*Eretmochelys imbricata*)	Endangered, U.S. Endangered Species Act; critically endangered, IUCN Red List
Loggerhead turtle (*Caretta caretta*)	Threatened, U.S. Endangered Species Act; Northwest Atlantic subpopulation considered least concern, IUCN Red List
Cuban gnatcatcher (*Polioptila lembeyei*)	Endemic, decreasing population size
Cuban tody (*Todus multicolor*)	Endemic, decreasing population size
Plain pigeon (*Columba inornata*)	Near threatened, IUCN Red List
White-crowned pigeon (*Patagioenas leucocephala*)	Near threatened, IUCN Red List
West Indian manatee (*Trichechus manatus*)	Endangered, U.S. Endangered Species Act and IUCN Red List
Queen conch (*Strombus gigas*)	CITES Appendix II (not threatened with extinction, but requiring trade control; harvest is illegal in Florida)
Pillar coral (*Dendrogyra cylindrus*)	Threatened, U.S. Endangered Species Act; vulnerable, IUCN Red List
Elkhorn coral (*Acropora palmata*)	Threatened, U.S. Endangered Species Act; critically endangered, IUCN Red List
Staghorn coral (*Acropora cervicornis*)	Threatened, U.S. Endangered Species Act; critically endangered, IUCN Red List
Rough cactus coral (*Mycetophyllia ferox*)	Threatened, U.S. Endangered Species Act; vulnerable, IUCN Red List

The Cuban iguana (*Cyclura nubila*) is common at Gitmo, where it is protected by the Endangered Species Act (Fig. 11.4). There are approximately 2000–3000 iguanas on the base, which could represent more than 5% of the total population in the country. Mainland iguanas outside of the base are declining at more than 1% per year. The most important habitat for the iguanas on the base is coastal limestone terrace, where they may be found at densities of more than 5 iguanas per hectare (Alberts et al. 2001). The density of iguanas along the coast of Gitmo is lower than in other undisturbed sites in Cuba, perhaps because of the many feral cats on the base, which are known to prey on juvenile iguanas (Gerber and Iverson 2000).

Fig. 11.4 The Cuban ground iguana is one of several endangered species at Gitmo. More than 5% of the known population is found on the base. (Drawing by Guantánamo courtroom sketch artist Janet Hamlin, commissioned by artist David Birkin for his project *Cyclura nubila* 2014)

In 2003, attorney Tom Wilner attempted to convince the U.S. Supreme Court to take the case of 12 Kuwaiti detainees being held in isolation at Guantánamo without access to a lawyer. He noted that when a Cuban iguana crosses the perimeter fence onto the base it becomes subject to U.S. law, and military personnel face fines of up to $10,000 for harming the animals. If the courts extended jurisdiction to include the iguanas while denying the detainees due process, Wilner argued, they would be providing more safeguards for the reptiles than for humans (Honigsberg 2009). The Supreme Court agreed to hear the case and later ruled in favor of the detainees. Even though Cuba held sovereignty over the land, U.S. jurisdiction over the base guaranteed the foreign nationals the same rights as if they were within the nation's borders.

The Cuban boa (*Epicrates angulifer*) is listed as near threatened by the IUCN. Widely persecuted in Cuba, especially near agricultural areas, it is considered a threat to domestic poultry and often killed on sight (Day and Tolson 1996; Matamoros et al. 1997). The base acts as a refuge for the species, as it is not hunted or persecuted at Gitmo. Cuban boas are active in spring and summer; they play an important role in regulating mammal populations, particularly hutias and bats (Alberts et al. 2001). Free from persecution, snakes grow larger on the base (up to 5 meters) and have bigger clutch sizes than in the rest of the country (Snider 2011).

The clearing of land in 1995 for Haitian and Cuban migrants—and the later use of the base for terrorism suspects—might have caused a decrease in the number of snakes, potentially reducing predation pressure on hutias and increasing their numbers.

The naval base is an important nesting area for the endangered green turtle (*Chelonia mydas*) and critically endangered hawksbill turtle (*Eretmochelys imbricata*). Blue, Windmill, and Pebble beaches on the windward side and AMC beach on the leeward side are among the most important nesting areas at the base. If Gitmo does becomes a peace park, these beaches—which account for 87% of the known nesting activity—should be managed to protect sea turtles (Alberts et al. 2001).

As the name of the nearby town of Caimanera suggests, crocodiles were once common in the area. The American crocodile (*Crocoldylus acutus*) was reportedly extirpated from Guantánamo Bay by the U.S. Navy by 1919 (King et al. 1982). Crocodiles were not observed during surveys conducted in the late 1990s, though naval station residents report seeing them (Sedaghatkish and Roca 1999).

11.3.3 Mammals

Mammalian diversity in Cuba is much lower than it is for other groups such as birds and fish. Of the 77 mammal species recorded on the island, about half are endemic and more than 80% have gone extinct (Woods and Eisenberg 1989). Bats have the highest diversity, with 33 extant and extinct species, followed by rodents, with 21, of which only 4 remain (Woods 1989). The largest native mammal is the Cuban hutia (*Capromys pilorides*), an endemic rodent that is uncommon in most of the country. Only found in the West Indies, hutias are threatened by overharvesting, invasive species, and habitat modification (Witmer and Lowney 2007). Thirteen hutia species have gone extinct in Cuba in recent times.

The Guantánamo Naval Base is exceptional, with widespread and abundant populations of *C. pilorides*, the only confirmed native terrestrial species found on the base. (Eight species of bats were also identified by the 1999 Rapid Ecological Assessment (Sedaghatkish and Roca 1999)). The Cuban hutia is primarily nocturnal, spending days in trees, and feeding on vegetation at night. Impacts from these high hutia densities include damage to landscaping, native vegetation, and vehicles; gnawing through cables; and the accumulation of feces in residential areas (Witmer and Lowney 2007). Since 2000, control efforts have included night spotlight shooting and occasional day shooting, though densities remain high, even in areas of intense population control. Future control efforts on the base should be mindful that the species is rare outside of Gitmo—mammologist William Kilpatrick noted that he saw more hutias during a couple of days on the base than during a month in the Caribbean—and it is likely important to the diet of Cuban boas (Alberts et al. 2001).

The area might provide a critical refuge for the West Indian manatee (*Trichechus manatus*) (Sedaghatkish and Roca 1999). Manatees were historically abundant in Cuba, especially in estuaries, river mouths, and freshwater springs, but their populations have declined after intensive hunting. They have been protected in Cuba since 1901, though poaching and fisheries entanglements continue to threaten the species. Manatees are expected to occur throughout the year at Gitmo, which provides potential foraging and resting areas, along with freshwater sources. A recent environmental DNA and tracking study of the naval station confirmed the presence of manatees in the region, with about 15–30 individuals using the bay (Hunter et al. 2018). The base maintains no-wake zones in high-use areas for manatees (Department of the Navy 2006).

11.3.4 Birds

A rapid ecological assessment of the base identified 101 species of birds, including 8 species that are endemic to Cuba. Several of these birds were considered of conservation concern by the Rapid Ecological Assessment, including the Cuban gnatcatcher (*Polioptila lembeyei*), Cuban tody (*Todus multicolor*), and plain pigeon (*Columba inornata*). All resident endemic bird species were found to be abundant on the naval station, with the exception of the Cuban vireo. The Cuban grassquit (*Tiaris canora*), which is common but declining near human settlements, has been observed in scrub, Phyllostylon-cactus forest, and mangroves on the base. Mangroves and tropical forests, globally threatened and harboring the greatest species richness of birds, are of highest conservation priority for this group at Gitmo (Sedaghatkish and Roca 1999).

11.3.5 Marine Species

Guantánamo is the largest and most important bay for marine life along the southeast coast of Cuba, which is comprised largely of rocky narrow cliffs and terraces. The mangrove shores, coral reefs, and seagrass beds on the base likely serve as critical recruitment and nursery sites for reef fauna. Although Santiago Bay, about 60 km to the west, could play a similar role, the city of Santiago has a population of more than 430,000; such a large population is likely to degrade marine resources through coastal development, overharvesting, and pollution. The marine communities in Guantánamo are better conserved because of its smaller human population and the restricted access to the base. Its relative isolation has helped preserve native ecosystems, such as coral reefs and seagrasses. Snappers and other fishes have been protected by no-take zones, such as Cuzco Beach, or areas that are off limits to divers because of military rules. Kittery Beach, for example, is closed because of its proximity to the detention camp. Yet, as an active naval base, Guantánamo has also affected the marine environment, especially through military activities and the over-harvesting of species such as queen conch (*Strombus gigas*). If the bay and surrounding uplands were protected, as a peace park and no-take area, many native species would likely flourish.

The naval station's coral reefs comprise a fringing reef system with well-developed spur-and-groove formations. According to researchers who conducted a rapid assessment in the 1990s, the predominance of corals on the base was surprising in light of a low abundance of herbivores and potential disease outbreaks on acroporid corals (Chiappone et al. 2001). A survey of the coral reef community found 72 benthic species of the phylum Cnidaria, including 44 stony coral taxa, octocorals, and gorgonians. Several species, now listed as threatened under the U.S. Endangered Species Act, were found on the base, such as the pillar coral (*Dendrogyra cylindrus*), elkhorn coral (*Acropora palmata*), staghorn coral (*Acropora cervicornus*), and rough cactus coral (*Mycetophyllia ferox*). A 2014 study of Guantánamo's reefs found heathy corals and gorgonians on the oceanic side of the base, but reefs exposed to the river plume in Guantánamo Bay, downstream from urban and agricultural areas in Cuba, were declining (Risk et al. 2014). International cooperation will be required to protect reefs exposed to upstream runoff.

Dive surveys recorded 92 fish species belonging to 29 families (Sedaghatkish and Roca 1999). This species richness is higher than for other nearby reefs and accounts for about a fourth of the total reef fish recorded in Cuba. Fishing in the waters on and near the base have been closed to Cubans since the early 1960s, but recreational harvest by U.S. military members and civilians has been occurring for decades. (A visitor map from the 1960s shows an entry point for the GTMO Reef Raiders on the windward side of the base.) One of the most important needs on the base is the enforcement of regulations protecting reef fish, queen conch, spiny lobster, and other species that are recreationally harvested (Sedaghatkish and Roca 1999).

11.3.6 Ecosystem Services

The ecosystem services provided by the Guantánamo Naval Base have yet to be quantified, but there are many benefits that likely flow from the bay, and these could be enhanced if the area is designated as a peace park and research station. The relatively intact mangroves, sea-grass beds, and coral reefs are expected to provide many of the services that they do in other areas, including the provision of finfish and shellfish, ocean recreation for military and staff on the base, flood control, and storm protection. It is essential that these interrelated ecosystems are protected and the interactions in this tropical seascape are maintained.

Among the many services provided by mangroves, the forests have extraordinarily high rates of primary productivity; they can sequester five times as much carbon as comparable land-based systems, and they are source of oceanic carbon

(Alongi 2002). Preserving and restoring them at Gitmo would help center the base in the fight to reduce climate change. Mangroves also serve as important nurseries for fish and crustaceans. In Cuba, mangroves are considered a valuable source for timber (especially buttonwood, *Conocarpus erectus*) and charcoal. They are also used as a source for honey, oysters, and shrimp (Sedaghatkish and Roca 1999).

The ecosystem services provided by seagrasses provide an estimated $1.9 trillion per year in the form of nutrient cycling globally; an order of magnitude enhancement of coral reef fish productivity; a habitat for thousands of fish, bird, and invertebrate species; and a major food source for endangered manatees and green turtles (Waycott et al. 2009). Seagrass meadows can support commercial fisheries worth as much as $3500 per hectare per year and subsistence fisheries for surrounding communities. Globally significant for the sequestration of carbon, seagrass meadows also play important roles as filters, improving water quality. Seagrasses and the organisms living in their leaves, for example, retain, filter, and actively capture suspended particles (Orth et al. 2006).

Coral reefs are essential to tropical fisheries, coastal protection, tourism, and recreation. They have high levels of biodiversity and are important for the migration of adult fish and invertebrates. These animals influence the productivity of seagrass beds, through grazing and nutrient export (Moberg and Folke 1999).

The ocean itself has been restorative for Americans and detainees. Swimming, scuba diving, and fishing are common forms of recreation for the military and civilians. One noncommissioned officer told me that the wildlife on the base helped him cope with his service. He became PADI-certified during his tour "inside the wire," and diving among the corals, octopuses, and barracuda helped spark a life-long passion for the marine environment. For the detainees, even a glimpse of the sea felt like freedom. After preparations for a coming hurricane prompted the removal of green tarps that blocked the ocean, one detainee recalled: "We all faced one direction: toward the sea… and the detainees started making art about the sea… I could see the detainees put their dreams, feelings, hopes and lives in them… the sea means freedom no one can control or own, freedom for everyone" (Adayfi 2017). After the detainees' work was exhibited, the US government declared the work US property subject to destruction.

One sustainable scenario for the future of the area would be to allow small-scale tourism in addition to research and artistic activities, with a focus on the natural and cultural history of the base. This transition would be easier for Gitmo than many other former militarized zones. Unlike the proposed Emerald Triangle Protected Forest Complex in Thailand, Laos, and Cambodia, which is riddled with landmines, the mines planted by the United States along the perimeters during the height of the Cold War have been removed. The Guantánamo Peace Park could also extend the network of protected areas in Cuba, including the Parque Nacionál de Gran Piedra and Parque Nacionál de Baconao, extensive natural areas that reach the outskirts of Santiago de Cuba. A coastal hiking trail extending from Cuba's second largest city to Guantánamo would help tie this area together.

11.3.7 Conservation Concerns

The future of Guantánamo as a thriving marine and terrestrial ecosystem depends on managing the landscape and seascape, with an awareness of past and future human activities. On land, there are several plant species, such as *Caribea littoralis* and the cactus *Pereskia zinniiflora*, that are rare and in need of conservation efforts at Gitmo. In general, the dry tropical forests and cactus scrub habitats found on the base are highly valuable and threatened throughout much of the Caribbean. They should be targeted habitats for conservation. The establishment of a no-take marine protected area in the restricted waters within and surrounding the base could serve as a refuge for reef fish and a source for nearby communities where harvest is allowed, and the area would also offer a refuge for many endangered species, such as sea turtles and manatees.

Several species on the base are reportedly under stress or overharvested. A 1996 survey found that recreational fishing is the most popular entertainment for military and civilian personnel (Bustamente et al. 2000). The spiny lobster (*Panulirus argus*), queen conch, and many reef fish populations are severely exploited and will require better management. The granadillo tree (*Brya ebenus*), which grows in cactus scrub and forest, has been overexploited for carving and craftsmanship at Gitmo and other areas of Cuba, resulting in severely reduced populations. Limits on harvesting this species has been recommended (Sedaghatkish and Roca 1999).

Invasive species are also an issue on the base. Lionfish (*Pterois* spp.) are common in Cuba and likely in Guantánamo, which despite its political isolation is not immune to marine invasions. Nonnative mammals include cats, dogs, goats, and white-tailed deer. Although there have been some efforts to remove these species, they remain on the base; feral cats are probably the biggest threat to the size and health of the iguana population (Alberts et al. 2001), and they presumably have an effect on other native species, such as Cuban boas, birds, and mammals. Such exotic species should be controlled and ideally eradicated from the area.

11.4 Future Scenarios

Guantánamo's future remains uncertain. More than 674 detainees have been released without charge, many after spending years in solitary confinement (Human Rights Watch 2016). At the time of this writing, 40 captives are in the Detention Center Zone, and visions for the base remain divided. Some in the United States think that Gitmo should remain open, as a legitimate and convenient place to keep suspected terrorists, and proceed with business as usual. In addition to incarcerating international terrorism suspects, the station supports other missions, including regional counterdrug operations, maritime migration interdiction, search and rescue, and humanitarian assistance, Other Americans – and much of the international community – strongly support closing the base and returning the land to Cuba.

The Cuban government and people appear united in the opinion that the base is a part of their national territory, held against their will. Many Cubans feel it should be returned as soon as possible, with no strings attached, an idea that is supported by the Community of Latin American States; CELAC has declared that the closing of the base should be part of the normalization process between the two nations (*Havana Times* 2015).

Transforming Gitmo into an international peace park would protect and restore the area's biodiversity, convert the active area into a research center, and provide ecosystem services to southeastern Cuba. Why would the Cuban government and its people accept anything less than an unconditional return of the land to Cuba? A jointly operated research facility and peace park would give global recognition to the country's conservation efforts. Cuba has many excellent scientists and a strong scientific tradition, but it has very little funding, especially for conservation. The nation has a relatively small research and development budget, with Cuban scientists earning about $36 per month (Stone 2015). As James Kraska, a professor at the U.S. Naval War College, and I proposed in the journal *Science*, the Guantánamo research park would provide financial support, up-to-date facilities for ecological and environmental work, and an opportunity to build capacity and train Cuban scientists and students (Roman and Kraska 2016). A poignant move would be to focus the interchange on residents from the surrounding eastern provinces. Far from Havana and the popular beaches of Varadero to the north, economic opportunities are fewer in the southeastern part of the country.

At the height of the Cold War, the U.S. placed more than 50,000 landmines along the perimeters of the base. Cuba planted cactuses, and landmines, on its side of the fence to discourage defectors. In 1996, President Bill Clinton ordered the demining of Gitmo, and landmines were removed by 1999. But the Cactus Curtain remains. The new relationship would reestablish communication between Gitmo and nearby communities—before the Cuban Revolution, thousands of local laborers earned their livelihoods on the wharfs, and in the machine shops and warehouse of the base—while avoiding the historical pitfalls of prostitution and military barriers. The Northeast Gate, long marking the division between Cuba and the U.S. compound, could provide a symbol for the transformation of the base, with Cubans returning to Guantánamo through the checkpoint after decades of exclusion (Fig. 11.5). Tearing down the 17-mile perimeter fence would be an act of ecological restoration and a symbolic gesture. The two countries could manage the area's wildlife jointly, rather than in isolation.

11.4.1 Research Center

Guantánamo could be central in expanding scientific exchange and promoting environmental cooperation between the United States and Cuba. Part of the developed leeward side of the base could become a Woods Hole of the Caribbean, housing research and educational facilities dedicated to addressing biodiversity loss, climate

Fig. 11.5 (**a**) The Northeast Gate of the Guantánamo Bay Naval Base (JTF Guantanamo file photo). (**b**) The gate could provide the first opening between Cuba and the U.S. naval base: an installation of photos of species native to the Guantánamo Bay region, with free-standing panels extending from the base checkpoint through the Cuban and U.S. borders (artwork courtesy Dave Hampton)

change, and ocean conservation. Laboratories for molecular genetics and geographic information systems could be built alongside videoconference rooms, and even art, music, and design studios, hosting scientists, scholars, and artists from Cuba, the United States, and around the world. The base could also provide facilities for captive breeding for endangered and endemic species such as Cuban crocodiles, corals, and bats.

Such a transition would be beneficial to the U.S. The military and detention facilities are economically, politically, and ecologically costly. Converting the base into a research station would extend a long tradition of U.S. naval support for scientific research and operational oceanography. The marine barracks, Joint Task Force headquarters, and other facilities could be refurbished, rather than torn down, to achieve conservation goals. The new facilities should strive to be carbon neutral, designed to reduce ecological damage to the surrounding marine and terrestrial ecosystems. Installing new renewable power systems could build on the precedent set by four 80-meter 950-kilowatt wind turbines completed in 2005, which now generate 2–3% of the base's power. (The turbines were planned before the terrorist attacks of September 11, 2001, when the base's energy demands were much lower.) The opening of the border would allow the flow of materials from the Guantánamo Province and other points in Cuba, rather than having most everything, including labor, flown in.

The research center would also be beneficial to Cuba and its neighbors. As a home for biodiversity and marine studies, Guantánamo could help stave the extinction crisis and work to protect the Caribbean from the increased degradation of coral reefs and marine fishes. The U.S. and Cuba could work together to restore native species and fight noxious invasives, such as lionfish (*Pterois* spp.), African catfish (*Clarius gariepinus*), and marabou (*Dichrostachys cinerea*). They could work with neighboring countries, such as Haiti, Jamaica, the Dominican Republic, and other Caribbean and Latin American countries, to restore their own coastal ecosystems.

11.4.2 Peace Park

During his 2016 visit to Cuba, Obama declared, "the embargo's going to end" (Liptak 2016). Later in the year, he eased travel regulations between the two countries. On August 31, Jet Blue became the first U.S. passenger airline to complete a commercial flight to Cuba in 50 years, from Fort Lauderdale to Santa Clara. There are now regular flights from New York, Miami, Houston, and other U.S. cities, although tourism remains restricted under the embargo. It is likely that resumed trade between the two countries will increase industrial agriculture and tourist development, after decades of slow growth and conservation in the archipelago.

The protection of Guantánamo as a peace park could help offset some of the ecological consequences of normalization between the United States and Cuba, offering an alternative path to rapid development. Gitmo could deindustrialize and demilitarize, joining the growing number of peace parks, such as the land along the corridor of the former Iron Curtain. The new European Green Belt could transform the continent and help species such as brown bears, imperial eagles, and lynx recover (Terry et al. 2006; Havlick, Chap. 9, this volume). As part of their peace agreement in 1999, Israel and Jordan established the Red Sea Marine Peace Park, an area under joint management where Israeli and Jordanian scientists and students come together to learn and protect shared coral reefs. After the U.S. military decommissioned Fort Clayton in Panama, the base was transformed into Ciudad de Saber (City of Knowledge), a government-sponsored complex that has attracted international scholars and the United Nations Development Program. Such international parks and decommissioned bases are signs that humans can work together, even after conflicts, and protect other species that share our planet.

Careful planning is important. As Saleem Ali notes in this book, peace can come at a cost to conservation (see Chap. 8 in this book). When the Cordillera del Condor Corridor was demilitarized, there was increased pressure to exploit the area. Guantánamo could experience similar pressures, with its ocean location and coastal infrastructure. As with other decommissioned military areas, the transformation of Gitmo should occur with the input and cooperation of local stakeholders—in this case, the people of the surrounding Guantánamo Province. Essential to this vision is understanding the spillover effect of ecosystem services beyond the base's borders. Research on the economic and ecological value of preserving the mangroves, sea grasses, and reefs for the nearby communities of Caimanera and Boquerón, and for the base, will help make the case for protecting the base and its surrounding waters. On a practical level, there is a need for updated botanical surveys, vegetation, and marine maps, most of which were completed before 9/11.

The restricted waters of the base should be converted to a no-take marine-protected area, and research conducted at Guantánamo could examine the impact of this change on endangered species and fisheries in Cuba. Protecting the coral reefs, sea turtle nesting beaches, and habitat for vulnerable birds and reptiles will ensure the future of these species. A high priority should be reintroducing critically endangered species. One potential candidate is the Cuban crocodile (*Crocodylus rhombifer*). Given that the American crocodile was extirpated from Guantánamo Bay about a century ago, the risk of hybridization between the two species would be lower here than in areas such as the Zapata Swamp (Milián-García et al. 2015). Providing an area for in situ and ex situ conservation around the bay could provide a much-needed refuge for this species, which is currently limited to just two locations, though whether there is enough habitat to support this species would have to be examined. Hutias, an important food source, are common; freshwater is not.

Isolation has played an important role in protecting some of Guantánamo's ecosystems, but there is no doubt that the military has had a large impact on the bay and its surrounding uplands. Ecological restoration will be an essential part of the future of the base, and managers could benefit from the experience of other military areas. As Machado and Hupy note (see Chap. 5 in this book), the battlefield of Verdun was described as a moonscape after World War I. The French forest ministry undertook a massive effort to restore the area, and it is now home to a wide diversity of wildlife and plants, with former shell holes and trenches providing unique ecological niches.

As in many U.S. military bases around the world, the local commanders compensate for isolation with a pumped-up Americanism. The base has two outdoor movie theaters, six dozen fast-food restaurants, and a miniature and nine-hole golf course (Toobin 2008). As the United States comes under increased pressure from Cuba and its allies in the Americas to close the base, it is essential to consider how these facilities can be repurposed to meet Cuban, Caribbean, and pan-American needs. In repurposing the base, it is important that we don't forget its past. The Guantánamo Public Memory Project, now housed at Columbia University and with extensive information available on line (www.gitmomemory.org), would be a logical partner in creating a site that recognizes, commemorates, and transcends the base's history.

Pope Francis and the Vatican played an essential role in establishing diplomatic relations between Cuba and the United States in 2014. They could help advance a peaceful future for Guantánamo. In the first papal encyclical on the environment, Pope Francis called for an ecological conversion and the widespread protection of biodiversity, remarking on the decline of coral reefs, "Who turned the wonderworld of the seas into underwater cemeteries bereft of color and life?" (Pope Francis 2015). We did, of course, through overfishing, deforestation, pollution, and burning fossil fuels. And we can turn it around.

11.5 Proposal Response

There have been several proposals for the postmilitary future of Guantánamo. These include converting the base into a hub for humanitarian relief work (Stavridis 2015) or a biomedical center for diseases associated with poverty in the Americas (Hotez 2008). Among the responses to our proposal for a peace park and research center, Admiral James Stavridis, the former head of the United States Southern Command, which oversees Guantánamo, called the proposal plausible. I considered this a rave. "Once something is plausible in a semi-democratic society," writer Adam Gopnik (2015) has noted, "it has a natural momentum toward becoming real." Senator Brian Schatz of Hawaii noted, "Guantánamo doesn't make us safer and costs millions of dollars and ought to be eventually closed." In contrast, Senator James Inhofe of Oklahoma responded, "That's the dumbest thing I ever heard. Why would we talk about a marine lab when we're trying to save American lives?" (Chemnick 2016).

Perhaps unsurprisingly, conservationists and scientists in the U.S. and Cuba have expressed support for the proposal, in the media and through personal communications. Author Elizabeth Kolbert (2016) remarked in *The New Yorker*: "The beauty of the latest proposal… is that it turns Guantánamo's historical liability—its isolation—into an asset… the notion of transforming the site of one of the world's most notorious prisons into a 'peace park' has an undeniable charm. As no less of an expert on reconciliation than Nelson Mandela once put it, the concept of the peace park 'can be embraced by all.'" Part of Robben Island, where Mandela and other ANC supporters were imprisoned, is now a protected penguin colony. Franklin

D. Roosevelt, who visited Gitmo twice as president, might have had similar thoughts, when he wrote to his Secretary of State, Cordell Hull: "Conservation is a basis for permanent peace." It was an idea presented to him by his friend Gifford Pinchot, whose work established the conservation and managed use of America's national forests (Pinchot 1947).

What is the likelihood that this proposal will become a reality? When I first presented it in 2015, the Obama administration had about 18 months to go, and the president had made it clear that he wanted the prison closed. But Obama's attempts had stalled, and the detainees remained. At the same time, Cuba insisted that the land should be returned to them with no strings attached, revealing little about its postbase aspirations. In discussions with Cubans during my visit to the country, there was strong support for the idea. Essential in the proposal was that the peace park was the first step in returning the land to Cuba. José Luis Perelló Cabrera, a professor in the faculty of tourism at the University of Havana wrote in an email that he considered the proposal "un magnífico proyecto," and was hopeful that there would be related opportunities for sustainable tourism. A Cuban official wishing to remain anonymous noted that it was a feasible idea, adding that "there are political complexities that cannot be ignored, for Cuba [the] return of this territory is a matter of sovereignty… and is not expected to accept a mediated way." The official thought that a two-step process, including an unconditional termination of the lease by the United States followed by the establishment of an area of shared administration, with Cuba retaining the land rights, could work.

Since that time, the Trump administration has shown little enthusiasm for improving relations with Cuba. And although the number of detainees has declined under Trump, despite his campaign promises to fill it back up, there are no public discussions of closing the base. I thought of the Cuban metaphor for long-term vision, *luces largas*, or high beams. The transformation of Gitmo will not happen overnight.

Kramer (2013) has noted that Gitmo has been cast, "as station and school, leverage and message, weapon and prison… essential to the United States' position in the world." The Guantánamo Peace Park and Research Center would add a new role: it would help salve memories of Cold War conflict, Cuban refugees, and 9/11 detainees. The park would serve as an exemplar, encouraging people across the world to convert military bases and conflict zones into areas of creativity and cooperation, while protecting the planet's biodiversity. The next generation should know Guantánamo as an inspiration, not a stain.

Acknowledgements I thank Todd R. Lookingbill and Peter D. Smallwood for encouraging me to submit this chapter. Many of the ideas were developed with James Kraska for our 2016 article, "Reboot Gitmo for U.S. Cuba research diplomacy," in *Science*. Influential early reviewers include Dan Whittle, Paul Kramer, and my steadfast editor, Debora Greger. Dave Hampton created images and helped envision the transition from base to park. David Birkin, Gillian Galford, Gillian Pirolli, and Sam Grubinger helped with figures, and David Evans provided background information. The Gund Institute for Environment, Paul Lattanzio, and Allison and Winthrop Pescosolido provided financial support.

References

Adayfi, M. (2017). In our prison on the sea. *The New York Times*, 15 September.

Alberts, A. C., Grant, T. D., Gerber, G. P., Comer, K. E., Tolson, P. J., Lemm, J. M., & Boyer, D. (2001). *Critical reptile species management on the U.S. Naval Base, Guantanamo Bay, Cuba*. Report to the United States Navy for Project No. 62470-00-5219.

Alongi, D. M. (2002). Present state and future of the world's mangrove forests. *Environmental Conservation, 29*, 331–349.

Binder, D. (1977, 30 August). Naval and civilian aides see Guantanamo base as declining asset. *New York Times*.

Bruck, C. (2016). Why Obama has failed to close Guantánamo. *The New Yorker*, 8 August.

Bustamente, G., Chiappone, M., Kelly, J. C., Lowe, A., & Sealey, K. S. (2000). Fish and fisheries in Guantánamo Bay, Cuba: Recommendations for their protection. *Proceedings of the 51st Gulf and Caribbean Fisheries Institute, 50*, 242–257.

Chemnick, J. (2016). The plan to shut Gitmo and turn it into a climate research lab. *Climatewire*, 18 March.

Chiappone, M., Sullivan-Sealy, K., Bustamante, G., & Tschirky, J. (2001). A rapid assessment of coral reef community structure and diversity patterns at Naval Station Guantánamo Bay, Cuba. *Bulletin of Marine Science, 69*, 373–394.

Cohn, J. P. (1996). New defenders of wildlife. *Bioscience, 46*, 11–14.

Cushman, Jr. J. H. (1994). U.S. force and Haitian refugees: A nervous wait. *New York Times*, 22 July.

Day, M., & Tolson, P. J. (1996). Chilabothrus angulifer. *The IUCN Red List of Threatened Species* 1996: e.T7815A12852846.

Department of the Navy. (2006). *Seasonality and distribution of marine life at U.S. Naval Station Guantanamo Bay (GTMO) and in the Guantanamo Operating Area (OPEREA)*, Norfolk, VA.

Fink, S. (2016). Where even nightmares are classified: Psychiatric care at Guantánamo. *New York Times*, 12 November.

Gerber, G. P., & Iverson, J. B. (2000). Turks and Caicos iguana, Cyclura carinata carinata. In A. C. Alberts (Ed.), *West Indian iguanas: Status survey and conservation action plan* (pp. 15–18). Gland: IUCN – the World Conservation Union.

Gopnik, A. (2015). Trollope trending. *The New Yorker*, 4 May.

Havana Times. (2015, 29 August). *CELAC calls for US-Cuba dialogue to include return of Guantánamo*.

Holpuch, A. (2015). US says future of Guantánamo Bay is not on the table in Cuba talks. *The Guardian*, 4 February.

Honigsberg, P. J. (2009). *Our nation unhinged: The human consequences of the war on terror*. Berkeley: University of California Press.

Hotez, P. J. (2008). Reinventing Guantánamo: From detainee facility to center for research on neglected diseases of poverty in the Americas. *PLoS Neglected Tropical Diseases, 2*, e201. https://doi.org/10.1371/journal.pntd.0000201.

Human Rights Watch. (2016). *Guantanamo, facts and figures*. https://www.hrw.org/video-photos/interactive/2017/03/30/guantanamo-facts-and-figures

Hunter, M. E., Meigs-Friend, G., Ferrante, J. A., Kamla, C. T., Dorazio, R. M., Diage, L. K., Luna, F., Lanyon, J. M., & Read, J. P. (2018). Surveys of environmental DNA (eDNA): A new approach to estimate occurrence in vulnerable manatee populations. *Endangered Species Research, 35*, 101–111.

King, F. W., Campbell, H. W., & Moler, P. E. (1982). Review of the status of the American crocodile. In *Proceedings of the 5th working meeting of the crocodile specialist group of the species survival commission of the international union for conservation of nature and natural resources convened at the Florida State Museum, Gainesville, Florida, U.S.A*, 12 to 16 August 1980.

Kolbert, E. (2016). Guantánamo: From prison to marine conservation peace park? *The New Yorker*, 17 March.
Kramer, P. (2013). A useful corner of the world: Guantánamo. *The New Yorker*, 30 July.
Lehnert, M. (2015). I helped create Gitmo. Now I want it shut down. *Politico*, 11 January.
Liptak, K. (2016). Obama tells Raul Castro: Cuban embargo is going to end. *CNN*, 21 March.
Lodge, H. C. (1895). Our blundering foreign policy. *Forum*, 8–17 March.
Matamoros, Y., Dulon, E. P., & Seal, U. (1997). Cuban conservation assessment and management plan report. *Conservation Breeding Specialist Group Newsletter, 8*(1), 14–15.
Milián-García, Y., Ramos-Targarona, R., Pérez-Fleitas, E., Sosa-Rodríguez, G., Guerra-Manchena, L., Alonso-Tabet, M., Espinosa-López, G., & Rusello, M. A. (2015). Genetic evidence of hybridization between the critically endangered Cuban crocodile and the American crocodile: Implications for population history and in situ/ex situ conservation. *Heredity, 114*, 272–280.
Miroff, N. (2015). Why the U.S. base at Cuba's Guantanamo Bay is probably doomed. *Washington Post*, 15 May.
Moberg, F., & Folke, C. (1999). Ecological goods and services of coral reef ecosystems. *Ecological Economics, 29*, 215–233.
Moore, M. (1990). Navy puts 94 bases on hit list. *Washington Post*, 25 April.
New York Times. (2018, May). *The Guantánamo Docket*. https://www.nytimes.com/interactive/projects/guantanamo
Newman, M. J. H., Paredes, G., Sala, E., & Jackson, J. B. C. (2006). Structure of Caribbean coral reef communities across a large gradient of fish biomass. *Ecology Letters, 9*, 1216–1227.
Orth, R. J., Carruthers, T. J. B., Dennison, W. C., et al. (2006). A global crisis for seagrass ecosystems. *Bioscience, 56*, 987–996.
Pinchot, G. (1947). *Breaking new ground*. New York: Harcourt Brace.
Pope Francis. (2015). *Laudato Si': Encyclic letter on care for our common home*. Rome: Vatican Press.
Risk, M. J., Burchell, M., Brunton, D. A., & McCord, M. R. (2014). Health of the coral reefs at the US Navy Base, Guantánamo Bay, Cuba: A preliminary report based on isotopic records from gorgonians. *Marine Pollution Bulletin, 83*, 282–289.
Rodríguez Milán, Y. (2016). *Scanning Guantanamo: Caimanera*. On Cuba: https://oncubamagazinecom/especiales/scanning-guantanamo/
Roman, J. (2018). The ecology and conservation of Cuba's coastal and marine ecosystems. *Bulletin of Marine Science, 94*, 149–169.
Roman, J., & Kraska, J. (2016). Reboot Gitmo for U.S.-Cuba research diplomacy. *Science, 351*, 1258–1260.
Rosenberg, C. (2016). Guards and staff outnumber captives 33 to 1 at Guantánamo prison. *Miami Herald*, 25 August.
Schwab, S. I. M. (2009). *Guantánamo, USA: The untold history of America's Cuban outpost*. Lawrence: University Press of Kansas.
Sedaghatkish, G., & Roca, E. (Eds.). (1999). *Rapid ecological assessment: U.S. Naval Station Guantánamo Bay Cuba*. Arlington: The Nature Conservancy.
Snider, A. (2011). Outside Guantanamo's Prisons, 'it's really a biologist's dream.' *New York Times*, June 17.
Stavridis, J. (2015). Reimagining Guantanamo Bay. *Huffington Post*, 9 March.
Stone, R. (2015). In from the cold. *Science, 348*, 746–751.
Terry, A., Ullrich, K., & Riecken, U. (2006). *The green belt of Europe—From vision to reality*. Glandt/Cambridge: IUCN.
Toobin, J. (2008). Camp justice. *The New Yorker*, 14 April.
Waycott, M., et al. (2009). Accelerating loss of seagrasses across the globe threatens coastal ecosystems. *Proceedings of the National Academy of Sciences, 106*, 12377–12381.
Whittle, D. J., & Rey Santos, O. (2006). Protecting Cuba's environment: Efforts to design and implement effective environmental laws and policies in Cuba. *Cuban Studies, 37*, 73–103.

Witmer, G. W., & Lowney, M. (2007). Population biology and monitoring of the Cuban hutia at Guantanamo Bay, Cuba. *Mammalia, 71*, 115–121.
Woods, C. A. (1989). The biogeography of West Indian rodents. In C. A. Woods (Ed.), *Biogeography of the West Indies: Past, present, and future* (pp. 741–798). Gainesville: Sandhill Crane Press.
Woods, C. A., & Eisenberg, J. F. (1989). The land mammals of Madagascar and the Greater Antilles: Comparison and analysis. In C. A. Woods (Ed.), *Biogeography of the West Indies: Past, present, and future* (pp. 799–826). Gainesville: Sandhill Crane Press.

Part IV
Synthesis

Chapter 12
Battlefields and Borderlands: The Past, Present and Future of Collateral Values

Peter D. Smallwood and Todd R. Lookingbill

Abstract Collateral values are the improvements and/or conservation of the natural capital of landscapes upon which societies conduct war and similar conflicts. This book presents case studies of actual or potential collateral values. The study of collateral values thus falls within the new field of Warfare Ecology, as introduced by Machlis and Hanson (Bioscience 58(8):729–736, 2008). Machlis and Hanson propose a "taxonomy" of warfare ecology effects, based on whether the activities that cause the effect are part of the preparations for war, the actual waging of war, or the post-conflict recoveries. In this chapter, we analyze each of the case studies presented in this book in light of the Machlis-Hanson taxonomy, noting that the landscape under study may be defined as a unit during one phase, and acquire its collateral values in a later phase. The evolution of warfare suggests that current and future wars are unlikely to define battlefields suitable for the development of collateral values. However, older battlefields with significant collateral values may be more likely to resist pressures for economic development, due to the historical and memorial significance of those landscapes. In contrast, tensions along borders and/or border disputes may continue to define new landscapes that can acquire collateral values. As protected areas of all types are under great pressure, we hope the recognition of collateral values may help conserve them.

12.1 Collateral Values

Natural capital is the value of the natural resources in a given landscape. It comprises the value of nonrenewable, extractable resources, and renewable resources, including the ecosystems and biodiversity inhabiting the landscape. In this sense,

P. D. Smallwood (✉)
Department of Biology, University of Richmond, Richmond, VA, USA
e-mail: psmallwo@richmond.edu

T. R. Lookingbill
Department of Geography and the Environment, University of Richmond, Richmond, VA, USA

T. R. Lookingbill, P. D. Smallwood (eds.), *Collateral Values*, Landscape Series 25,
https://doi.org/10.1007/978-3-030-18991-4_12

natural capital is a part of a country's portfolio of wealth. It is very difficult to measure, but the consensus is that natural capital is a substantial fraction of global wealth. Lange et al. (2018) estimate 9% of global wealth is in natural capital, but for low-income countries it represents nearly half their wealth.

If natural capital is a part of a country's portfolio of wealth, ecosystem services are analogous to the earnings from the ecosystems in that portfolio: the harvestable food and fiber, pollination, and cleansing of water and air, to name a few. While the value of ecosystem services is also hard to measure, two things stand out. First, the value is very large: by some estimates the value of ecosystem services globally is more than twice that of global gross domestic product (Costanza et al. 2014). Second, because of the precipitous decline in the renewable resources of natural capital, the value of ecosystem services has begun to decline (Costanza et al. 2014; ELD Initiative 2015; Shepherd et al. 2016). The United States Agency for International Development (the world's largest donor for promoting economic development in less-developed countries) recognizes the critical importance of ecosystem services. It provides guidance and support for estimating the value of ecosystem services, including the effects of proposed development projects in cost/benefit analyses of the proposals (Kashi et al. 2018).

The effects of war and similar conflicts are overwhelmingy negative for both natural capital and the ecosystem services they provide, through a wide range of mechanisms. These include direct destruction; motivating unsustainable exploitation to fund the conflict; and putting large populations of people to flight across landscapes that cannot adequately support them (Hanson et al. 2009; Gaynor et al. 2016; Daskin and Pringle 2018). This book in no way disputes those harms. However, it is also true that there are cases – however rare – where war preserves ecosystems and the services they provide, or protects the landscape well enough to allow it to regenerate its natural capital (Hanson 2018). We call this a collateral value (with thanks to Gary Machlis for the term). This book presents case studies of such collateral values created by conflicts between human societies, from many places across the globe (see Fig. 1.1). Given the rapid destruction of our natural capital, we cannot afford to ignore any cases where it is conserved, even those that come from war.

12.2 The Past and Present of Collateral Values

Machlis and Hanson (2008) introduced the term warfare ecology, and proposed a framework for conceptualizing the effects of war and similar conflicts on landscapes (elaborated in Machlis et al. 2011). They divided these effects into those caused by preparations for conflict, those created while waging war, and those caused during post-conflict activities. Below, we consider the cases in this book in light of the Machlis-Hanson framework, noting that the landscape may be defined as a battlefield unit during one phase, and acquire its collateral values in a later phase.

Part I of this book is an exploration of the collateral values created by historic battlefields. In a sense, they are all similar: the battlefield landscapes considered here were defined during the conflicts themselves. Their natural capital developed later in the post-conflict period, once these lands were set aside from development as memorials to battles and those who fought. One might argue that forts built by the American government during the conflict between native Americans and European American settlers who displaced them (Julian, Chap. 3) were created during the preparation phase of conflicts. However, that conflict was a long, ongoing affair. It began long before the forts were built and continued long afterwards. It thus seems more reasonable to consider these forts as created during the active phase of the conflict.

The border regions in part II are more variable in this analysis. For the Korean DMZ and the Cordillera del Condor protected areas between Ecuador and Peru, the creation of those borders marked the transition from warfighting to post-conflict activities, albeit in Korea it has been more of an extended, uneasy ceasefire. Much of the natural capital in the Korean DMZ developed by accident in the post-conflict period (Brady, Chap. 7). In contrast, the Cordillera del Condor, which was created in part to preserve biodiversity, has not lived up to its promise for sustainably managing the natural capital of the border region (Ali, Chap. 8).

The landscape of the Iron Curtain between the Warsaw Pact of countries allied with the USSR and the American-led NATO alliance was defined during the cold war (Havlick, Chap. 9). However, despite numerous proxy wars elsewhere on the globe, the two alliances never engaged in active fighting across that border. The same can be said of Wakhan buffer zone between the great powers of the nineteenth century, the Russian empire and the British Raj in India (Smallwood and Shank, Chap. 10). In that respect, one could argue that these border zones were defined during the preparations for conflict, and succeeded in preventing open warfare in the border region. Since the end of the cold war, Europe has been developing the natural capital of its Iron Curtain border. In 2014, Afghanistan declared the Wakhan border region to be its second national park; we hope this designation will help better preserve and sustainably manage the natural capital of that region.

Finally, there is the U.S. Naval Base in Cuba, Guantanamo (Roman, Chap. 11). If Cuba was an outpost for the Warsaw pact, then Guantanamo Naval base became a fort for America on the border of that outpost (albeit it was originally established long before the cold war). Like the Iron Curtain, it was created during the preparation phase of conflict. Roman's proposal for the development and preservation of the area's natural capital by declaring it a biodiversity park is a very hopeful view of the potential for environmental peacemaking.

12.3 The Future of Collateral Values

The future of the kinds of collateral values described here depend upon both protecting what we already have and on the production of additional landscapes of collateral values. Protecting the landscapes we already have is no small matter; Jones et al. (2018) estimate that one third of the world's protected areas are under intense human pressures. We hope that protected areas that originated as collateral values from conflicts have a better chance of resisting those pressures, in part because of their origin. The desire to honor their histories and memorialize the human suffering on those landscapes may buttress efforts to preserve them as effective protected areas.

We think it is unlikely that many new landscapes of collateral values will be created from battlefields of current and future conflicts. The nature of warfare has changed. Battles are no longer fought with massive armies clashing in large, well-defined battlefields as typified by the America Civil War battlefield of Antietam (Lookingbill et al., Chap. 6) or the World War I battle at Verdun (Machado and Hupy, Chap. 5). The battles of today are smaller, and most often take place in cities, as typified by recent battles with the Islamic State in Iraq and Syria (Maurer 2018). These battlefields are rather unsuitable for the establishment of natural landscapes that could provide ecosystem services: the survivors of these conflicts strive to rebuild their cities. Remote natural landscapes of forest and mountains may continue to serve as retreats for insurgent forces, and battles may continue to take place in these landscapes when their adversaries discover their locations. However, those battles will be small in scale. Therefore, they are also unlikely to be good candidates for the establishment of protected areas large enough to provide significant ecosystem services. The era of massed armies of hundreds or even tens of thousands of fighters assembled for set piece battles across a well-defined battlefield seems to be over (Latiff 2017; Wright and Thompson 2017). As noted military historian Andrew Bacevich (e.g., Bacevich 2016) commented, "The range, accuracy, and lethality of modern weapons have made massed conventional armies obsolete" (Bacevich, pers. comm. to PDS, 2016). Military historians Douglas Macgregor (Macgregor 2016) and Roger Cirillo (Cirillo 2004) concur (pers. comm. to PDS, 2018). Thus if we are to develop new natural capital from battlefields, it will be from old battlefields coming under new protections (Sibilia et al., Chap. 3).

However, there are many border disputes between parties in conflict, in many parts of the world. Walker (2011) reviewed 13 of the most dangerous border disputes of the time; as of this writing, none of them have been resolved. Many of these borders cut through landscapes that provide valuable ecosystem services: between Sudan and South Sudan; Cambodia and Thailand; Angola and the Democratic Republic of Congo; Pakistan and India; and Afghanistan and Pakistan. The latter is particularly interesting, as it is not a particular section of this border: Afghanistan disputes the entire length of the border (see Smallwood and Shank, Chap. 10). There are many other borders disputed besides those on Walker's list. China has border disputes with India, Pakistan, Vietnam, and others. In many cases, these disputes are driven partly by national pride and associated politics. These may provide opportunities for environmental peacemaking based on better management of the border region's natural capital. That could provide a face-saving way for the parties to agree to the border (Ali, Chap. 8; Roman, Chap. 11). However, we must take to heart the dangers. McNeely (2003) provides cases where peace came with notable increases in unsustainable exploitation of natural capital. Ali (Chap. 8) shows that even when peace is achieved explicitly by means of environmental peacemaking, sustainable management of the landscape may fall by the wayside.

Even settled borders may provide opportunities. The border between Iraq and Iran is not in dispute (at least, not at this writing), but substantial stretches have remained undeveloped and relatively unexploited due to the dangers of old minefields. This may be a contributing factor to the persistence of native wild goat and sheep, and even Persian leopard (*Panthera pardus tulliana*) in the border region (Schwartzstein 2014; Avgan et al. 2016). There have been efforts to create biodiversity parks along the border (Gies 2018), but current tensions between Iraqi Kurdistan and Iran have led the Iraqi wildlife biologists to shift their conservation efforts away from the border for now.

On September 11, 2001, terrorists hijacked four commercial airlines in the United States, turning them into weapons for mass casualties. Two of them were crashed into the iconic Twin Towers of the World Trade Center in New York City, and one was crashed into the Pentagon just outside of Washington, D.C. The fourth flight, United Airlines #93, did not reach its target, presumed to be the U.S. Capitol Building. A group of passengers attacked the hijackers in an attempt to take control of the flight, and in the ensuing struggle, UAL flight 93 crashed into the countryside in rural southern Pennsylvania (9/11 Commission 2011).

The crash was located on a site of active coal strip mining. In the aftermath of the crash, it was decided to cease mining activity and establish a memorial to those who died on flight 93. The Flight 93 National Memorial now includes a visitor center, an observation platform, a memorial wall with the names of the crew and passengers of the flight, and at this writing, 2200 hectares (~8.5 square miles) of land, now protected from development (Fig. 12.1). Remediation of pollution from acidic mine drainage has been a challenge, but has also created a remarkable partnership of families of passengers and crew aboard the flight, local people, the National Park Service, and other federal organizations (Emili et al. 2016). They are creating new natural capital on the site of this tragedy.

If collateral values can be developed even from such a senseless act of terrorism, surely we can wrest such values from other landscapes of war. We hope the examples in this book inspire readers to support efforts to create and preserve the natural capital found on old battlefields and old and new borderlands of conflict.

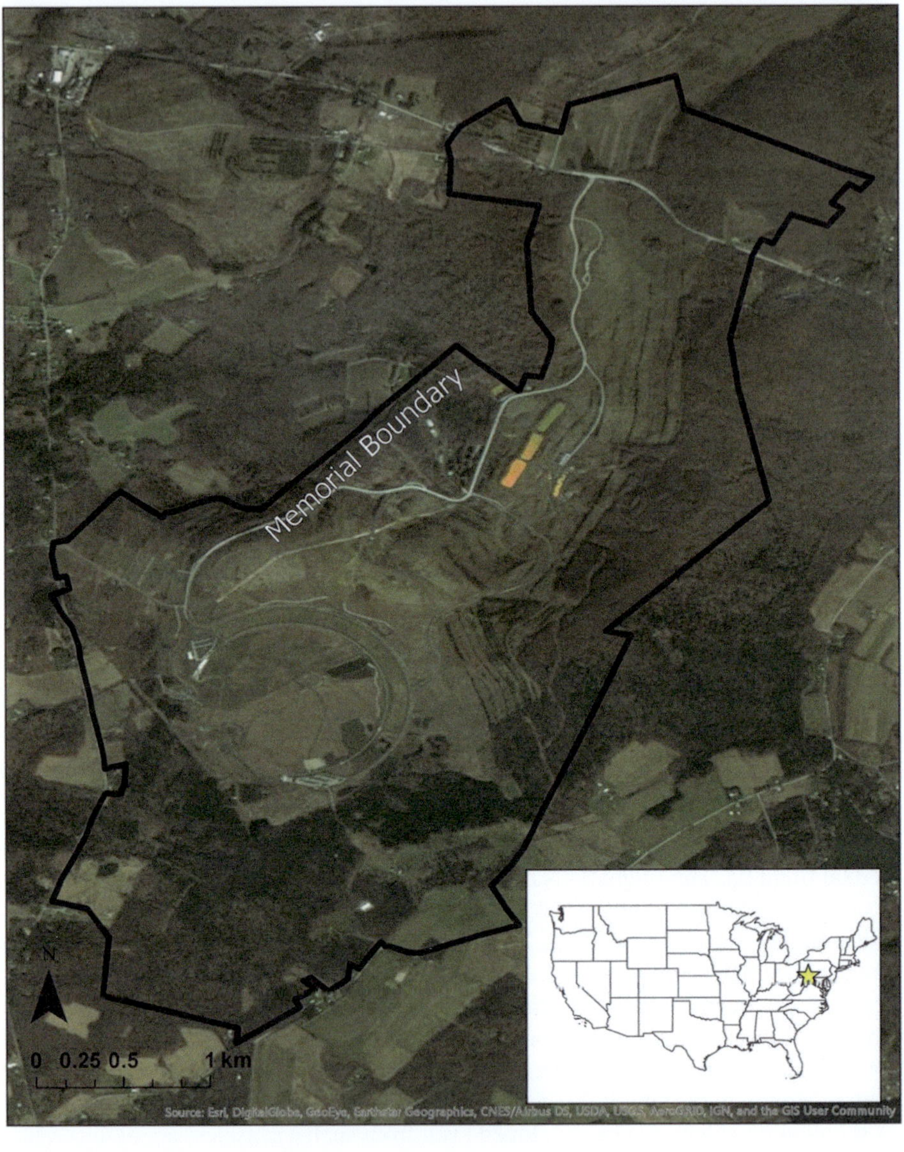

Fig. 12.1 Flight 93 National Memorial. The creation of this Memorial Park has involved protection of existing forests, remediation of lands afflicted by acidic coal mine drainage, and reforestation, on 2200 hectares of land

References

9/11 Commission. (2011). *The 9/11 commission report: Final report of the national commission on terrorist attacks upon the United States*. Government Printing Office.

Avgan, B., Raza, H., Barzani, M., & Breitenmoser, U. (2016). Do recent leopard Panthera pardus records from northern Iraq and South-Eastern Turkey reveal an unknown population nucleus in the region? *Zoology in the Middle East, 62*(2), 95–104.

Bacevich, A. J. (2016). *America's war for the greater Middle East: A military history*. New York: Random House.

Cirillo, R. (2004). *Ardennes-Alsace. The campaigns of World War II* (A World War II commemorative series, Vol. 72, No. 26). US Government Printing Office.

Costanza, R., de Groot, R., Sutton, P., Van der Ploeg, S., Anderson, S. J., Kubiszewski, I., Farber, S., & Turner, R. K. (2014). Changes in the global value of ecosystem services. *Global Environmental Change, 26*, 152–158.

Daskin, J. H., & Pringle, R. M. (2018). Warfare and wildlife declines in Africa's protected areas. *Nature, 553*, 328.

ELD Initiative. (2015). *The value of land: Prosperous lands and positive rewards through sustainable land management*. http://www.eld-initiative.org/fileadmin/pdf/ELD-main-report_en_10_web_72dpi.pdf. Accessed 30 Oct 2018.

Emili, L. A., Pizarchik, J., & Mahan, C. G. (2016). Sustainable remediation of legacy mine drainage: A case study of the Flight 93 National Memorial. *Environmental Management, 57*(3), 660–670.

Gaynor, K. M., Fiorella, K. J., Gregory, G. H., Kurz, D. J., Seto, K. L., Withey, L. S., & Brashares, J. S. (2016). War and wildlife: Linking armed conflict to conservation. *Frontiers in Ecology and the Environment, 14*, 533–542.

Gies, E. (2018). Patience, peace, and Persian leopards. *BioGraphic*. https://www.biographic.com/posts/sto/patience-peace-and-persian-leopards. Accessed 20 Oct 2018.

Hanson, T. (2018). Biodiversity conservation and armed conflict: A warfare ecology perspective. *Annals of the New York Academy of Sciences, 1429*, 50–65. https://doi.org/10.1111/nyas.13689.

Hanson, T., Brooks, T. M., Da Fonseca, G. A., Hoffmann, M., Lamoreux, J. F., Machlis, G., Mittermeier, C. G., Mittermeier, R. A., & Pilgrim, J. D. (2009). Warfare in biodiversity hotspots. *Conservation Biology, 23*(3), 578–587.

Jones, K. R., Venter, O., Fuller, R. A., Allan, J. R., Maxwell, S. L., Negret, P. J., & Watson, J. E. (2018). One-third of global protected land is under intense human pressure. *Science, 360*, 788–791.

Kashi, B., Simpson, D., Simone, C., Higgins, M., Manion, M., & Bruner, A. (2018). *Integrating ecosystem values into cost-benefit analysis: Recommendations for USAID and practitioners*. United States Agency for International Development (USAID). https://rmportal.net/biodiversityconservation-gateway/resources/projects/bridge/integrating-ecosystem-values-cost-benefit-analysis. Accessed 18 Oct 2018.

Lange, G. M., Wodon, Q., & Carey, K. (Eds.). (2018). *The changing wealth of nations 2018: Building a sustainable future*. Washington, DC: The World Bank.

Latiff, R. H. (2017). *Future war: Preparing for the new global battlefield*. New York: Knopf.

Macgregor, D. (2016). *Margin of victory: Five battles that changed the face of modern war*. Annapolis: Naval Institute Press.

Machlis, G. E., & Hanson, T. (2008). Warfare ecology. *Bioscience, 58*(8), 729–736.

Machlis, G. E., Hanson, T., Špirić, Z., & McKendry, J. E. (Eds.). (2011). *Warfare ecology: A new synthesis*. Dordrecht: Springer.

Maurer, T. (2018). ISIS's warfare functions: A systematized review of a proto-state's conventional conduct of combat operations. *Small Wars & Insurgencies, 29*(2), 229–244.

McNeely, J. A. (2003). Conserving forest biodiversity in times of violent conflict. *Oryx, 37*(2), 142–152.

Schwartzstein, P. (2014). For leopards in Iran and Iraq, land mines are a surprising refuge. *National Geographic*. https://news.nationalgeographic.com/news/2014/12/141219-persian-leopard-iran-iraq-land-mine/. Accessed 30 Oct 2018.

Shepherd, E., Milner-Gulland, E. J., Knight, A. T., Ling, M. A., Darrah, S., van Soesbergen, A., & Burgess, N. D. (2016). Status and trends in global ecosystem services and natural capital: Assessing progress toward Aichi biodiversity target 14. *Conservation Letters, 9*, 429–437.

Walker, P. (2011). *The world's most dangerous borders*. Foreign Policy [Online]. https://foreignpolicy.com/2011/06/24/the-worlds-most-dangerous-borders/. Accessed on 26 Sept 2014.

Wright, M. R., & Thompson, C. G. (2017). *Are brigade combat teams relevant for the future operating environment?* Army War College publication #3514 http://publications.armywarcollege.edu/pubs/3514.pdf. Accessed 30 Oct 2018.

Index

T. R. Lookingbill, P. D. Smallwood (eds.), *Collateral Values*, Landscape Series 25,
https://doi.org/10.1007/978-3-030-18991-4

MIX
Papier aus verantwortungsvollen Quellen
Paper from responsible sources
FSC® C105338

If you have any concerns about our products,
you can contact us on
ProductSafety@springernature.com

In case Publisher is established outside the EU,
the EU authorized representative is:
Springer Nature Customer Service Center GmbH
Europaplatz 3, 69115 Heidelberg, Germany

Printed by Libri Plureos GmbH
in Hamburg, Germany